Textbook of Pomology

Textbook of Pomology

Edited by **Edgar Crombie**

SYRAWOOD
PUBLISHING HOUSE

New York

Published by Syrawood Publishing House,
750 Third Avenue, 9th Floor,
New York, NY 10017, USA
www.syrawoodpublishinghouse.com

Textbook of Pomology
Edited by Edgar Crombie

International Standard Book Number: 978-1-68286-136-3 (Hardback)

The publisher's policy is to use permanent paper from mills that operate a sustainable forestry policy. Furthermore, the publisher ensures that the text paper and cover boards used have met acceptable environmental accreditation standards.

Trademark Notice: Registered trademark of products or corporate names are used only for explanation and identification without intent to infringe.

Printed in the United States of America.

Contents

Preface

The world is advancing at a fast pace like never before. Therefore, the need is to keep up with the latest developments. This book was an idea that came to fruition when the specialists in the area realized the need to coordinate together and document essential themes in the subject. That's when I was requested to be the editor. Editing this book has been an honour as it brings together diverse authors researching on different streams of the field. The book collates essential materials contributed by veterans in the area which can be utilized by students and researchers alike.

Pomology is an interdisciplinary field that incorporates concepts from botany and agricultural science, and focuses upon cultivation of fruit trees and fruits. The book includes development, cultivation and physiological studies of fruit trees. It focuses upon planting systems and fruit production, processing of fruits, etc. The purpose of this book is to provide a comprehensive overview of present status of tree-fruit cultivation and measures to enhance the quality of fruits. It emphasises on the significance of pomology in understanding and improving tree-fruit productivity.

Each chapter is a sole-standing publication that reflects each author´s interpretation. Thus, the book displays a multi-facetted picture of our current understanding of applications and diverse aspects of the field. I would like to thank the contributors of this book and my family for their endless support.

Editor

High performance liquid chromatography-diode array detector (HPLC-DAD) detection of *trans*-resveratrol: Evolution during ripening in grape berry skins

Angelo Maria Giuffrè

Università degli Studi Mediterranea di Reggio Calabria – Dipartimento di Agraria. E-mail: amgiuffre@unirc.it.

Trans-resveratrol (3, 5, 4' – trihydroxy-trans-stilbene) is a stilbene naturally present in a number of plant families. Berry skins of grape (*Vitis vinifera* L.), Alicante, Black Malvasia, Nerello and Prunesta cultivars were analyzed at weekly intervals for *trans*-resveratrol production at five ripening stages during the last five weeks of maturation, from August 26th to September 23th. Analysis was carried out using the high performance liquid chromatography-diode array detector (HPLC-DAD) technique after fractioning of *trans*-resveratrol through a 500 mg C_{18} column (Solid Phase Isolation – SPI technique). A continuous decrease in *trans*-resveratrol content in all cultivars was observed during ripening. Alicante had the highest *trans*-resveratrol content, showing a decrease of 46% from the first to last sampling, while Black Malvasia produced the lowest amount of *trans*-resveratrol, 4 to 5 times less than Alicante at each sampling date.

Key words: Antioxidants, grape, phenols, stilbenes, *trans*-resveratrol, high performance liquid chromatograph (HPLC).

INTRODUCTION

The increase in food quality information has led to a cultural evolution among consumers. Food quality can be defined as the capacity of a product to answer consumer needs and can be characterized by numerous aspects: chemical, physical, microbiological, nutritional, hedonistic and economical. One of the most important aspects of food quality is related to nutraceutical and functional activity. The term "functional food" originated in Japan in the 1980s, and is defined as "any food or ingredient that has a positive impact on an individual's health, physical performance or state of mind in addition to its nutritive value" (Hardy, 2000; Goldberg, 1994). The increasing interest for nutraceuticals and functional foods is also related to the awareness that chemical pharmaceutics are always expensive, superfluous and sometimes unsafe and have dubious benefits (Bagchi, 2006). Fruits are among the most important foods that naturally contain functional biomolecules like polyphenols. Trans-

resveratrol is a polyphenolic component of the stilbenes group, (3, 5, 4' - trihydroxy-trans-stilbene) and its structure and metabolism have been widely studied (Guiso et al., 2002; Deak and Falk, 2003; Halls and Yu, 2008; Szekeres et al., 2010). Its formation is caused by a response of plant tissues to external stress (Frémont, 2000; Hammerschmidt, 2004) and the berry skins of red grape contain this minor but important substance. In several studies, trans-resveratrol, a cardioprotective component, has been associated with many beneficial effects on human health (Frémont, 2000; Dawn, 2007) and cardiovascular benefits have also been observed among the Greek population (Kopp, 1998). An important role was also found in diabetes with insulin regulation (Szkudelska and Szkudelski, 2010) and beneficial effects have also been observed for anti-inflammatory arthritis with experiments carried out on animals (Elmali et al., 2007). Beneficial effects have been described for skeletal muscles, increasing glucose transport, restraining protein degradation, improving strength and endurance and protecting from oxidative injury (Dirks Naylor, 2009). In addition, studies have shown that trans-resveratrol, as a natural antioxidant, plays an important role in chemoprevention (Schneider et al., 2000; Savouret and Quesne, 2002; Dong, 2003; Goswami and Das, 2009). The trans-resveratrol concentration in grape skins and wines is influenced by a number of factors including the geographical area of production, climateage of the wine, grape cultivar, fungal infection and ultraviolet irradiation (Roggero and Garcia-Parrilla, 1995; Roggero, 1996; Roldán et al., 2003; Moreno-Labanda et al., 2004).In Northern Italian red wines, Recioto (sweet) and Amarone (dry) produced from the same grape cultivar in the same geographical area and with the same technical proce-dures by processing dried grapes recorded a 0.05 to 0.40 and 0.05 to 0.80 mg/L trans-resveratrol content for Recioto and Amarone, respectively (Celotti et al., 1996). In Czech red wines from the Bohemian and Moravian regions, a trans-resveratrol content of between 0.92 and 6.25 mg/L was observed (Kolouchová-Hanzlíková et al., 2004). In French red wines, the resveratrol concentration in young and nine-month old wines was analyzed with the highest content generally found in the young wines (Roggero and Archier, 1994). In Hungarian wines, a maximum of 10.4 mg/L was recorded with a relatively higher content in the warm "Villany" Region (Southern Hungary) and a lower content in wines produced in the cold and humid "Eger" Region (Montsko et al., 2010).

Despite a wide interest in grape cultivation and wine production in the Calabria Region (Southern Italy), little information has been found regarding the chemical characteristics and the correct time of harvest of the grape cultivars of this region, discriminating between industrial and technological maturity. Industrial maturity time is characterized by a higher concentration of sugars in the fruit and therefore leads to a higher theoretical degree of alcohol in the wine while technological maturity is reached when the grapes have the optimum chemical composition for obtaining a wine with specific characteristics (Robredo et al., 1991; Jin et al., 2009). Although, several studies have focused on the trans-resveratrol concentration in red wines, relatively few have examined the trans-resveratrol content in berry skins, but no data has been found on the trans-resveratrol content in the berry skins of Calabrian grape cultivars. The aim of this paper was to study the evolution of the trans-resveratrol content in berry skins of 4 red grape cultivars which are widely cultivated on the Tyrrhenian side of the Province of Reggio Calabria, South West Calabria (Southern Italy) during ripening, in order to define the moment of maximum accumulation of this antioxidant. For trans-resveratrol quantification, a solid-phase isolation (SPI) coupled with a HPLC analysis was studied in UV-visible by diode array detector (DAD). No previous data regarding this antioxidant exists for this geographical area.

MATERIALS AND METHODS

Chemicals

Trans-resveratrol standard was purchased from Sigma (Milan, Italy) and a C_{18} Sep-Pak separation column was purchased from Millipore (Milan, Italy). All other reagents were purchased from Merck, Darmstadt (Germany).

Sampling

Four red grape cultivars, Alicante, Black Malvasia, Nerello and Prunesta were grown in vineyards at 80 meters above sea level in the Bagnara Calabra-Scilla area, a coastal location in the province of Reggio Calabria (Southern Italy). The area is characterized by a favourable climate for grape cultivation with rainfall also occurring in summer from the middle of August at regular intervals and especially at night. In this contest, Black Malvasia, Nerello and Prunesta are autochthonous cultivars, Alicante is allochthonous. Temperatures are mild and there are no sudden changes between day and night. Twenty-five grape plants per cultivar, 18 years old in 2009, with similar growth and production characteristics were selected in the vineyard. Plants were selected and labeled in mono-cultivar groves with the same microclimatic conditions. Each mono-cultivar grove was at least 700 meters from the others. Plants were well managed, uniform in size and had no nutrient deficiency or pest damage and were spaced in adjacent rows 1.2 m apart and 1.5 m apart within rows. For all cultivars, cordons were supported by two wire mounted on a trellis at 1.0 m above the vineyard floor. No irrigation was applied to the area during the course of this study. All cultivars were fertilized with 60 kg N ha^{-1}, 5 kg P ha^{-1} and 50 kg ha^{-1} per year. Briefly, no differences there were in applied measures or agrochemicals or climate between individual vineyards. Five samplings were carried out for each cultivar with grapes harvested on the following weekly basis: August 26th, September 2th, 9th, 16th and 23th of the 2009 to 2010 crop year. The fifth sample was conducted at industrial ripeness (Robredo et al., 1991; Jin et al., 2009). Berries were harvested when fully ripe (that is, at the moment when they possessed the required industrial maturity) (Robredo et al., 1991; Di Stefano and Cravero, 1991). Two hundred berries/cultivar were randomly hand-picked from plants and separated into three lots: small (0.00 to 1.00 g), medium (1.01 to 2.00 g), and large (2.01 to 3.00 g) according to their weight. From the three lots, ten sub-lots of 10 berries were formed, each containing small, medium and large berries. At this point, the three most similar sub-lots by weight were chosen to carry out analysis in three replicates. Berries were neither washed nor cleaned before peeling.

Extraction of trans-resveratrol from berry skin

Berries of the three most similar sub-lots were longitudinally cut with a bistoury into halves and the skin was removed within 4 h of harvest. Extraction was conducted according to the Di Stefano and Cravero method (1991) modified as follows: the skins of ten berries were quickly placed in a 25 ml extracting buffer solution (pH=3.2), from a 1 L buffer solution consisting of 200 ml de-ionized water, 5 g tartaric acid, 120 ml ethanol (95% type), 2 g sodium metabisulfite, 22 ml of a 1 N sodium hydroxide solution and de-ionized water up to the 1 L volume. Phenolic compounds were extracted from the skin in darkness for 24 h by means of the above-mentioned extracting buffer solution at a temperature of 25°C. The extracting solution of the four cultivars was separated from the skin by decanting and was then stored in a freezer (-10°C) until analysis.

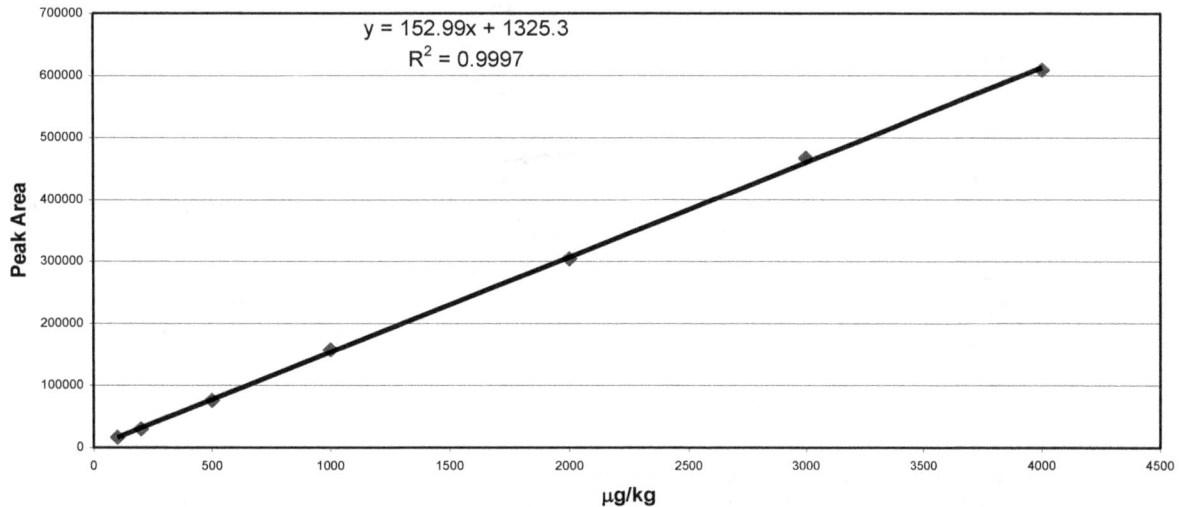

$$y = 152.99x + 1325.3$$
$$R^2 = 0.9997$$

Figure 1. Linear response of 100 to 4,000 µg/kg for *trans*-resveratrol as external standard, after analysis by HPLC-DAD.

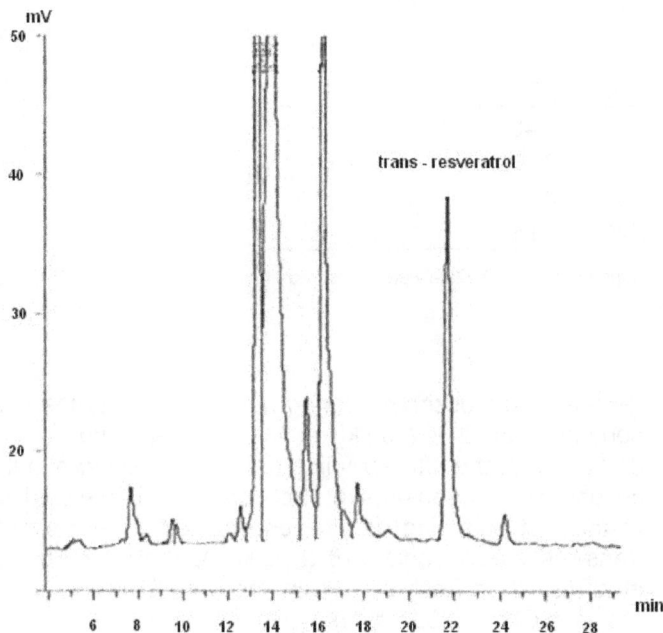

Figure 2. HPLC-DAD chromatogram showing *trans*-resveratrol in grape berry skin.

Solid-phase isolation of trans-resveratrol

The Mattivi (1993) and Kallithraka et al. (2001) methods, adapted for grape skin, were used for the solid-phase isolation of trans-resveratrol. A 5 ml quantity of phenolic extract was adjusted to pH 7 with a 1N NaOH solution. The neutralized extract was quantitatively transferred into a 10 ml volumetric flask and filled up to volume with de-ionized water. A 5 ml volume was placed in a 500 mG C_{18} column.

After adsorbing the sample was washed with 3 ml of a pH 7 buffer solution and 5 ml of de-ionized water, respectively. The eluted washing solution was discarded. Trans-resveratrol was eluted with 40 ml of ethyl acetate. After eluting, the sample was dried by means of a rotary evaporator and a flow of nitrogen. At this point, the dried sample was dissolved in 1 ml of methanol (HPLC grade) in order to prepare the injecting solution.

Chromatographic analysis

Trans-resveratrol was identified by comparison of retention times with standard and by enrichment of skin extracts with authentic sample. For quantitative analysis and response factor calculation, the external standard method was used with trans-resveratrol as the authentic sample. Figure 1 shows the linearity of the response correlation coefficient (R^2) for a range of standards from 100 to 4,000 µg/L which was found to be 0.9997.

The chromatographic detection of trans-resveratrol was conducted using the Roldán et al. (2003) and Kolouchovà-Hanzlìkovà et al. (2004) methods modified by using a Knauer Instrument, consisting of a Smartline Pump 1000 and a Smartline UV detector 2600, equipped with a Lichroshper C_{18} 100 mm × 5 µm pre-column and a ODS Hypersil 200 mm × 2.1 mm × 5 µm separation column (Figure 2). Eluent A, H_3PO_4 10^{-3} M; eluent B, CH_3CN. Program: linear gradient from 0% A to 60% A in 35 min; flow 0.8 ml/min; operating temperature 30°C; detection 310 nm; injection volume 20 µl.

Statistical analysis

"Statistica" software was used for statistical analysis (ANOVA One-way) and Duncan's multiple range tests ($p \leq 0.05$) was used for comparison of means. Excel software was used to build the Figures 1 and 3 and to calculate the SD.

RESULTS AND DISCUSSION

Data of the sample analysis of berries at fifth sampling about physico-chemical characteristics are shown in Table 1 as mean ± standard deviation of ten replicates. The highest °Brix and the lowest TA were found in Black Malvasia. The pH value was similar for all cultivars. Nerello had the lowest °Brix and the highest TA.

Generally was found that the higher the °Brix, the lower the TA. Trans-resveratrol content was expressed as µg/kg

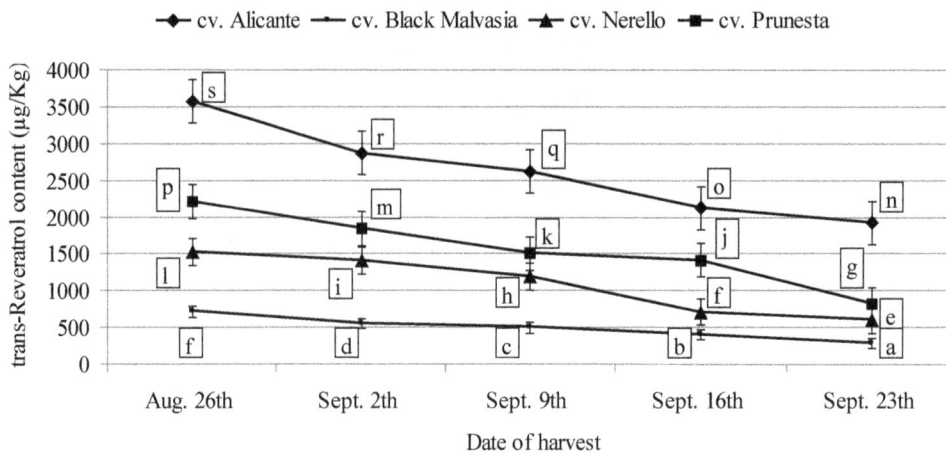

Figure 3. HPLC-DAD analyses of *trans*-resveratrol on fresh berry skins during grape ripening. The results are mean values of three independent experiments ± SD. Data followed by different letters are significantly different by Duncan's multiple range test (P≤0.05).

Table 1. Titratable acidiy (TA) as grams tartaric acids equivalents per liter, °Brix and pH.

Cultivar	TA (g/L) ± SD	°Brix ± SD	pH ± SD
Alicante	7.0 ± 0.64	22.9 ± 0.40	3.5 ± 0.02
Prunesta	8.1 ± 0.70	22.0 ± 0.45	3.7 ± 0.03
Nerello	8.3 ± 0.72	21.8 ± 0.57	3.6 ± 0.02
Black Malvasia	6.9 ± 0.58	23.0 ± 0.51	3.5 ± 0.04

The results are mean values of ten independent experiments ± SD. Data were measured at the fifth sampling.

of the studied stilbene contained in fresh berry skin. Analysis of variance showed a high significant difference between the four cultivars and dates of harvest with a variation coefficient of 2.08% and Duncan's test shows 19 separated homogeneous groups (Figure 3). Analysis of trans-resveratrol evidenced a decreasing trend for all cultivars from the first to the fifth sampling. Of all the cultivars, Alicante showed by far the highest trans-resveratrol content in all sampling dates, ranging from 3,573 µg/kg recorded on August 26th to 1,917 µg/kg recorded after 5 weeks at the industrial ripeness grade, resulting in a decrease of 46% during the five sampling dates. Similar data (2,980 µg/kg) were found by Li et al. (2008) in ripe fresh berry skin of Takasuma cultivar grown in the Institute of Botany of the Chinese Academy of Science. In contrast, Black Malvasia showed the lowest content in all harvest dates, 715, 553, 498, 400 and 286 µg/kg, measured on August 26th, September 2th, September 9th, September 16th and September 23th, respectively, decreasing by 60%. The same decrease (60%) was found in the Nerello cultivar, the third highest for total trans-resveratrol content, varying from 1,521 µg/kg on August 26th, to 607 µg/kg on September 23th. In Prunesta berry skins the highest trans-resveratrol

decrease was observed during ripening, 63% of the total content, from 2,209 µg/kg at the first sampling date to 812 µg/kg at the fifth sampling date. This cultivar had the second highest trans-resveratrol content of those studied. In the last sampling date, Prunesta had a very similar trans-resveratrol content (812 µg/kg) to that found by Liu et al. (2013) in fresh berry skins of Saint-Emilion cultivar grown in China (880 µg/kg).

The trans-resveratrol content of berry skins of Black Malvasia grown in South West Calabria (Southern Italy) in the first week of September (500 to 550 µg/kg) was similar to that of ripe fresh berry skin in Tano red cultivar (600 µg/kg) grown in the Institute of Botany of the Chinese Academy of Science (Li et al., 2008). Other Authors have found a 3, 5, 4´ – trihydroxy-trans-stilbene content varying between 32,000 µg/kg (Tempranillo cultivar) to 245,000 µg/kg (Bobal cultivar) measured on dry weight of berry skins (Navarro et al., 2008). Pascual-Martí et al. (2001) have found, in dried grape skin, a 3,5, 4´ – trihydroxy-trans-stilbene content ranging between 48,000 µg/kg (Cabernet cv) and 172,000 µg/kg (Tempranillo cv).

All Calabrian grape cultivars showed a trans-resveratrol content in fresh berry skins lower than Merlot,

Cabernet Sauvignon, Colorino del Valdarno e Montepulciano cultivars grown in Tuscany (Central Italy), (Iacopini et al., 2008).

Esna-Ashari et al. (2008) reported a trans-resveratrol content in red grape cultivars ranging between a not detectable amount and 5.52 mg/kg calculated on whole berry; more importantly, it appeared that the second major trans-resveratrol content was found in a cultivar named Italia (5.48 mg/kg) out of a pool of 147 cultivars grown in Iran.

Conclusion

Trans-resveratrol has been found to be an important functional bio-molecule with beneficial effects on human health as a cardioprotective substance. The aims of this work was to verify the efficacy of the SPI-HPLC-DAD technique applied to this analysis and to quantify the trans-resveratrol content in red skin of four grape cultivars grown in South West Calabria (Southern Italy) during ripening. The SPI-HPLC-DAD technique applied to red berry skin analysis was successful in extracting and quantifying trans-resveratrol. Analysis of trans-resveratrol in the berry skin of red grapes cultivated in the South West Calabria (Southern Italy) evidenced a decreasing trend for all the studied cultivars during ripening. Significant differences were found in the trans-resveratrol content among four cultivars grown in the same area. These differences proved to be interesting elements for the characterization of cultivars without climatic or agronomic interference. A higher degree of grape ripeness leads to a lower trans-resveratrol content. Alicante (that is, the allochthonous cultivar) was found to be with the highest trans-resveratrol content among those studied. As all the plants were cultivated with identical agronomical and climatic conditions and as an identical sanitary state was observed, the different results for trans-resveratrol content are mainly due to the genetic characteristics of each cultivar and to the relationship between cultivar and environment (that is, soil and climate).

REFERENCES

Bagchi D (2006). Nutraceuticals and functional foods regulations in the United States and around the world. Toxicology 221:1-3.

Celotti E, Ferrarini R, Zironi R, Conte LS (1996). Resveratrol content of some wines obtained from dried Valpolicella grapes: Recioto and Amarone. J. Chromatogr. A. 730:47-52.

Dawn B (2007). Resveratrol: ready for prime time? J. Mole. Cell. Cardiol. 42:484-486.

Deak M, Falk H (2003). On the chemistry of Resveratrol diastereomers. Monatshefte für Chemie 134:883-888.

Di Stefano R, Cravero MC (1991). Metodi per lo studio dei polifenoli dell'uva. Rivista di Viticoltura e di Enologia 2:37-45.

Dirks Naylor AJ (2009). Cellular effects of resveratrol in skeletal muscle. Life Sci.84:637-640.

Dong Z (2003). Molecular mechanism of the chemopreventive effect of

resveratrol. Mutat. Res. 523–524:145–150.

Elmali N, Baysal O, Harma A, Esenkaya I, Mizrak B (2007). Effects of Resveratrol in Inflammatory arthritis. Inflammation 30:1-2.

Esna-Ashari M, Gholami M, Zolfigol MA, Shiri M, Mahmoodi-Pour A, Hesari M (2008). Analysis of trans-resveratrol in Iranian grape cultivars by LC. Chromatographia 67:1017-1020.

Frémont L (2000). Biological effects of Resveratrol. Life Sci. 66:663-673.

Goldberg I (1994). Functional foods, designer foods, pharmafoods, nutraceuticals. London: Chapman & Hall. (Please provide page number).

Goswami SK, Das DK (2009). Resveratrol and chemoprevention. Cancer Lett. 284:1-6.

Guiso M, Marra C, Farina A (2002). A new efficient resveratrol synthesis. Tetrahedron Lett. 43: 597-598.

Halls C, Yu O (2008). Potential for metabolic engineering of resveratrol biosynthesis. Trends in Biotechnol.26: 77-81.

Hammerschmidt R (2004). The metabolic fate of Resveratrol: key to resistance in Grape? Physiol. Mole. Plant Pathol. 65:269-270.

Hardy G (2000). Nutraceuticals and functional foods: introduction and meaning. Nutrition 16:688-689.

Iacopini P, Baldi M, Storchi P, Sebastiani L (2008). Catechin, epicatechin, quercetin, rutin and resveratrol in red grape: Content, in vitro antioxidant activity and interaction. J. Food Composit. Anal. 21:589-598.

Jin ZM, He JJ, Bi HQ, Cui XY, Duan CQ (2009). Phenolic compounds profiles in berry skins from nine red wine grape cultivars in Northwest China. Mole. 14:4922-4935.

Kallithraka S, Arvanitoyannis I, El-Zajouli A, Kefalas P (2001). The application of an improved method for trans-resveratrol to determine the origin of Greek red wines. Food Chem. 75:355-363.

Kolouchová-Hanzlíková I, Melzoch K, Filip V, Šmidrkal J (2004). Rapid method for resveratrol determination by HPLC with electrochemical and UV detections in wines. Food Chem. 87:151-158.

Kopp P (1998). Resveratrol, a phytoestrogen found in red wine. A possible explanation for the conundrum of the "French paradox". Eur. J. Endocrinol. 138:619-620.

Li X, Zheng X, Yan S, Li S (2008). Effects of salicylic acid (SA), ultraviolet radiation (UV-B and UV-C) on trans-resveratrol inducement in the skin of harvested grape berries. Front. Agric. China 2:77-81.

Liu C, Wang L, Wang J, Wu B, Liu W, Fan P, Liang Z, Li S (2013). Resveratrols in Vitis berry skins and leaves: Their extraction and analysis by HPLC. Food Chem. 136:643-649.

Mattivi F (1993). Solid phase extraction of trans-resveratrol from wines for HPLC analysis. Z Lebensm Unters Forsch 196:522-525.

Montsko G, Ohmacht R, Mark L (2010). Trans-resveratrol and trans-Piceid content of Hungarian wines. Chromatographia Suppl. 71:S121-S124.

Moreno-Labanda JF, Mallavia R, Pérez-Fons L, Lizama V, Saura D, Micol V (2004). Determination of piceid and resveratrol in Spanish wines deriving from Monastrell (Vitis vinifera L.) Grape variety. J. Agric. Food Chem. 52:5396-5403.

Navarro S, León M, Roca-Pérez L, Boluda R, García-Ferriz L, Pérez-Bermúdez P, Gavidia I (2008). Characterisation of Bobal and Crujidera grape cultivars, in comparison with Tempranillo and Cabernet Sauvignon: Evolution of leaf macronutrients and berry composition during grape ripening. Food Chem. 108:182-190.

Pascual-Martí MC, Salvador A, Chafer A, Berna A (2001). Supercritical fluid extraction of resveratrol from grape skin of Vitis vinifera and determination by HPLC. Talanta 54:735-740.

Robredo LM, Junquera B, Gonzales-Sanjose ML, Barron LJR (1991). Biochemical events during ripening of grape berries. Italian J. Food Sci. 3:173-180.

Roggero JP, Archier P (1994). Dosage du resvératrol et de l'un de ses glycosides dans les vins. Sci. des Aliments 14:99-107.

Roggero JP, Garcia-Parrilla C (1995). Effects of ultraviolet irradiation on resveratrol and changes in resveratrol and various of its derivatives in the skins of ripening grapes. Sci. des Aliments 15:411-422.

Roggero JP (1996). Évolution des teneurs en resvératrol et en picéid des vins en cours de fermentation ou de vieillissement. Comparaison des cépages grenache et mourvèdre. Sci.des Aliments 16:631-642.

Roldán A, Palacios V, Caro I, Pérez L (2003). Resveratrol content of Palomino fino Grapes: influence of vintage and fungal infection. J. Agric. Food Chem. 51:1464-1468.

Savouret JF, Quesne M (2002). Resveratrol and cancer: a review. Biomedicine and Pharmacotherapy 56: 84-87.

Schneider Y, Vincent F, Duranton B, Badolo L, Gossé F, Bergmann C, Seiler N, Raul F (2000). Anti-proliferative effect of resveratrol, a natural component of grapes and wine, on human colonic cancer cells. Cancer lett. 158:85-91.

Szekeres T, Fritzer-Szekeres M, Saiko P, Jäger W (2010). Resveratrol and Resveratrol analogues-structure-activity relationship. Pharmaceut. Res. 27:1042-1048.

Szkudelska K, Szkudelski T (2010). Resveratrol, obesity and diabetes. Eur. J. Pharmacol. 635:1-8.

2

Responses of sapodilla fruit (*Manilkara zapota* [L.] P. Royen) to postharvest treatment with 1-methylcyclopropene

Victor Manuel Moo-Huchin[1], Iván Alfredo Estrada-Mota[1], Raciel Javier Estrada-Leon[1], Elizabeth Ortiz-Vazquez[2], Jorge Pino Alea[4], Adriana Quintanar-Guzman[3], Luis Fernando Cuevas-Glory[2] and Enrique Sauri-Duch[2]

[1]Laboratorio de Instrumentación Analítica. Instituto Tecnológico Superior de Calkiní. Av. Ah-Canul, C.P. 24900, Calkiní, Campeche, México.
[2]Laboratorio de Ciencia y Tecnología de Alimentos. División de Estudios de Posgrado e Investigación. Instituto Tecnológico de Mérida, C.P. 97118, km 5 Mérida-Progreso, Mérida, Yucatán, México.
[3]Adriana Consulting Services inc, Consultant Food Science and Technology, P. O. Box 6762, Siloam Springs, AR 72761, U.S.A.
[4]Instituto de Investigaciones para la Industria Alimentaria, Carretera al Guatao km 3½, La Habana, CP 19200, Cuba.

The effect of 1-methylcyclopropene (1-MCP) on ripening and chilling injury in sapodilla fruits was investigated. Sapodilla fruits were treated with four different concentrations of 1-MCP (0.0, 0.2, 0.5 or 1.0 µL/L) and two exposure times (12 or 24 h) in sealed chambers under different temperatures (15 and 25°C). Following the previous treatment, fruits were stored at 25°C with 85 to 95% relative humidity (RH) for ripening assessment. Subsequently, we evaluated the effect of 1-MCP (1.0 µL/L for 24 h at 25°C) on chilling injury when fruits were stored at 6°C, and matured afterwards at 25°C. 1-MCP treatment delayed the ripening of sapodilla fruits (from 4 to 11 days). Ethylene and carbon dioxide production were reduced and delayed significantly ($P < 0.05$) by 1-MCP treatment. In general, all quality characteristics of fruits were maintained. Sapodilla fruit stored at 6°C for 3, 10 and 14 days developed chilling injury. These chilling injury symptoms were reduced by 1-MCP treatment.

Key words: 1-methylcyclopropene (1-MCP), sapodilla, chilling injury, postharvest.

INTRODUCTION

Sapodilla tree (*Manilkara zapota* (L.) P. Royen) is native to tropical America and very widespread in southern Mexico. The fruit production is mainly consumed locally. However, this fruit could be exported either in fresh or processed form, due to its sweet taste and unique aroma, highly appreciated by consumers (Balerdi and Shaw, 1998;

Ma et al., 2003; Sauri-Duch et al., 2010). Worldwide, one of the major producers of sapodilla fruit is India, with a production of 1,346,000 tons (Indian Horticulture Database, 2010). In Mexico, the production has been estimated at 20,000 tons produced principally in the States of Campeche, Yucatan and Veracruz (SAGARPA,

2011). Sapodilla fruits are climacteric (Lakshminarayana, 1979); its ripening is rapid and is characterized by a significant increase in respiration and ethylene production, which makes this fruit highly perishable, and consequently very difficult to preserve and commercialize. The fruits ripen between 3 to 7 days after harvesting at 25°C and although it can be stored under refrigeration (15°C) (Broughton and Wong, 1979), the use of low temperatures is restricted as it can cause a physiological dysfunction referred to as "chilling injury" which leads to deterioration and loss of quality (Saltveit, 2002). Because sapodilla fruits are highly perishable, it became necessary to study its conservation, handling and postharvest methods, which may extend its shelf life in order to promote sales abroad (Ganjyal et al., 2003; Sauri-Duch et al., 2010).

To prolong the postharvest life of fruits, a promising solution is the use of chemical compounds in order to delay the ripening process and maintain quality. In this regard, 1-methylcyclopropene (1-MCP) is a compound that blocks the ethylene receptor sites, avoiding the physiological actions of this phytohormone by delaying ripening and senescence (Sisler and Serek, 1997). This chemical compound has been studied in various fruits and vegetables, finding a significant decrease in the production and action of ethylene. The effective concentrations used vary between 0.0025 and 1.0 µL/L at temperatures between 20 and 25°C for 12 or 24 h (Blankenship and Dole, 2003). The action of 1-MCP is affected by various factors, such as the species and treated variety, treatment conditions and the stage of fruit maturity or level of ethylene production at the moment of treatment (Blankenship and Dole, 2003).

The effect of 1-MCP has been evaluated in a broad variety of fruits in order to delay the ripening process, prolong shelf life and maintain quality, such as in papaya (Hofman et al., 2001; Manenoi et al., 2007), avocado (Adkins et al., 2005), pear (Liu et al., 2005; Calvo and Sozzi, 2009), plum (Manganaris et al., 2008; Luo et al., 2009), apple (Kashimura et al., 2010) and sapodilla (Qiuping et al., 2006; Kunyamee et al., 2008). It has also been reported that treatment with 1-MCP reduces symptoms of damage by chilling, such as internal browning, rot and superficial scald in pineapple, pear and apple (Selvarajah et al., 2001; Argenta et al., 2003; Sabban-Amin et al., 2011). The evidence supports that 1-MCP treatment could potentially represent a promising technology to extend the postharvest life of sapodilla fruits and to maintain quality for longer periods, however, the suitable conditions (doses, treatment duration and temperature) for commercial use of 1-MCP on sapodilla are far from being standardized, and there are several concerns regarding quality and appearance of the treated fruit. As far as we know, this is the first time 1-MCP is reported for reducing symptoms of damage by chilling on sapodilla fruits. Therefore, the objective of this work was to evaluate the influence of different treatments with 1-MCP on the ripening and chilling injury of sapodilla fruits, thereby increasing their commercial and economic viability.

MATERIALS AND METHODS

Harvested sapodilla fruits were collected in an orchard located in the rural community of "Cansahcab", in the State of Yucatan, Mexico.

This study was conducted in two steps: The first experiment was carried out in order to evaluate the influence of various treatments with 1-MCP on the quality of sapodilla fruit during ripening; the second experiment was performed in order to evaluate the effect of treatment with 1-MCP on the onset of symptoms of chilling injury.

First experiment

640 sapodilla fruits were harvested in February of 2009. The fruits were identified according to quality based on the stage of physiological maturity and determined by the absence of latex (Sulladmath and Reddy, 1990). The fruits were weighed and measured, representing an average of 200 g, 10 cm of length, 176 N of firmness and 0.27 g of malic acid/100 g of fresh pulp at the time of harvest.

Treatment with 1-MCP

The fruits were divided into 32 lots of 40 fruits; each lot received a single treatment. The doses applied of 1-MCP in each treatment were as follows: 0 (control), 0.2, 0.5 or 1.0 µL/L of 1-MCP for 12 and 24 h. Treatments were applied in closed containers (0.07 m^3) at 15 and 25°C. Each treatment was performed twice. The commercial product SmartFresh® (Rohm and Haas, USA) in powder form (0.14% of the active ingredient), was the source of 1-MCP. The doses were calculated based on fruit weight and volume of the container, taking into account that 1.6 g of powder provides 1.0 µL/L of 1-MCP at 1.0 m^3. The chemical compound was dissolved in a pre-weighed flask with 25 mL of distilled water at 40°C (Akbudak et al., 2009). Subsequently, each flask was placed alongside the fruit, inside a sealed container. The period between harvest and the initiation of treatment was two days. During this period the fruits were stored at 25°C.

After treatments, the fruits of each lot were removed from the containers and maintained, until ripening, at 25°C and 85 to 95% RH. During the ripening process, certain parameters were measured as follows: Respiration rate, production of ethylene, percentage of mature fruits and percentage of weight loss. When the fruits reached ripeness-consumption stage (this being characterized by the time the fruits were soft to the touch), five fruits per treatment were removed from the containers in order to measure whole fruit firmness, titratable acidity, total soluble solids, reducing sugars and color of the pulp.

Respiration and ethylene production rates

Respiration and ethylene production were measured daily using the same fruits of each treatment. Three fruits from each replication were sealed for 2 h at 25°C in 2 L plastic containers prior to sampling. A 2 mL gas sample was withdrawn by a syringe through a rubber septum and analyzed by a gas chromatograph (Varian Star model 3400, Walnut Creek, CA, USA). Carbon dioxide and ethylene were determined using a thermal conductivity detector (TCD) and flame ionization detector (FID) respectively, with a Porapak Q column. Injector and detector temperatures were both set at 250°C, and an isothermal program was run at 30°C. Helium was used as the carrier gas at a flow rate of 1 ml/min. Based on

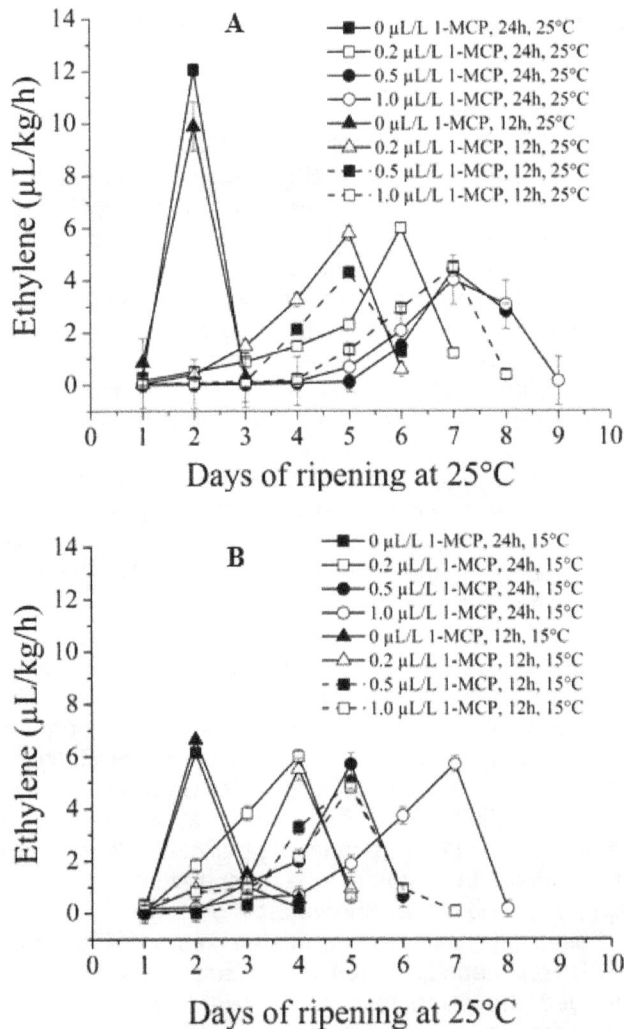

Figure 1. Ethylene production in sapodilla fruits treated with varying concentrations of 1-MCP with subsequent fruit ripening at 25°C. Average value of three determinations ± standard deviation. (A) Treated fruit at 25°C; (B) Treated fruit at 15°C.

areas of standard gases, concentrations of carbon dioxide and ethylene were calculated. Respiration rate was expressed as mL/kg/h while ethylene production rate was expressed as µL/kg/h.

Ripening and physicochemical characteristics

In order to determine the degree of ripeness in which the fruits reach the best edible quality, at each day of evaluation five fruits from each treatment were evaluated by a non-trained panel. Panelists were volunteers experienced in fruit tasting and selected from among the staff that work at the institute.

The accumulated weight losses were measured in percentage with respect to the initial weight of the fruits; ten individually weighed fruits were used each day after treatment. The measurement was made with a digital balance (OHAUS, Adventure Pro AV3102, USA) and results were expressed as percentage.

For determination of soluble solids content (SSC), titratable acidity (TA) and reducing sugar, five ripe fruits per replication were homogenized and the homogenates filtered through a cheese cloth to obtain clear juice. SSC was determined by a digital refractometer

(Model PR-1, Atago, Tokyo, Japan) and expressed as °Brix. TA was determined by titrating 5 mL of juice with 0.1 N NaOH, to pH 8.1 and expressed as grams of malic acid per 100 g. Reducing sugars were determined following the colorimetric method described by Nelson (1944) and Somogyi (1952) and expressed as grams of glucose per 100 g of pulp.

Pulp color was measured with a Minolta portable colorimeter CR-200 (Minolta Co; Ltd., Osaka, Japan). Two measurements in the equatorial area of the pulp fruit were carried out, using ten fruits from each treatment. The parameters 'L*', 'a*' and 'b*' were measured and the final results were expressed as hue angle (h= arctan[b*/a*]) (McGuire, 1992).

The fruit firmness was determined on whole, unpeeled and full-ripe stage fruit using an Instron Universal Testing Instrument (Model 4422, Canton, MA, USA) fitted with a flat-plate probe (5 cm diameter) and 50 kg load cell. The force was recorded at 5 mm deformation and was determined at two equidistant points on the equatorial region of each fruit. Ten fruits per treatment were used. The mean values of the firmness were expressed as Newton (N).

This experiment demonstrated that 1.0 µL/L treatment of 1-MCP at 25°C for 12 or 24 h, significantly delays the fruit ripening (4 to 11 days); however, the lower production of ethylene and carbon dioxide in the fruit was achieved at 24 h (Figures 1 and 2). This criterion was taken into account for choosing 1.0 µL/L treatment of 1-MCP, 25°C and 24 h, for the second experiment.

Second experiment

Based on the results of the first experiment and in order to evaluate the effect of 1-MCP in the chilling injury during ripening under refrigeration conditions of sapodilla fruit, a second experiment was designed. Sapodilla fruits were randomly divided into 8 lots of 30 fruits each with two replications; four of these lots were randomly selected as control (0 µL/L of 1-MCP for 24 h at 25°C) and four treated with 1-MCP (1.0 µL/L for 24 h at 25°C). Following treatment, the fruits were stored at 6°C, 80-90% RH for 3, 10 and 14 days and subsequent to each period of cooling the fruits of each batch were matured at 25°C for 10 days and 50 to 60% RH. During ripening, the symptoms of chilling injury were visually evaluated in the pulp. The fruits were cut longitudinally in order to identify symptoms of tissue browning and watery areas in the pulp, lignification of vascular bundles and irregular softening of fruit (Téllez et al., 2009). Data were reported as percentage of fruits that had shown the indicated symptoms.

Statistical analysis

Statgraphics® Plus, ver. 2.1 (Manugistic, Inc., Rockville, Md., USA) was utilized for the analysis of variance (ANOVA) between treatments. When the significant mean was obtained, comparison test was performed by using Tukey ($p<0.05$) (Montgomery, 2005). Analysis was carried out in triplicate.

RESULTS AND DISCUSSION

Effect of 1-MCP in ethylene production rates

Accordingly to the results obtained, the maximum ethylene production rates achieved by the fruits treated with 1-MCP (0.2, 0.5 and 1.0 µL/L), at 25°C and two exposure times evaluated, with a subsequent ripening 25°C), were significantly lower (4.0-6.0 µL/kg/h) (Figure 1A) when compared to the untreated fruits (0 µL/L of

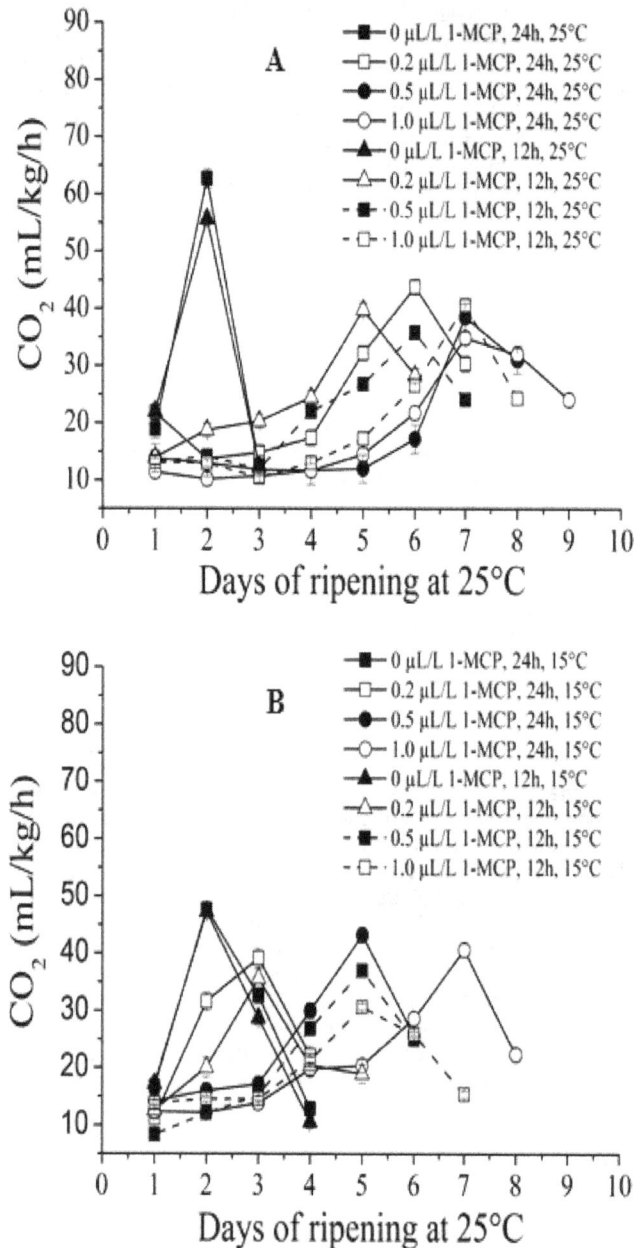

Figure 2. Carbon dioxide production in sapodilla fruits treated with varying concentrations of 1-MCP with subsequent fruit ripening at 25°C. Average value of three determinations ± standard deviation. (A) Treated fruit at 25°C; (B) Treated fruit at 15°C.

of the binding site for 1-MCP (Blankenship and Dole, 2003). Also, fruits treated with 1-MCP at 25 and 15°C achieved climacteric peak between 4 and 7 days, whereas normally untreated fruits reach this peak at 2 days. This result indicates that 1-MCP treatment at two temperatures delayed the ethylene peak in sapodilla fruit.

Several studies also have shown that 1-MCP delays the onset of ethylene peak such as in banana (Zhang et al., 2006), persimmon (Luo, 2007) and Hami melon (Li et al., 2011). It has also been shown that 1-MCP not only delayed the ethylene peak but also caused a decrease in ethylene production; this behavior has been observed in other fruits such as in pineapples (Selvarajah et al., 2001), kiwi (Boquete et al., 2004) and fresh-cut melon (Guo et al., 2011). There were no significant statistical differences due to the dose, exposure time and temperature in the treated fruits. Furthermore untreated fruits, stored in closed containers at 25°C and transferred for ripening at the same condition (25°C), showed an increase in the production of ethylene (9.9 μL/kg/h for 12 h and 12.0 μL/kg/h for 24 h), unlike the untreated fruits, stored and matured at 15°C (6.6 μL/kg/h for 12 h and 6.1 μL/kg/h for 24 h). This indicates that storing fruits in closed containers at temperatures lower than room temperature, leads to a decrease in production of ethylene. In relation to this fact, Bron et al. (2005) observed that guava fruit stored at a relatively low temperature (11°C) produced significantly less ethylene than guava fruit stored at a temperature above 21°C, approximating room temperature.

Based on these results, post-harvest application of 1-MCP on sapodilla fruit significantly decreased and delayed the climacteric peak, retarding ripening when compared to untreated fruits.

Effect of 1-MCP on respiration rates

1-MCP treatment (Figure 2A) (0.2, 0.5 and 1.0 μL/L) at 25°C and two exposure times with the subsequent ripening of sapodilla fruits (25°C) showed a significant reduction in the maximum respiratory rate (30.7 - 43.8 mL/kg/h carbon dioxide) when compared to the untreated fruits (0 μL/L of 1-MCP) (47.3 - 62.7 mL/kg/h carbon dioxide). No statistically significant differences were found due to doses of 1-MCP, temperature and exposure time between treated sapodilla. A reduction of maximum respiratory rate (30.7-37.0 mL/kg/h carbon dioxide) was observed in treated fruits exposed to 15°C for 12 h (Figure 2B) and matured afterwards (25°C), when compared to untreated fruits (47.3 mL/kg/h carbon dioxide). However, this effect was not observed in fruits treated for 24 h, which suggests that treatment time influences respiration rate. The action of 1-MCP delayed in all treatments the increase of the respiratory peak (3-7 days) when compared to untreated fruits (two days).

The reduction in respiration rate in fruits and vegetables due to the action of 1-MCP has also been

1-MCP) (9.9-12.0 μL/kg/h). There is evidence indicating that this effect may be due the 1-MCP is an inhibitor of ethylene perception in plant tissues; it is thought that it binds irreversibly to ethylene receptors (Sisler et al., 1996), reducing the production of ethylene in treated fruits (Moya-León et al., 2004). According to figure 1B, the maximum ethylene production was similar in control and 1-MCP treated fruits at 15°C. This behavior may be because the lower temperatures might decrease the affinity

Table 1. Ripening percentage, time to ripen and firmness when sapodilla fruits reached their ripening at 25°C after treatment with 1-MCP.

Doses of 1-MCP (µL/L)	Treatment temperature (°C)	Treatment time (h)	Ripening percentage* (%)	Time to ripen (days)	Firmness* (N)
0	15	12	91.6	7	11.85
0	25	24	91.6	4	11.85
0	25	12	91.6	4	11.85
0	15	24	91.6	7	11.95
0.2	15	12	91.6	8	11.85
0.2	25	24	91.6	10	11.95
0.2	25	12	91.6	7	11.85
0.2	15	24	91.6	8	11.85
0.5	15	12	91.6	8	11.85
0.5	25	24	91.6	10	11.85
0.5	25	12	95.8	7	11.85
0.5	15	24	91.6	9	11.85
1.0	15	12	95.8	9	11.95
1.0	25	24	91.6	11	11.85
1.0	25	12	91.6	11	11.85
1.0	15	24	95.8	9	11.85

*There was no significant difference at P<0.05 (Tukey test). The values are averages of three determinations.

observed in several other studies, such as in tomato (Choi and Huber, 2008), banana (Zhang et al., 2006) and persimmon (Luo, 2007). It has been suggested that changes in the respiratory rate in fruits is due to the action of ethylene (Golding et al., 1998), which causes the deterioration of horticultural products; consequently the delay in respiration rate could increase or decrease the shelf life of fruits (Perera et al., 2003).

These results indicates that application of 1-MCP on sapodilla fruit after harvest decreased and delayed maximum respiratory rate when compared to untreated fruit, which represent a noteworthy effect in delaying fruit ripening.

Effect of 1-MCP on ripening of sapodilla fruit

Fruits treated with 1-MCP took longer to reach an edible maturity (7-11 days), providing an increase in sapodilla fruit shelf life at 25°C. No significant changes were found in the percentage of maturation in treated fruits (91.6-95.8%) and non-treated fruits (91.6%) (Table 1). In this sense, it is known that 1-MCP blocks ethylene action and delays fruit ripening (Blankenship and Dole, 2003). Posteriorly following the inhibitory action of ethylene the fruits ripen. This effect could be explained on the basis that new ethylene receptors are constantly formed in the fruit during storage (Liu et al., 2005). The effectiveness of 1-MCP in delaying fruit ripening has also been observed in several studies, such as in banana (Pelayo et al., 2003), plums (Salvador et al., 2003), papaya (Moya-León et al., 2004), pears (Liu et al., 2005) and sapodilla fruit

(Morais et al., 2006).

It should be noted that treated and untreated fruits demonstrated similar firmness when mature for consumption (11.85-11.95 N) with no significant difference between them. This data corresponds to the firmness when sapodilla fruit normally matures (data not shown). In this regard, Morais et al. (2006) found that when sapodilla fruits treated with 1-MCP reach ripeness, firmness values are similar to control fruits without significant differences between them. With respect to this matter, evidence suggests that the characteristic firmness value of a mature fruit is achieved because the ethylene stimulates the activity of enzymes that degrade the cell wall (Abeles et al., 1992), which explains why the treated sapodilla fruit had a normal maturation.

It is worth mentioning that when treated and untreated fruits reached their maturity, the fruits showed a normal pulp with the characteristics appreciated by consumers, such as color brownish-orange, soft and smooth texture, juicy and sweet taste.

On the other hand, fruits exposed to 1-MCP at the evaluated temperatures (25 and 15°C) (Figure 3A and 3B) showed an increase in weight loss when exposed to ripen at 25°C. Furthermore, fruits treated at 25°C, showed more weight loss than those treated at 15°C.

In relation to time of treatment, fruits at 25°C, exposed to 0.2 and 0.5 µL/L of 1-MCP for 24 h, showed a loss of weight (15.7-17.0%) at maturity, significantly greater than those treated for 12 h (10.2-12.0%), indicating that longer treatment times causes greater weight loss. In this sense, Calvo (2004) suggested that the effect of 1-MCP in weight loss has not been entirely elucidated and Porat et

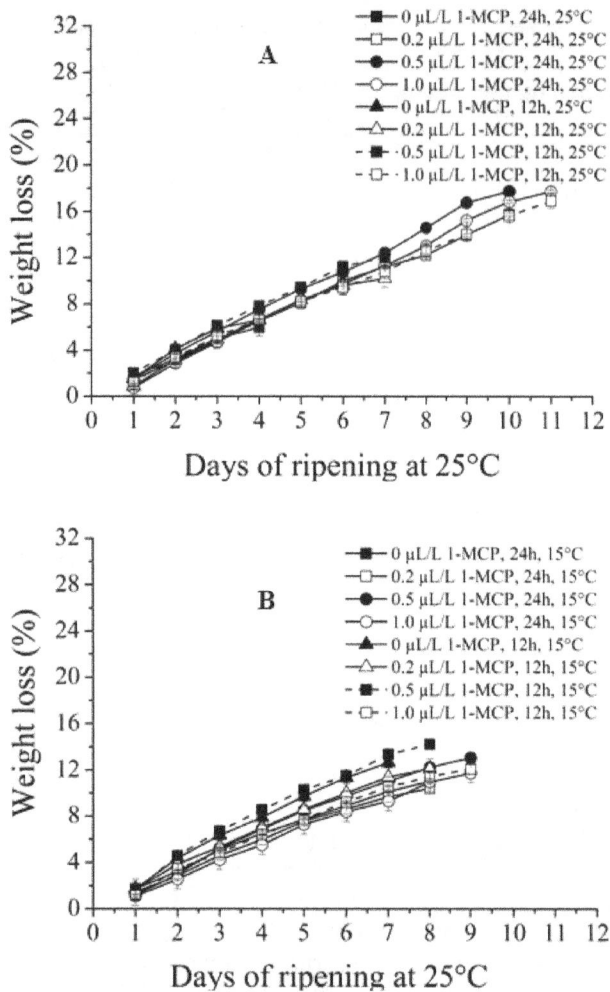

Figure 3. Weight loss (%) in sapodilla fruits treated with varying concentrations of 1-MCP with subsequent fruit ripening at 25°C. Average value of three determinations ± standard deviation. (A) Treated fruit at 25°C; (B) Treated fruit at 15°C.

lightness (50.0-50.8 L*) (Table 2).

The total soluble solids content of treated fruit with 0.5 and 1.0 μL/L of 1-MCP, at the times and temperatures evaluated, was significantly higher (16.6-17.1 °Brix) when compared to untreated fruit (15.8 °Brix), except the dose of 0.5 μL/L at 15°C for 12 h (16.3 °Brix). This increase in the content of soluble solids due to the effect of 1-MCP has been also observed in apples (Fan et al., 1999) and papaya (Hofman et al., 2001). However, Watkins et al. (2000) observed that 1-MCP treatment did not influence the sugar content in apples, which indicates that exposing fruits to 1-MCP does not affect the metabolism of sugars (Salvador et al., 2003). These results indicate that the effect of 1-MCP on the total soluble solids content is variable. Consequently, it has been suggested that this effect in treated fruits could be attributed to low respiration rates; however, it may also depend on the type of crop and storage conditions (Watkins et al., 2000).

Other studies have demonstrated that 1-MCP does not affect the development of acidity in fruits such as plums (Menniti et al., 2004), peach and apple (Mir et al., 2001; Salvador et al., 2003). In addition, significant differences in color of treated apricots were not detected (Fan et al., 2000; Botondi et al., 2003).

The results obtained in this study indicate that postharvest application of 1-MCP on sapodilla fruit only produced a significant effect when high doses were used, which increased total soluble solids (°Brix), without modifying the other physicochemical characteristics of the pulp. The similarity in the results between mature fruits treated and untreated indicates that although 1-MCP delays the physiological ripening of sapodilla fruit, the characteristics of a mature fruit are practically unmodified, producing fruits with highly valued characteristics of quality.

al. (1999) found that 1-MCP had no effect on weight loss in fruits such as orange, whereas Cao et al. (2012) found that 1-MCP reduced weight loss in green bell peppers.

One of the most important findings, after analyzing the results, is that treatments with 1.0 μL/L of 1-MCP at 25°C for 12 or 24 h significantly increased the time to ripen of sapodilla fruit up to 11 days, while fruits not treated under the same conditions had a time to ripen of 4 days only.

Effect of 1-MCP on the physicochemical properties of mature sapodilla fruit

The physicochemical characteristics of fruits at maturity at 25°C following treatment did not demonstrate significant differences in titratable acidity (0.12-0.13 g of malic acid/100 g), reducing sugars (9.9-10.0 g/100 g), pulp color measured as hue angle (79.5-80.3 °hue) and

Effect of 1-MCP on the development of chilling injury symptoms

Fruits treated and untreated with 1-MCP stored at 6°C for short periods of time (3 days) showed chilling injury (6.67% of fruits) (Figure 4). The damages found were as follows: Irregular pulp softening, hardening and browning of vascular tissue in various areas of the pulp. The fruits treated with 1-MCP (6.67 and 20% for 10 and 14 days, respectively) and refrigerated for a longer period of time thereafter (10 and 14 days) at 6°C showed a significant reduction of chilling injury symptoms when reached maturity, compared to refrigerated (6°C) and untreated fruits (20 and 30% for 10 and 14 days), which demonstrates the effectiveness of 1-MCP in reducing chilling injury in sapodilla fruit. It has also been suggested that chilling injury symptoms are less, as 1-MCP causes a lower production of ethylene in treated fruits, such as found in apple, avocado and pineapple (Watkins, 2006).

Table 2. Effect of 1-MCP in the physicochemical characteristics of ripe sapodilla fruits.

Doses of 1-MCP (µL/L)	Treatment temperature (°C)	Treatment time (h)	Titratable acidity (g malic acid/100 g)	Reducing sugars (g/100 g)	Soluble solids content (°Brix)	Color (°hue)	Lightness (L*)
0	15	12	0.13[a]	10.0[a]	15.8[ab]	79.9[a]	50.0[a]
0	25	24	0.13[a]	9.9[a]	15.8[ab]	80.0[a]	50.8[a]
0	25	12	0.12[a]	9.9[a]	15.8[ab]	80.0[a]	50.1[a]
0	15	24	0.12[a]	9.9[a]	15.8[ab]	79.9[a]	50.0[a]
0.2	15	12	0.12[a]	10.0[a]	15.5[a]	79.6[a]	50.2[a]
0.2	25	24	0.12[a]	9.9[a]	15.7[ab]	80.0[a]	50.3[a]
0.2	25	12	0.13[a]	9.9[a]	15.5[a]	80.3[a]	50.5[a]
0.2	15	24	0.12[a]	9.9[a]	15.7[ab]	79.7[a]	50.2[a]
0.5	15	12	0.12[a]	9.9[a]	16.3[bc]	79.6[a]	50.7[a]
0.5	25	24	0.13[a]	10.0[a]	16.7[cd]	80.0[a]	50.7[a]
0.5	25	12	0.13[a]	9.9[a]	16.6[cd]	80.1[a]	50.2[a]
0.5	15	24	0.13[a]	10.0[a]	16.6[cd]	79.7[a]	50.6[a]
1.0	15	12	0.13[a]	10.0[a]	16.9[cd]	79.5[a]	50.8[a]
1.0	25	24	0.13[a]	10.0[a]	17.0[d]	79.7[a]	50.5[a]
1.0	25	12	0.13[a]	10.0[a]	17.1[d]	79.9[a]	50.2[a]
1.0	15	24	0.13[a]	10.0[a]	16.8[cd]	79.7[a]	50.8[a]

[a, b, c, d]Different letters in column show significant difference at $P<0.05$ (Tukey test). The values are averages of three determinations.

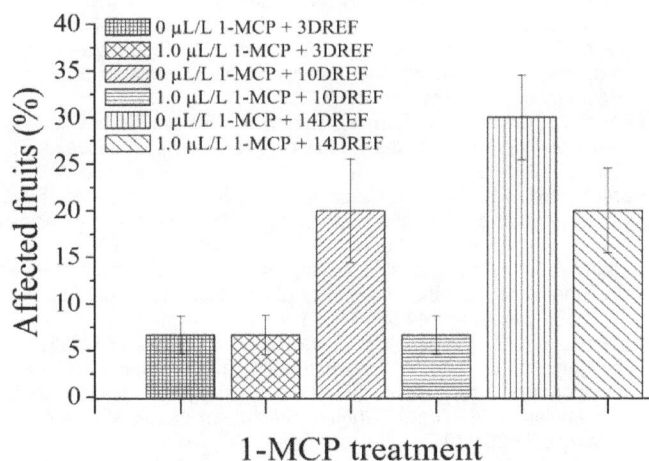

Figure 4. Chilling injury severity in fruits of sapodilla treated with 1-MCP, stored at 6°C for 3, 10 and 14 days and ripens at 25°C. Average value of three determinations ± standard deviation. DREF: Days of refrigeration.

Conclusions

This study demonstrates that using 1-MCP significantly reduces and delays maximum production of ethylene and carbon dioxide in sapodilla fruits. As a result of the effect of 1-MCP, fruit ripening was delayed by an average of 4 to 11 days, which represents a significant 7 day increase in shelf life at room temperature (25°C). Also, fruits treated with 1-MCP showed a normal ripeness, with no significant differences in quality parameters between treated and untreated fruits; excepting for some treatments where a slight increase of total soluble solids content and weight loss were found. According to the results, it is possible to infer that the use of 1-MCP is a suitable alternative to extend the shelf life of sapodilla fruits, which in turn, could greatly facilitate its commercialization. In this regard, the recommended doses of 1-MCP in sapodilla fruits are as follows: 1.0 µL/L of 1-MCP at 25°C for 12 or 24 h of treatment. In addition, treatment with 1-MCP could reduce the symptoms of chilling injury when sapodilla fruits are going to be stored at 6°C for 10 and 14 days, which represents a beneficial effect of 1-MCP in sapodilla fruit conservation and ensuing positive economic implications.

ACKNOWLEDGMENTS

The authors express their gratitude to the General Directorate of Higher Education and Technology and Foundation Pablo Garcia of the State of Campeche for the support provided.

REFERENCES

Abeles FB, Morgan PW, Saltveit JR ME (1992). Ethylene in plant biology. San Diego, California: Academic Press. pp. 63-222

Adkins MF, Hofman PJ, Stubbings BA, Macnish AJ (2005). Manipulating avocado fruit ripening with 1-methylcyclopropene. Postharvest Biol. Technol. 35:33-42.

Akbudak B, Ozer MH, Erturk U, Cavusoglu S (2009). Response of 1-methylcyclopropene treated "Granny smith" apple fruit to air and controlled atmosphere storage conditions. J. Food Quality 32:18-33.

Argenta LC, Fan XT, Mattheis JP (2003). Influence of 1-methylcyclopropene on ripening, storage life, and volatile production by 'd'Anjou' cv. pear fruit. J. Agric. Food Chem. 51:3858-3864.

Balerdi CF, Shaw PE (1998). Sapodilla, sapote and related fruit. In: Shaw PE, Chan HT, Nagy S (Eds) Tropical and subtropical fruits, Auburndale: AgScience pp. 78-136.

Blankenship SM, Dole JM (2003). 1-Methylcyclopropene: a review. Postharvest Biol. Technol. 28:1-25.

Boquete EJ, Trinchero GD, Fraschina AA, Vilella F, Sozzi GO (2004). Ripening of "Hayward" kiwifruit treated with 1-methylcyclopropene after cold storage. Postharvest Biol. Technol. 32:57-65.

Botondi R, Desantis D, Bellincontro A, Vizovitis K, Mencarelli F (2003). Influence of ethylene inhibition by 1-methylcyclopropene on apricot quality, volatile production, and glycosidase activity of low and high aroma varieties of apricots. J. Agric. Food Chem. 51:1189-1200.

Bron IU, Ribeiro RV, Cavalini FC, Jacomino AP, Trevisan MJ (2005). Temperature-related changes in respiration and Q_{10} coefficient of guava. Sci. Agric. (Piracicaba, Braz.) 62:458-463.

Broughton WH, Wong HC (1979). Storage conditions and ripening of chiku fruits (Achras zapota). Sci. Hortic. 10:377-385.

Calvo G (2004). Efecto del 1-metilciclopropeno en peras cv. Williams cosechadas con dos estados de madurez. RIA, Rev. Investig. Agropec. 33: 3-26.

Calvo G, Sozzi G (2009). Effectiveness of 1-MCP treatments on "Bartlett" pears as influenced by the cooling method and the bin material. Postharvest Biol. Technol. 51:49-55.

Cao S, Yang Z, Zheng Y (2012). Effect of 1-methylcyclopropene on senescence and quality maintenance of green bell pepper fruit during storage at 20 °C. Postharvest Biol. Technol. 70:1-6.

Choi ST, Huber DJ (2008). Influence of aqueous 1-methylcyclopropene concentration, immersion duration, and solution longevity on the postharvest ripening of breaker-turning tomato (Solanum lycopersicum L.) fruit. Postharvest Biol. Technol. 49: 147-154.

Fan X, Argenta L, Mattheis JP (2000). Inhibition of ethylene action by 1-methylcyclopropene prolongs storage life of apricots. Postharvest Biol. Technol. 20:135-142.

Fan X, Blankenship SM, Mattheis JP (1999). 1-Methylcyclopropene inhibits apple ripening. J. Am. Soc. Hort. Sci. 124:690-695.

Ganjyal GM, Hanna MA, Devadattam DSK (2003). Processing of zapota (sapodilla): Drying. J. Food Sci. 68:517-520.

Golding J, Shearer D, Wyllie S, Mcglasson W (1998). Application of 1-MCP to identify ethylene-dependent ripening processes in mature banana fruit. Postharvest Biol. Technol. 14:87-98.

Guo Q, Cheng L, Wang J, Che F, Zhang P, Wu B (2011). Quality characteristics of fresh-cut 'Hami' melon treated with 1-methylcyclopropene. Afr. J. Biotechnol. 10:18200-18209.

Hofman PJ, Jobin DM, Meiburg GF, Macnish AJ, Joyce DC (2001). Ripening and quality responses of avocado, custard apple, mango, and papaya fruit to 1-methylcyclopropene. Aust. J. Exp. Agr. 41: 567-572.

Indian Horticulture Database (2010). National Horticulture Board. Ministry of Agriculture, Government of India 85, Institutional Area, sector 18, Gurgaon-122015 India. pp. 118-123.

Kashimura Y, Hayama H, Ito A (2010). Infiltration of 1-methylcyclopropene under low pressure can reduce the treatment time required to maintain apple and Japanese pear quality during storage. Postharvest Biol. Technol. 57:14-18.

Kunyamee S, Ketsa S, Imsabai W, Van Doorn W (2008). The transcript abundance of an expansin gene in ripe sapodilla (Manilkara zapota) fruit is negatively regulated by ethylene. Functional Plant Biol. 35:1205-1211.

Lakshminarayana S (1979). Sapodilla and pickly pear. In: Magy S, Shaw PE (Eds) Tropical and subtropical fruits, AVI Publishing Connecticut Inc. Wesport.

Li XW, Jin P, Wang J, Zhu X, Yang HY, Zheng YH (2011). 1-methylcyclopropene delays postharvest ripening and reduces decay in Hami melon. J. Food Quality 34:119-125.

Liu H, Jiang W, Zhou L, Wang B, Luo Y (2005). The effects of 1-methylcyclopropene on peach fruit (Prunus persica L. Cv. Jiubao) ripening and disease resistance. Int. J. Food Sci. Tech. 40:1-7.

Luo Z (2007). Effect of 1-methylcyclopropene on ripening of postharvest

persimmon (Diospyros Kaki L.) fruit. LWT–Food Sci. Technol. 40:285-291.

Luo Z, Xie J, Xu T, Zhang L (2009). Delay ripening of "Qingnai" plum (Prunus salicina Lindl.) with 1-methylcyclopropene. Plant Sci. 177:705-709.

Ma J, Luo XD, Protiva P, Yang H, Ma CY, Basile MJ, Weinstein IB, Kennelly EJ (2003). Bioactive novel polyphenols from the fruit of Manilkara zapota (sapodilla). J. Nat. Prod. 66:983-986.

Manenoi A, Bayogan ER, Thumdee S, Paull RE (2007). Utility of 1-methylcyclopropene as a papaya postharvest treatment. Postharvest.Biol. Technol. 44:55-62.

Manganaris GA, Crisosto CH, Bremer V, Holcroft D (2008). Novel 1-methylcyclopropene immersion formulation extends shelf life of advanced maturity 'Joanna Red' plums (Prunus salicina Lindell). Postharvest. Biol. Technol. 47:429-433.

Mcguire RG (1992). Reporting of objective color measurements. HortScience 27:1254-1255.

Menniti AM, Gregori R, Donati (2004). 1-methylcyclopropene retards postharvest softening of plums. Postharvest Biol. Technol. 31:269-275.

Mir NA, Curell E, Khan N, Whitaker M, Beaudry RM (2001). Harvest maturity, storage temperature, and 1-MCP application frequency alter firmness retention and chlorophyll fluorescence of "Redchief delicious" apples. J. Amer. Soc. Hort. Sci. 126:618-624.

Montgomery D (2005). Diseño y análisis de experimentos. Segunda edición, LimusaWiley, México. pp. 194-201

Morais PLD, Oliveira LC, Alves RE, Alves JD, Paiva A (2006). Amadurecimiento de sapoti (Manilkara zapota L.) submetidoao 1-metilciclopropeno. Rev. Bras. Frutic. (Jaboticabal) 28:369-373.

Moya-León MA, Moya M, Herrera R (2004). Ripening of mountain papaya (Vasconcellea pubescens) and ethylene dependence of some ripening events. Postharvest Biol. Technol. 34:211-218.

Nelson NJ (1944). A photometric adaptation of the Somogyi method for the determination of the glucose. J. Biol. Chem. 153:375-380.

Pelayo C, Vilas BE, Benichou M, Kader A (2003).Variability in responses of partially ripe bananas to 1-methylcyclopropene. Postharvest. Biol. Technol. 28:75-85.

Perera CO, Balchin L, Baldwin E, Stanley R, Tian M (2003). Effect of 1-Methylcyclopropene on the Quality of Fresh-cut Apple Slices. J. Food Sci. 68:1910-1914.

Porat R, Weiss B, Cohen L, Daus A, Goren R, Droby S (1999). Effects of ethylene and 1-methylcyclopropene on the postharvest qualities of shamouti oranges. Postharvest Biol. Technol.15: 155-163.

Qiuping Z, Wenshui X, Jiang Y (2006). Effects of 1-Methylcyclopropene treatments on ripening and quality of harvested sapodilla fruit. Food Technol. Biotechnol. 44:535-539.

Sabban-Amin R, Feygenberg O, Belausov E, Pesis E (2011). Low oxygen and 1-MCP pretreatments delay superficial scald development by reducing reactive oxygen species (ROS) accumulation in stored 'Granny Smith' apples. Postharvest Biol. Technol. 62:295-304.

Sagarpa (2011). Sistema de información agroalimentaria de consulta (SIACON). Centro de Estadística Agropecuaria (CEA). Versión 1.1. D.F., México.

Saltveit ME (2002). The rate of ion leakage from chilling-sensitive tissues does not immediately increase upon exposure to chilling injury. Postharvest Biol. Technol. 26:295-304.

Salvador A, Cuquerella J, Martínez-Jávega JM (2003). 1-MCP treatment prolongs postharvest life of santa rosa plums. J. Food Sci. 68:1504-1510.

Sauri-Duch E, Centurión-Yah AR, Vargas-Vargas L (2010). Alternative tropical fruits in order to increment the offer to European market. Acta. Hort. 864:305-316.

Selvarajah S, Bauchot AD, Jhon P (2001). Internal browning in cold stored pineapples is suppressed by a postharvest application of 1-methylcyclopropene. Postharvest. Biol. Technol. 23:167-170.

Sisler EC, Dupille E, Serek M (1996). Effect of 1-methylcyclopropene and methylenecyclopropane on ethylene binding and ethylene action on cut carnations. Plant Growth Regul. 18:79-86.

Sisler EC, Serek M (1997). Inhibitors of ethylene responses in plants at the receptor level: Recent developments. Plant Physiol. 100:577-582.

Somogy M (1952). Notes on sugar determinations. J. Biol. Chem. 195:

19-23.

Sulladmath UV, Reddy NMA (1990). Sapota. In: Bose TK, Mitra SK (Eds) Fruits: Tropical and Subtropical, Naya Prokash, Calcutta, India.

Téllez PP, Saucedo VC, Arévalo GML, Valle GS (2009). Ripening of mamey fruits (*Pouteria sapota* Jacq.) treated with 1-methylcyclopropene and refrigerated storage. CyTA-J. Food 7:45-51.

Watkins CB (2006). The use of 1-methylcyclopropene (1-MCP) on fruits and vegetables. Biotechnol. Adv. 24: 389-409.

Watkins CB, Nock JF, Whitaker BD (2000). Response of early, mid and late season apple cultivars to postharvest application of 1-methylcyclopropene under air and controlled atmosphere storage conditions. Postharvest Biol. Technol. 19:17-32.

Zhang M, Jiang Y, Jiang W, Liu X (2006). Regulation of ethylene synthesis of harvested banana fruit by 1-methylcyclopropene. Food Technol. Biotechnol. 44:111-115.

3

Examination of some physiological and biochemical changes based on ripening in fruits of different types of apricots

Zehra Tugba Abacı[1] and **Bayram Murat Asma[2]**

[1]Department of Food Engineering, Faculty of Engineering, Ardahan University, 75000 Ardahan-Turkey.
[2]Department of Biology, Faculty of Science and Literature, Inonu University, 44069 Malatya-Turkey.

This study was performed to determine some of the physiological and biochemical changes that occurred during ripening period in fruits of different types of apricots. In the fruits of six types of apricots (Hasanbey, Canino, Turfanda Eskimalatya, Hacihaliloglu, Özal, and Levent) collected during green, mature green and ripe periods, amounts of total soluble solids (TSS) (Brix°), titratable acidity, chlorophyll a (Cha), chlorophyll b (Chb) and total chlorophylls (Ch) were determined. During ripening, the highest and lowest increase in TSS occurred in apricot types called 'Hacihaliloglu' and 'Turfanda Eskimalatya', respectively. In all three ripening periods, it was found that 'Hacihaliloglu' had the lowest acid content. During ripening, decreases in amounts of Cha, Chb and Ch were observed. Differences between apricot types in terms of decrease in chlorophyll amounts was detected and the highest difference occurred in apricot type called 'Turfanda Eskimalatya.'

Key words: Apricot, ripening, Brix, chlorophyll.

INTRODUCTION

Although commercial apricot production in the world involves quite extensive areas including Asia, Europe and America, worldwide apricot production is extremely low. According to the results of the World Agriculture Organization, worldwide fresh apricot production varies between 3 to 3.5 million tons (Anonymous, 2011). Turkey is the first producer of fresh and dried apricot, followed by Iran, Pakistan, Uzbekistan, France, Italy and Spain. In Turkey, apricot production is about 650,000 tons (Anonymous, 2010).

Like in many other fruit types, fruit development in apricot starts with flowering. After pollination and fertilization, developments starting in ovule spread to other tissues. Apricot fruit has double-sigmoid growth curve. In the fruit, there are three different developmental stages that are first fast, then slow and fast again lastly

(Karacali, 1990).

Ripening of fruit includes a serial complex biochemical, physiological and structural changes such as starch hydrolysis, degradation of chlorophyll, production of carotenoid, anthocyanin and phenolic substance, accumulation of sugar and organic acid, modifications in structures and compositions of cell wall polysaccharides, color, taste and texture changes (Speirs and Brady, 1991; Giovannoni, 2001, 2004; Gulao and Olivera, 2008; Borsani et al., 2009).

First observable sign of ripening is the color change due to degradation of chlorophyll (Seymour et al., 1993). During ripening, while chlorophyll amount in fruit tissues decreases quickly, amount of carotenoid increases (Merzlyak et al., 1999). Chloroplasts found in green fruits are converted into chromoplasts by degradation of chlorophylls and carotenoid synthesis during progress of ripening (Hortensteiner, 2006). While surface color of fruit is initially green, it starts to be yellow with ripening due to degradation of chlorophyll (Bureau et al., 2009).

Fruit size, titratable acidity and total soluble solids (TSS) content depend on type, environment and cultivation conditions (Kingston, 1992). Changes in TSS are quite important for fruit taste development. In most fruits, ripening and fruit quality are determined by sugar content (Villanueva et al., 2004).

High acid content generally reduces fruit quality; however, intermediate acid concentration causes tastier fruits (Silva et al., 2004). Different organic acids are found in different fruit types. For example, citric and malic acids are basic organic acids found in cirtus and melon, and apple and loquat, respectively (Yamaki, 1989; Flores et al., 2001; Chen et al., 2009). Although titratable acidity of fruits decreases during different ripening periods, their pH and TSS increase (Jiménez et al., 2011).

In this study, amounts of TSS, titratable acidity, chlorophyll a (Cha), chlorophyll b (Chb) and total chlorophyll (Ch) were examined in six apricot types whose fruit samples were collected during green, ripe-green and ripe periods.

MATERIALS AND METHODS

Plant material

The apricot samples used in this study were taken from apricot collection garden found in Malatya Inonu University, Apricot Research and Application Center. Green and mature green fruit samples were collected after 30 and 60 days before flowering, respectively, and ripe fruit samples were collected before harvest from early-ripening apricot types; Canino, Turfanda Eskimalatya, Hasanbey and Hacihaliloglu. By considering fruit developmental periods, green and mature green fruit samples were collected after 30 and 90 days before flowering, respectively, and ripe fruit samples were taken before harvest from late-ripening apricot types; Levent and Ozal.

Total soluble solid content (TSS)

TSS in fruits was measured of the juice obtained from the pulp of 10 fruits by digital Brix refractometer (Asma and Ozturk, 2005).

Titratable acidity

10 ml juice was completed to 100 ml with distilled water and titrated by 0.1 N NaOH until pH 7.0. Titration results were calculated in terms of malic acid (Cemeroglu, 1992).

Pigment analysis

Extraction and purification of pigments from fruit samples were performed by De-Kok and Graham (1989) method. 1 g of each sample was homogenized by grinding with 500 cc acetone for 5 min and left in shaking incubator for 30 min. After that, they were stored at +4°C for 24 h. After filtering the samples taken out from

refrigerator and addition of 1/5 volume distilled water, they were left in shaking incubator for 15 min and centrifuged at 300 rpm for 10 min. Absorbance of supernatants was measured at 662 and 645 nm for Cha, Chb and total Ch and calculated by using following standard equation.

Statistical analysis

Statistical analysis was performed using SPSS 10.0 software. Duncan's test (1955) was used for significance control ($P < 0.05$) following variance analysis.

RESULTS AND DISCUSSION

Total soluble solid content (TSS, °Brix)

TSS contents of apricot types were found statistically similar in green and mature green periods ($P < 0.05$) (Figure 1). While the highest TSS was found in 'Hacihaliloglu' and 'Ozal' (1.9%), the lowest TSS was found in 'Levent' (1.2%) in green period, slight increase was observed in mature green period, the highest TSS was found in 'Canino' (3.0%) and the lowest TSS was observed in 'Ozal' and 'Hacihaliloglu' (2.5%) ($P < 0.05$). The fastest increase in TSS occurred in mature green period and it was found that the highest and the lowest rates of TSS were observed in 'Hacihaliloglu' (22.0%) and 'Turfanda Eskimalatya' (14.1%), respectively. During fruit ripening, the highest TSS increase was found in 'Hacihaliloglu' and the lowest increase was observed in 'Turfanda Eskimalatya'.

In the studies carried out with different fruit types, increase in TSS was reported during fruit ripening period (Wu et al., 2005; Karlidag and Bolat, 2007; Prinsi et al., 2011). Jiménez et al. (2011) stated that TSS in gulupa (*Passiflora edulis*) fruits increased from 13.5 to 17.4% during progress of ripening.

Titratable acidity

While titratable acid content in fruits is the highest in mature green period, it has the lowest content in ripe period (Figure 2). Statistically significant differences between acidity were not observed in green fruit samples collected from different apricot types after 30 days before flowering ($P < 0.05$). In this period, while 'Canino' (2.88%) and 'Hacihaliloglu' (1.94%) had the highest and the lowest acid contents, respectively, 'Canino' (3.11%) and 'Turfanda Eskimalatya' (2.97%) had the highest content and 'Hacihaliloglu' (1.99%) had the lowest acid content in mature green period. A significant decrease in acid contents of fruits was observed in ripe period ($P < 0.05$) and it was found that 'Canino' (1.16%) and 'Levent' (1.15%) had the highest acid contents and 'Hacihaliloglu' (0.45%) had the lowest content.

Nunes et al. (2009) reported that the acid content in fruits of guava (*Psidium guajava*) and plum (*Prunus*

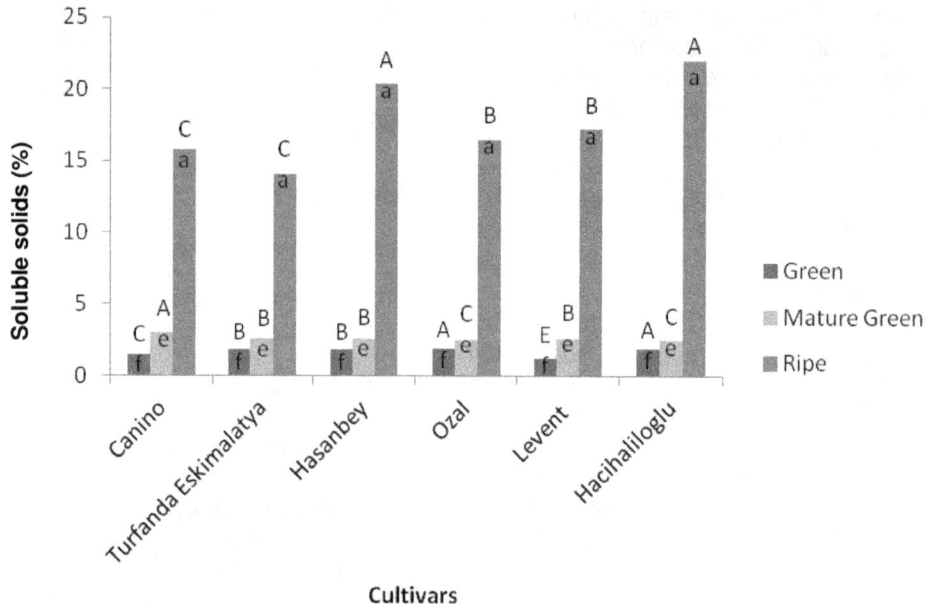

Figure 1. Soluble solids of different cultivars during the same ripening stages (capital letters) and during different ripening stages (small letters). Data followed by different letters are significiantly different from each other (P < 0.05) according to Duncan's test.

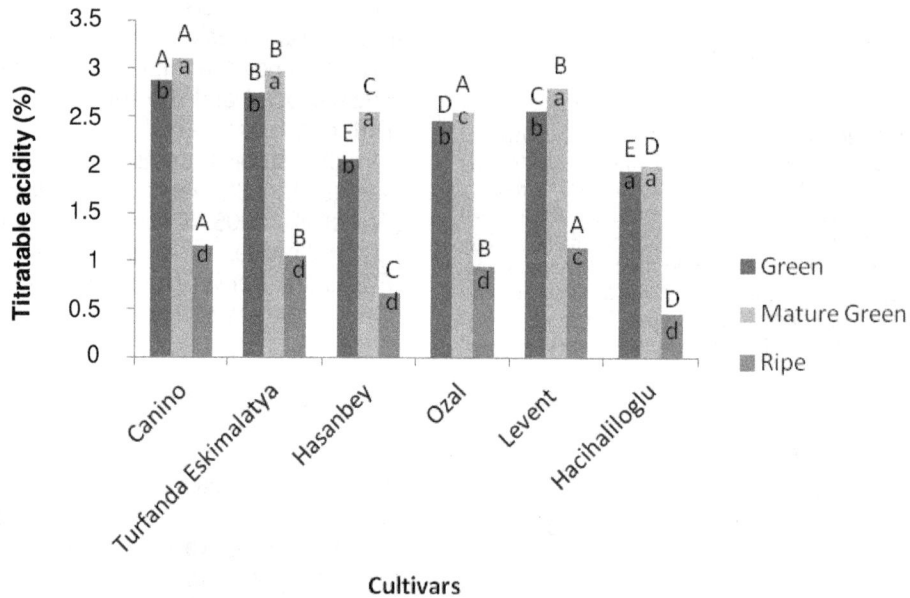

Figure 2. Titratable acidity of different cultivars during the same ripening stages (capital letters) and during different ripening stages (small letters). Data followed by different letters are significiantly different from each other (P < 0.05) according to Duncan's test.

domestica) initially increased significantly and then decreased during ripening period. Similarly, it is stated that acid content of gulupa (*P. edulis*) fruits which is 4.86% in green period, decreases to 2.51% in ripe period and the source of this decrease is organic acid consumption due to increased respiration rate in fruit during ripening period (Jiménez et al., 2011).

Pigment analysis

It was found that Cha, Chb and Ch contents in all apricot

Table 1. Mean Cha, Chb and ch contents of different cultivars during the same ripening stages (*unripe, #half ripe, °ripe) and during different ripening stages (letters on right).

Apricot cultivar	Ripening stage	Chlorophyll a (µg/g)	Chlorophyll b (µg/g)	Total chlorophyll (µg/g)
Hacihaliloglu	Green	*^{b}1.69 ± 0.003^a	*^{e}0.11 ± 0.006^a	*^{c}1.80 ± 0.003^a
	Mature green	#^{c}1.26 ± 0.003^b	#^{c}0.03 ± 0^b	#^{c}1.29 ± 0.003^b
	Ripe	°^{c}0.67 ± 0.005^d	°^{c}0.008 ± 0.01^d	°^{c}0.67 ± 0.01^d
Levent	Green	*^{d}1.24 ± 0.01^a	*^{b}0.89 ± 0.04^a	*^{b}2.13 ± 0.04^a
	Mature green	#^{c}1.23 ± 0.01^a	#^{b}0.23 ± 0.01^b	#^{b}1.46 ± 0.02^b
	Ripe	°^{a}1.15 ± 0.008^b	°^{a}0.06 ± 0.02^d	°^{a}1.21 ± 0.01^c
Hasanbey	Green	*^{a}1.78 ± 0.02^a	*^{c}0.44 ± 0.0003^a	*^{a}2.22 ± 0.02^a
	Mature green	#^{a}1.55 ± 0^b	#^{a}0.30 ± 0.01^b	#^{a}1.85 ± 0.01^b
	Ripe	°^{c}0.60 ± 0.02^d	°^{d}-0.07 ± 0.02^d	°^{d}0.53 ± 0.01^d
Turfanda Eskimalatya	Green	*^{d}1.33 ± 0.01^a	*^{a}1.16 ± 0.003^a	*^{d}2.49 ± 0.01^a
	Mature green	#^{b}1.37 ± 0.01^a	#^{c}0.08 ± 0.01^c	#^{b}1.45 ± 0.01^a
	Ripe	°^{d}0.56 ± 0.01^d	°^{d}-0.01 ± 0.003^d	°^{d}0.55 ± 0.01^d
Canino	Green	*^{c}1.46 ± 0.03^a	*^{b}0.81 ± 0.04^a	*^{a}2.27 ± 0.01^a
	Mature green	#^{d}1.09 ± 0.01^b	#^{a}0.37 ± 0.03^b	#^{b}1.46 ± 0.05^b
	Ripe	°^{d}0.58 ± 0.02^d	°^{b}0.01 ± 0.02^d	°^{c}0.59 ± 0.02^d
Ozal	Green	*^{b}1.60 ± 0.005^a	*^{d}0.27 ± 0.003^a	*^{c}1.87 ± 0.005^a
	Mature green	#^{c}1.22 ± 0.03^b	#^{d}-0.08 ± 0.03^d	#^{d}1.14 ± 0.01^b
	Ripe	°^{b}0.91 ± 0.006^c	°^{c}0.006 ± 0.02^c	°^{b}0.91 ± 0.02^d

Values are means ± standard deviation (SD) of three replications. Data followed by different letters are significantly different from each other ($P < 0.05$) according to Duncan's test.

types displayed statistically significant decrease with ripening (Table 1) ($P < 0.05$). While Cha, Chb and Ch contents of fruits were found highest in green period, it had the lowest content in ripe period. While the one having the highest chlorophyll content among apricot types was 'Hasanbey' (1.78 µg/g) in green period, 'Turfanda Eskimalatya' (0.56 µg/g) was the one having lowest content in ripe period. It was determined that 'Turfanda Eskimalatya' having 1.33 µg/g Cha, 1.16 µg/g Chb and 2.49 µg/g Ch content in green period was the one that lost the most chlorophyll content and so that displayed more observable color change with ripening. In ripe period, its Ch content decreased to 0.55 µg/g. The type showing the lowest decrease in Ch content with ripening was 'Levent'. While Ch content of 'Levent' was 2.13 µg/g in green period, it decreased to 1.21 µg/g in ripe period. The type 'Levent' has light yellow fruit color in ripe period.

In most studies carried out, it was reported that degradation of chlorophylls and formation of chromoplasts with ripening were observed which was similar to our results (Beltran et al., 2005; Iglesias et al., 2008). Cox et al. (2004) stated that decrease in Cha and Chb levels in fruits of Hass avocado (*Persea americana*) occurred during ripening and anthocyanin concentration increased. Researchers reported that Cha content which was 0.43 mg g^{-1} in one period of the study decreased to 0.36 mg g^{-1} in last ripening period and Ch content decreased from 0.63 to 0.57 mg g^{-1}. In addition, total anthocyanin amount increased from 150 to 524 mg kg^{-1}. It was found that these pigment changes were related with surface and inside fruit color changes.

In the study carried out about ripening and chlorophyll changes in 5 apple types (Antonovk, Zhigulevskoe, Granny Smith, Golden Delicious and Renet Simirenko), Merzlyak et al. (2003) found that while chlorophyll contents were 11 nmol/cm^2 in apple types in green period, it decreased to 0.2 nmol/cm^2 with ripening. In the study, it was reported that Granny Smith had the highest chlorophyll content and the content decreased about 3 times in Antonovka and Golden Delicious types with ripening and storage.

Conclusion

In this study, it was determined that some physiological and biochemical changes and the relationship between this change and ripeness stages showed differences between different apricot types during ripening period. The result showed that ripening apricot fruit is a process with stages well-differentiated in their physicochemical properties. During this process, TSS increased and titratable acid content decreased. This situation created the fruit characteristic taste. With ripening of fruits, degradation of Cha, Chb and Ch increased and fruit color displayed change from green to yellow and orange. It can be said that different varieties and different ripening stages have effects on these changes.

ACKNOWLEDGMENTS

This research was supported by a grant (BAP 2005/44) from Inonu University.

ABBREVIATIONS

TSS, Total soluble solids; **Cha,** chlorophyll a; **Chb,** chlorophyll b; **Ch,** total chlorophylls.

REFERENCES

Anonymous (2010). FAO Production Year Book, Rome, Italy.

Anonymous (2011). Agricultural Structure and Production. Turkish Statistical Institute, Turkey.

Asma BM, Ozturk K (2005). Analysis of morphological, pomological and yield characteristics of some apricot germplasm in Turkey. Gen. Res. Crop Evol. 52:305-313.

Beltran G, Aguilera MP, Rio CD, Sanchez S, Martinez L (2005). Influence of fruit process on the natural antioxidant content of Hojiblanca virgin olive oils. Food Chem. 89:207-215.

Borsani J, Budde CO, Porrini L, Lauxmann MA, Lombardo VA, Murray R, Andreo CS, Drincovich MF, Lara MV (2009). Carbon metabolism of peach fruit after harvest: changes in enzymes involved in organic acid and sugar level modifications. J. Exp. Bot. 60:1823-1837.

Bureau S, Renard MGC, Reich M, Ginies C, Audergon JM (2009). Change in anthocyanin concentrations in red apricot fruits during ripening. LWT - Food Sci. Technol. 42:372-377.

Cemeroglu B (1992). Fundamental Analysis methods for fruit and vegetable processing industry. Ankara, Turkey. P. 381.

Chen FX, Liu XH, Chen LS (2009). Developmental changes in pulp organic acid concentration and activities of acid-metabolising enzymes during the fruit development of two loquat (Eriobotrya japonica Lindl.) cultivars differing in fruit acidity. Food Chem. 114:657-664.

Cox KA, McGhie TK, White A, Woolf AB (2004). Skin colour and pigment Chl anges during ripening of "Hass" avacado fruit. Postharvest Biol. Tec. 31:287-294.

De-Kok L, Graham M (1989). Levels of pigments, soluble proteins, amino acids and sulfhydryl compounds in foliar tissue of Arabidopsis thaliana during dark induced and natural senesence. Plant Physiol. Biochem. 27:203-209.

Duncan DB (1955). Multiple range and multiple F Tests Biometrics. 11:1-14.

Flores FB, Martı´nez-Madrid MC, Sa´ nchez-Hidalgo FJ, Romojaro F (2001). Differential rind and pulp ripening of transgenic antisense ACC oxidase melon. Plant Physiol. Bioch. 39:37-43.

Giovannoni J (2001). Molecular regulation of fruit ripening. Ann. Rev. Plant Physiol. Plant Mol. Biol. 52:725-749.

Giovannoni JJ (2004). Genetic regulation of fruit development and ripening. Plant Cell. 16:170-180.

Gulao LF, Olivera CM (2008). Cell wall modification during fruit ripening: when a fruit is not the fruit. Trends Food Sci. Technol. 19: 4-25.

Hortensteiner S (2006). Chlorophyll degradation during senescence. Annu. Rev. Plant Biol. 57:55-77.

Iglesias I, Echeverria G, Soria Y (2008). Differences in fruit colour development, anthocyanin content, fruit quality and consumer acceptability of eight 'Gala' apple strains. Sci Hortic. 119:32-40.

Jiménez AM, Sierra CA, Rodríguez-Pulido FJ, González-Miret ML, Heredia FJ, Osorio C (2011). Physicochemical characterisation of gulupa (Passiflora edulis Sims. fo edulis) fruit from Colombia during the ripening. Food Res. Int. 44:1912-1918.

Karacali I (1990). Marketing and preservation of Horticultural Crops. Ege Uni. Agric. Fac. Pub. Bornova, İzmir (In Turkish) 494:413.

Karlıdag H, Bolat I (2007). Determination of the chemical and physical properties of some apricot cultivars growing at different altitudes in Malatya, Turkey. V. Nat. Hort. Congr. 1:782-785.

Kingston CM (1992). Maturity indices for apples and pears. Hort. Rev. 13:407-432.

Merzlyak MN, Gitelson AA, Chivkunova OB, Rakitin VY (1999). Non-destructive optical detection of pigment changes during leaf senescence and fruit ripening. Physiol. Plant. 106:135-141.

Merzlyak MN, Solovchenco AE, Gitelson AA (2003). Reflectance spectral features and non-destructive estimation of chlorophyll, carotenoid and anthocyanin content in apple fruit. Postharvest Biol. Tec. 27:197-211.

Nunes C, Santos C, Pinto G, Silva S, Lopes-da-Silva JA, Saraiva JA, Coimbra MA (2009). Effects of ripening on microstructure and texture of "Ameixa d'Elvas" candied plums. Food Chem. 115:1094-1101.

Prinsi B, Negri AS, Fedeli C, Morgutti S, Negrini N, Cocucci M, Espen L (2011). Peach fruit ripening: A proteomic comparative analysis of the mesocarp of two cultivars with different flesh firmness at two ripening stages. Phytochemistry 72:1251-1262.

Seymour GB, Taylor JE, Tucker GA (1993). Biochemistry of Fruit Ripening. Chapman and Hall, London.

Silva BM, Andrade PB, Goncalves AC, Seabra RM, Oliveira MB, Ferreira MA (2004). Influence of jam processing upon the contents of phenolics, organic acids and free amino acids in quince fruit (Cydonia oblonga Miller). Eur. Food Res. Technol. 218:385-389.

Speirs J, Brady CJ (1991). Modification of gene expression in ripening fruit. Aust. J. Plant Physiol. 18:519-532.

Villanueva MJ, Tenorio MD, Esteban MA, Mendoza MC (2004). Compositional changes during ripening of two cultivars of muskmelon fruits. Food Chem. 87:179-185.

Wu BH, Quilot B, Genard M, Kervella J, Li SH (2005). Chl anges in sugar and organic acid concentrations during fruit maturation in peaches, P. davidiana and hybrids as analyzed by principal component analysis. Sci. Hortic. 103:429-439.

Yamaki YT (1989). Organic acids in the juice of citrus fruits. J. Jpn. Soc. Hortic. Sci. 58:587-594.

4

Responses of Mmupudu (*Mimusops zeyheri*) indigenous fruit tree to three soil types

Phatu W. Mashela[1], Kgabo M. Pofu[2] and Bombiti Nzanza[1]

[1]School of Agricultural and Environmental Sciences, University of Limpopo, Private Bag X1106, Sovenga 0727, South Africa.
[2]Agricultural Research Council, VOPI, Private Bag X293, Pretoria, 0001, South Africa.

Mmupudu (*Mimusops zeyheri*) is an indigenous fruit tree to South Africa, with the potential of being domesticated for its aesthetic and nutritional attributes in rural and urban communities which were historically settled on heavy clay and loam soils, respectively. A pot study was conducted to investigate the performance of *M. zeyheri* seedlings under loam, clay and sandy soils. Relative to loam, clay increased leaf growth by 19 to 20% and 9 to 58% at 9 and 12 months after transplanting, respectively, while sand consistently reduced leaf growth by 10 to 88% and 21 to 49%, respectively. In conclusion, the positive performance of *M. zeyheri* on clay and loam soils enhanced its potential for domestication for rural and urban greening in South Africa.

Key words: Domestication, indigenous trees, moepel, vitamin C.

INTRODUCTION

Mmupudu (*Mimusops zeyheri*) fruit tree is indigenous to the northern parts of the Republic of South Africa and has high potential to serve in economic and nutritional projects in arid and semi-arid regions (Venter and Venter, 1996). Trees of *M. zeyheri* are evergreen and grow up to 15 m high, with non-aggressive lateral root systems and the tree species is both frost-hardy and drought-tolerant (Mashela and Mollel, 2001). Due to its high level of latex in above ground organs, the tree is relatively pest-free, except that ripe fruits are sensitive to an unidentified fruit-borer. Additionally, the tree is resistant to the southern root-knot (*Meloidogyne incognita*) nematode, which is highly injurious to a wide range of crops (Pofu et al., 2012). The first leaf flush occurs in late winter (July to August), while the second, along with flowers, starts during late spring through early summer, when fruits are ready for harvest. Due to a large number of flowers and the high retention of both flowers and fruits, alternate fruit

bearing is common in *M. zeyheri* trees. However, the intensity of alternate fruit bearing can be ameliorated through best management practices (Hartmann et al., 1988). Fruits are laterally borne, which implies that pruning is necessary for fruit production (Hartmann et al., 1988). Fruit of *M. zeyheri* have the highest vitamin C per unit among locally available fruits, both endemic and exotic (Venter and Venter, 1996). The tree is particularly important in the amelioration of the incidence of scurvy, which is prevalent in these parts of the country.

In Limpopo Province, South Africa, *M. zeyheri* had been identified by local communities as an indigenous fruit tree with the potential for domestication in arid areas (Mashela and Mollel, 2001). Micropropagation and cultivation protocols for the tree are already advanced (Maila, 2001). Most marginal communities in Limpopo Province were historically settled in areas with soils of low agricultural potential, where clay and sandy soils are predominant (Mashela and Mollel, 2001). Incidentally, commercial farming and urban communities are situated in high potential agricultural soils such as loam. Soil type has a strong influence on the productivity of crops (Hartmann et al., 1988), which may be direct through physical abrasion of soil particles on the root system and/or indirect through the influence of the soil on the availability of water and/or nutrient elements (Hartmann et al., 1988; Mashela et al., 1991). The growth potential of *M. zeyheri* on different soil types is not documented. The objective of this study, therefore, was to determine the influence of three soil types (loam, clay and sandy soils) on the growth potential of *M. zeyheri* seedlings under greenhouse conditions.

MATERIALS AND METHODS

Loam soil was collected from the University of Limpopo Experimental Farm (23°5310 S, 29°4415 E), clay soil from Tzaneen (23°5520 S, 30°0155 E) and sand from Magatle Village (24°2719 S, 29°2339 E). Soil samples collected from each soil type were sieved to pass through a 2-mm sieve, shade-dried and analyzed for soil particle distribution using the rapid method (Bouyoucos, 1961). Soil nutrients and pH for each soil texture were determined (Rhue and Kidder, 1983) and (1) loam soil comprised 31% sand, 59% clay and 10% slit, (2) clay soil 27%, (3) sand, 70% clay and 3% silt, while sand constituents were 98% sand, 0.5% clay and 1.5% silt. Soil pH(H$_2$O) on clay, loam and sand were 7.52, 7.27 and 7.2, respectively. Quantities of nutrient elements in various soil types were negligible and therefore, supplemented using fertilizers which included macronutrients and micronutrients. Each of the three soil types was steam-pasteurized (300°C, 1 h).

Seeds were planted in a 160-hole seedling tray potted with steam-pasteurized river sand. At 110 days after seeding, uniform seedlings were transplanted into 10 L plastic bags, each was filled to 9.5 L using separate soil types. Treatments, namely, clay, loam and sand, were arranged in a randomized complete block design, with 18 replications. Average day/night temperatures in winter (May to July) were 15/4°C, spring (August to October) 18/10°C, summer 25/17°C and autumn 20/11°C. Maximum greenhouse temperatures were regulated using thermostatically-activated fans, while minimum temperatures depended upon the greenhouse effect. Seedlings were fertilized at transplanting using 2 g 2:1:2 (43), which provided a total of 0.35 mg N, 0.35 mg K and 0.16 mg P, 0.9 mg Mg, 0.75 mg Fe, 0.075 mg Cu, 0.35 mg Zn, 1.0 mg B, 3.0 mg Mn and 0.07 mg Mo per ml water and thereafter, bimonthly. Dolomitic lime at 5 g/pot to provide Ca and Mg was applied once during potting. Four sets of Hadeco Moisture Meters (Hadeco Magic^R, New Delhi, India) were inserted to 20 cm depths in randomly selected pots of each treatment to monitor soil moisture tension. Plants were automatically irrigated with 500 ml tapwater as soon as at least 50% of the moisture meters have average readings below 2 units with irrigation interval depending on soil type.

Data were collected at 9 and 12 months after transplanting. Four leaves/plant selected for measurements were tagged in order to ensure that data were collected from the same leaves. Chlorophyll meter (SPAD 502) was used to measure chlorophyll content from the tagged leaves. Stem root-collar diameters were measured at 3 cm above the soil surface using a digital vernier caliper. Plant height, leaf length, leaf width and leaf petiole length were measured. Plants were severed at the soil surface, with root systems removed from pots, immersed in water to remove soil particles, blotted dry and both roots and shoots dried at 70°C for 72 h.

Data were analyzed using SAS software (SAS Institute Inc.,

2008). Data on number of leaves/plant were transformed using $\log_{10}(x + 1)$ prior to analysis of variance to standardize the variances (Gomez and Gomez, 1984). Data were subjected to analysis of variance and means separated using Fisher's least significant difference test. Loam soil was used as a standard and data from clay and sand expressed relative to the standard using the relationship:

Impact (%) = [(Mean performance on sand or clay soil/ Mean performance on loam soil – 1) × 100].

RESULTS AND DISCUSSION

Soil type had no effect on chlorophyll content, dry shoot mass and dry root mass, while sand reduced root/shoot ratios of *M. zeyheri* seedlings (data not shown). Relative to loam soil, clay had no effect on plant height at 9 and 12 months. However, relative to loam soil, sand reduced this variable by 88 and 49% at 9 and 12 months, respectively (Table 1). Clay had no effect on stem diameter at 9 months, but increased the variable by 9% at 12 months. Sand significantly reduced the variable by 22 and 21% at 9 and 12 months, respectively.

Relative to loam, clay soil increased leaf width by 19% at 9 months and leaf petiole at 9 and 12 months by 20 and 19%, respectively (Table 1). In contrast, sand consistently reduced various components of the leaf during both measurement periods. Dry shoot mass and root mass were not affected by soil type, which supported the view that *M. zeyheri* is a hardy plant (Venter and Venter, 1996). Using the two variables, it was previously demonstrated that *M. zeyheri* was resistant to the root-knot nematodes *Meloidogyne* species (Unpublished data), which are widespread in various parts of Limpopo Province (Kleynhans et al., 1996). Generally, *M. zeyheri*, like most indigenous plants in arid and semi-arid areas, is a slow grower, with dry shoot mass and root mass being poor indicators of growth (Maila, 2001). In this study, plant height, stem diameter and various components of the leaf were good indicators for assessing growth of *M. zeyheri* in response to soil type.

Relative to loam, clay soil had either no effect or increased the variables. Clay soil has high nutrient adsorption capabilities and high water holding capacities (Hartmann et al., 1988), all of which might have had an influence on the number of leaves and their growth potentials, which by 12 months had increased almost 50% more than at 9 months. Generally, most exotic fruit trees with either taproot or fibrous root systems, grow poorly on clay soil (Hartmann et al., 1988), probably due to their sensitivity to waterlogging conditions and soil-borne diseases, which invariably induce root and crown rots (Agrios, 2005). Poorly developed taproot systems and hardy fibrous root system of *M. zeyheri* appear to be well-suited for clay soils, which are widespread in marginal rural communities of Limpopo Province.

Sand consistently reduced the indicators which were used for assessing growth potential of *M. zeyheri*,

Table 1. Influence of clay and sand relative to loam on mean ± SE plant height, stem diameter, leaf length, leaf width, leaf petiole and the number of leaves on *M. zeyheri* seedlings at nine and twelve months after initiating the treatment (n = 54).

Variable	9 months after initiating the treatment					Twelve months after initiating the treatment				
	Loam	Clay	Impact (%)	Sand	Impact (%)	Loam	Clay	Impact (%)	Sand	Impact (%)
Plant height (cm)	6.8 ± 1.8^a	7.0 ± 1.2^b	3^{ns}	4.8 ± 1.2^b	-88*	13.0 ± 2.1^a	13.6 ± 3.0^a	5^{ns}	6.6 ± 0.7^b	-49*
Stem diameter (mm)	2.4 ± 0.5^a	2.6 ± 0.6^a	10^{ns}	1.8 ± 0.5^b	-22*	3.9 ± 0.7^b	4.3 ± 1.7^a	9*	3.1 ± 1.3^c	-21*
Leaf length (cm)	4.0 ± 0.6^a	3.8 ± 0.7^a	-5^{ns}	4.1 ± 0.4^a	2^{ns}	6.2 ± 1.39^a	5.8 ± 1.2^a	-6^{ns}	4.1 ± 1.1^b	-34*
Leaf width (cm)	1.9 ± 0.5^b	2.2 ± 0.6^a	19*	1.4 ± 0.1^c	-44*	2.6 ± 0.8^a	2.6 ± 0.5^a	2^{ns}	1.9 ± 0.6^b	-28*
Leaf petiole (cm)	3.4 ± 0.9^b	4.0 ± 2.0^a	20*	3.1 ± 1.4^c	-10*	4.0 ± 0.8^b	4.7 ± 1.3^a	19*	3.0 ± 1.0^c	-23*
Number of leaves	5.0 ± 0.9^a	4.8 ± 0.5^a	-5^{ns}	3.2 ± 1.1^b	-36*	9.1 ± 2.1^b	13.9 ± 1.0^a	58*	6.37 ± 1.3^c	-30*

Impact (%) = [(Mean performance on sand or clay soil/mean performance on loam soil − 1) × 100], where ^{ns} meant that the impact was not significant at P ≤ 0.05, while *meant it was significant at P ≤ 0.05. Row means within a sampling period followed by the same letter were not different (P ≤ 0.05) according to Fisher's least significant difference test.

suggesting the high susceptibility of the indicators to soil type. Although there was no reduction in either dry shoot mass and dry roots mass, the integrated reduction of the number of leaves and components of the leaf due to soil type would eventually result in poor overall plant productivity, since the leaf is the major source of carbohydrates required for growth. Mashela et al. (1991) demonstrated that sandy soil consistently suppressed growth of Alyceclover (*Alysicarpus vaginalis* L.), a leguminous forage crop with a fibrous root system in Florida, USA. In that study, it was argued that roots were the most sensitive to sand damage and that the damage was mainly due to physical abrasiveness of sand particles, although other factors such as water stress and deficiency of nutrient elements were not ruled out. The observed reduction in stem diameter due to sand in *M. zeyheri* seedlings is consistent with the increased flow of sucrose to the root system whenever plants are subjected to extrinsic factors that reduce the root/shoot ratio (Mafeo, 2005). The increased flow of carbohydrates in the form of sucrose is intended to assist the plant to increase growth of the root system in order to re-establish the normal root/shoot ratio (Mashela and Nthangeni, 2002).

Conclusion

Most rural communities in South Africa were historically settled in areas with heavy clay soils. Results of this study suggested that *M. zeyheri* trees would be suitable for soil conditions with both clay and loam soils, while areas with sandy soils would be unfavourable. Since urban and certain rural areas have predominantly loam and clay soils, respectively, prospects for domesticating *M. zeyheri* trees are good in terms of its performance on the two soil types.

ACKNOWLEDGEMENTS

The authors are grateful to the Land Bank Chair, University of Limpopo for financial support.

REFERENCES

Agrios GN (2005). Plant pathology, 5th ed. Academic Press: London. p. 952.

Bouyoucos GJ (1961) Hydrometer method improved for making particle size analysis of soils: Agron. J. 54:464-465.

Gomez KA, Gomez AA (1984). Statistical procedure for agricultural research, 2nd ed. Wiley: New York. p. 680.

Hartmann HT, Kester DE, Davies AM, Geneve RL (1988). Plant propagation: Principles and practices, 8th ed. Prentice Hall: New York. p. 896.

Kleynhans KPN, Van den Berg E, Swart A, Marais M, Buckley NH (1996). Plant nematodes in South Africa. Plant Protection Institute: Pretoria. p. 165.

Mafeo TP (2005). Propagation, fertilization and irrigation of Cucumis myriocarpus. MSc Dissertation, University of Limpopo, Sovenga, South Africa.

Maila YM (2001). In-vitro propagation of Mmupudu (Mimusops zeyheri) fruit tree. Master Dissertation, University of the North, Sovenga, South Africa.

Mashela PW, Nthangeni ME (2002). Osmolyte allocation in response to Tylenchulus semipenetrans infection, stem girdling and root pruning in citrus. J. Nematol. 34:273-277.

Mashela PW, Mollel N (2001). Farmer – identified and selected indigenous fruit tree with suitable attributes for smallholder farming systems in South Africa's Northern Province. S. Afr. J. Ext. 30:1-12.

Mashela PW, McSorley R, Duncan LW (1991) Correlation of Belonolaimus longicaudatus, Hoplolaimus galeatus and soil texture with yield of Alyceclover. Nematropica 21:177-184.

Pofu K, Mashela P, De Waele D (2012). Survival, flowering and productivity of watermelon (Citrullus lanatus) cultivars in inter-generic grafting on nematode-resistant Cucumis seedling rootstocks in Meloidogyne-infested fields. Int. J. Agric. Biol. 14:217-222.

Rhue RD, Kidder G (1983). Procedures used by IFAS extension

soil laboratory and interpretation of results. Circ. No. 595. Fl. Co-op. Ext. Serv., IFAS, Univ. Fl., Gainesville.

SAS Institute (2008). Statistical analysis systems computer package, Cary, New York, USA.

Venter F, Venter JA (1996). Making the most of indigenous trees. Briza Publications: Pretoria. p. 304.

5

Standardization of germination media for the endangered medicinal tree, bael (*Aegle marmelos*)

B. Venudevan[1], P. Srimathi[2], N. Natarajan[3] and R. M. Vijayakumar[4]

[1]Department of Seed Science and Technology, Tamil Nadu Agricultural University, Coimbatore-3, India.
[2]Seed Centre, Tamil Nadu Agricultural University, Coimbatore-3, India.
[3]Department of Nano Science and Technology, Tamil Nadu Agricultural University, Coimbatore-3, India.
[4]Department of Medicinal and Aromatic Crops, Tamil Nadu Agricultural University, Coimbatore-3, India.

Bael is an endangered medicinal tree, highly propagated through seeds. Either on trade or before sowing evaluation of seed germination is essential. International rules for seed testing recommended different techniques for seed quality evaluation. One such requirement is the selection of media for germination to assure actual germination as media provide proper platform for full expression of seedling emergence and growth. Hence, the studies were conducted at the Department of Seed Science and Technology, Tamil Nadu Agricultural University, Coimbatore, on standardization of media and methodologies for germination test under germination room conditions in line with ISTA for Bael seeds. The results revealed that, either river sand or paper media could be used for obtaining reproducible and complete expression of germination of seed. In river sand, in-sand method (seed are to be sown at depth of 2 cm) and in paper, between paper (Roll towel) method had better expression for germination (83 and 78%) respectively, recommending it as the best media for evaluation of germination percentage of Bael seeds in seed testing.

Key words: Bael, endangered, sand or paper media, seedling emergence, germination.

INTRODUCTION

Aegle marmelos (L.) Corr. is a medicinal tree which belongs to the family Rutaceae and its various parts are used in Ayurvedic and Siddha medicines to treat a variety of ailments. Bael is highly habitated to tropical and subtropical climate of India, Burma, Pakistan, Bangladesh, Sri Lanka, Northern Malaya, Java and Philippine (Islam et al., 1995). It is a medium sized tree having profuse dimorphic branches, alternate, trifoliate and deep green leaves, membranous leaflets, large sweet scented, greenish white flowers, large and globose fruits (Purohit and Vyas, 2005). Mazumder et al. (2006) revealed that, approximately 200 to 250 kg of fruits could be obtained per tree. All parts of the tree are highly useful in preparation of herbal medicines (Kala, 2006).

The roots are useful for treating diarrhea, dysentery and dyspepsia. The aqueous stem and root bark extracts are used as medicine for malaria, fever, jaundice, cancer, ulcers, urticaria and eczema (Nadkarni, 1954).The fruit and root of the plant have antiamoebic and hypoglycaemic activities (Ponnachan et al., 1993). Goel et al. (1997) revealed that, crop is rich with the alkaloids aegline, marmesin, marmin and marmelosin. Rana et al. (1997) revealed that, the seed is rich in luvangetin and pyranocoumarin compounds, which has antiulcer activity. They also revealed that, essential oil isolated from the leaf has antifungal activity. The foundation for revitalization of local health traditions (FRLHT), Banglore, India listed bael (*Aegle marmelos*) as rare, endangered

Table 1. Seeds sown in different media adopting different methodologies.

Media		Methodologies for germination test
Germination paper	✓	Top of the paper (TP)
	✓	Between the paper (Roll towel)
River sand	✓	sowing at 2 cm depth
Quartz sand	✓	sowing at 2 cm depth
Vermiculite	✓	sowing at 2 cm depth
Inclined plate		

and threatened (RET) species specifically endangered species. This underutilized tree is generally propagated through seeds. ISTA (1993) formulate, procedures for testing the physical, physiological and health status of seed, which differ from seed to seed and newer crops are added based on necessity. Among the seed quality characters, evaluation of germination is the prime and most important reliable character that explores the relative planting value of the seed lot. It should be evaluated in a correct media, to give accurate and reproducible results. The objective of the germination test is to express the maximum germination potential of seed which is the most important than any other quality parameters.

The use of standardized ideal techniques in the laboratory as prescribed by ISTA is warranted as it ensures that, results obtained for a given seed lot in one laboratory would be identical with those obtained from any other laboratory in the same or other countries (Willan, 1985). In the laboratory, the environmental conditions, including moisture, temperature, aeration and light, must not only be specific enough to indicate germination but also favorable for the development of the seedlings to a stage where interpretation as normal and abnormal types were possible. Medium plays an important role in germination testing, because seeds have characteristic requirements of moisture and oxygen for germination. Thus, minute seeds germinate well on top of paper (TP) rather than sand. Best suited medium depends upon the physical condition of the seeds. Generally, sand as medium is best suited for large sized seeds and paper for small sized seeds (Nawabahar, 2008). Hence, studies were initiated to standardize suitable media and methodology for evaluation of seed germination of Bael seeds at the Department of Seed Science and Technology, Tamil Nadu Agricultural University, Coimbatore, Tamil Nadu during 2012.

MATERIALS AND METHODS

Bael seeds were collected from different places of Coimbatore district (76°57 E, 11°8 N and 320 MSL) which are; Ram Nagar, Perur, Karamadai, and Saibaba colony. To evaluate the germination potential and seedling growth rate, the seeds were sown in different media adopting different methodologies as follows (Table 1 and Figure 1).

In each of the media/methodologies, 100 seeds in four replications were sown and kept under the germination room maintained at 25°C and 95 ± 2% RH. Data on daily germination was recorded in each of the media until no further germination was observed that is, up to 23 days of germination period. Then the germination test was terminated and the resultant seed and seedlings were categorized into normal, abnormal, and dead seeds. In randomly selected all normal seedlings, the seedlings were measured for their root and shoot length and the mean reported as root and shoot length per seedlings. The seedlings were dried for 48 h in an oven maintained at 85°C and weighed in a top pan balance and the mean dry matter per seedlings in milligram. Vigour index values, the totality expressions of seed quality characters, were also computed as per Abdul and Anderson (1973) adopting the following formula and the results reported as whole number.

Vigour index = Germination (%) × Total seedling length (cm)

The data gathered from various sources for all the seed quality parameters were subjected for analysis of variance (ANOVA) as per Panse and Sukhathme (1995) to determine the level of significance (5%).

RESULTS AND DISCUSSION

Highly significant variations were observed for the evaluated seed quality parameters obtained from different growing media, methodologies and locations (Table 2). The results revealed that, germination was initiated 11 days after sowing in river sand on adopting in-sand method while it took 12 days for initiation of germination in roll towel (RT) and modified RT method. The initiation of germination was observed between 13 to 15 days. The seed germination recorded based on normal seedlings was the highest germination (83%) in sand media sown at 2 cm depth and was followed by the paper; RT 78%, quartz sand 76%, inclined plate 73% and vermiculite 67%. The root 12.8 cm and shoot length 11.6 cm, dry matter production 47.9 mg, and vigour index (1866) also recorded the highest value in sand media sown at 2 cm depth and was followed by between the paper (roll towel) (Tables 3 and 4). It was also observed that, expression of seedlings as normal seedling was more in sand and paper media on adoption of modified RT method and the lowest germination percentage (56%), root (5.0 cm) and shoot length (5.4 cm), dry matter production (20.0 mg) and vigour index (504) recorded with top of the paper (TP) method expressing

Figure 1. Standardization of different germination media.

that, it could not be used for evaluation of germination of bael seeds, might be due to the insufficiency of water and space requirement for development of the seedlings as normal seedlings for this medium sized seeds.

Seeds sown in sand at 2 cm depth produced the maximum germination of 83% irrespective of seed sources. However, Zhe et al. (2009) reported that, seeds in quartz sand seemed to grow faster and the seedlings would die in shorter durations, but in the present study, though there is no death in quartz sand, germination was easier in river sand than quartz sand.

Based on easy availability and economic utility, for Bael the river sand could be used as the testing media, but in that too, in-sand method (sowing of the seed at 2 cm) should be adopted for obtaining reproducible results in the germination test. It was followed by the other media that is, between the paper (Roll towel) and quartz sand. According to Nawabahar (2008), the choice of media depends on the species being tested, as species seems to have a distinctive preference for a particular media. Minute seeds are best germinated on TP while large seeds are best germinated in sand media or RT. Towel paper is more commonly used for medium sized seed because it is easier to handle than sand and permits

Table 2. Standardization of germination media on days to first emergence and germination percentage in Bael (*Aegle marmelos*) seeds .

Media (M)	Days to first emergence					Germination (%)				
	Ram nagar	Perur	Karamadai	Saibaba colony	Mean	Ram nagar	Perur	Karamadai	Saibaba colony	Mean
Top of the paper (Petri plate)	16	17	13	15	15	54(47.29)	55(47.87)	58(49.60)	56(48.44)	56(48.44)
Between the paper (roll towel)	13	10	11	15	12	76(60.66)	79(62.72)	80(63.43)	78(62.02)	78(62.02)
Sand (2 cm depth)	12	10	11	12	11	80(63.43)	87(68.86)	81(64.15)	85(67.21)	83(65.65)
Vermiculate	15	14	12	14	14	68(55.55)	69(56.16)	66(54.33)	65(53.73)	67(54.94)
Quartz sand	14	15	11	13	13	74(59.34)	75(60.00)	78(62.02)	76(60.66)	76(60.66)
Inclined plate	15	11	13	15	14	72(58.05)	73(58.69)	71(57.41)	74(59.34)	73(58.69)
SEd	1.570	1.428	1.526	1.549		2.429	1.903	1.882	1.940	
CD (P = 0.05)	3.140	2.947	3.045	3.070		5.104	3.998	3.954	4.077	

Table 3. Standardization of germination media on root and shoot length in bael (*Aegle marmelos*).

Media (M)	Root length (cm)					Shoot length (cm)				
	Ram nagar	Perur	Karamadai	Saibaba colony	Mean	Ram nagar	Perur	Karamadai	Saibaba colony	Mean
Top of the paper (Petri plate)	4.7	5.7	2.7	6.7	5.0	4.6	6.6	2.6	7.6	5.4
Between the paper (roll towel)	10.4	11.4	8.4	12.4	10.7	9.2	11.2	7.2	12.2	10.0
Sand (2 cm depth)	12.5	13.5	10.5	14.5	12.8	10.8	12.8	8.8	13.8	11.6
Vermiculate	7.8	8.8	5.8	9.8	8.1	8.6	10.6	6.6	11.6	9.4
Quartz sand	8.5	9.5	6.5	10.5	8.8	9.1	11.1	7.1	12.1	9.9
Inclined plate	8.9	9.9	6.9	10.9	9.2	8.8	10.8	6.8	11.8	9.6
SEd	0.768	0.542	0.459	0.614		0.473	0.502	0.494	0.537	
CD (P = 0.05)	1.614	1.140	0.965	1.291		0.995	1.055	1.039	1.129	

easier and quicker development of seedlings. Large seeds could also be rolled in paper towel with limited seeds but is a very inconvenient method to test a large number of seeds at a time. Sand is not suitable for very small seeds but is widely used for large seeds especially for tree seeds that have longer germination period. Sand can also be sterilized easily and fungal development on sterilized sand is controlled better than paper media. It also provides good contact

between the seed and moisture as seed is pressed into the medium, the bigger particle size aids in aeration that favoured the production of normal seedling in higher number. The sand was found to be the best medium for improving the forest tree seed germination as expressed by Anber (2010).

These results were in supportive of the findings of Egharevba et al. (2005) in African walnut, *Plukenetia conophorum*, Bahuguna et al. (1987a,

b) in *Terminalia myricarpa* and *Adhatoda vasica*, Murugesan et al. (2008) in oil palm, Vilela and Ravette (2001) in species of Prosopis, Docker and Hubble (2008) in some Australian tree species and by Thapliyal and Rawat (1991) in *Alnus nitida* and *A. nepalensis* who, recommends sand medium for germination testing under the process of seed testing for expression as normal seedlings in germination room condition irrespective of the source seeds.

Table 4. Standardization of germination media on dry matter production and vigour index in bael (*Aegle marmelos*).

Media (M)	Dry matter production (mg seedlings^{-10})					Vigour index				
	Ram nagar	Perur	Karamadai	Saibaba colony	Mean	Ram nagar	Perur	Karamadai	Saibaba colony	Mean
Top of the paper (Petri plate)	19.9	20.3	19.5	20.2	20.0	502	507	497	508	504
Between the paper (roll towel)	44.2	44.6	43.8	44.5	44.3	1411	1416	1406	1417	1413
Sand (2 cm depth)	47.8	48.2	47.4	48.1	47.9	1864	1869	1859	1870	1866
Vermiculate	42.1	42.5	41.7	42.4	42.2	1115	1120	1110	1121	1117
Quartz sand	43.9	44.3	43.5	44.2	44.0	1302	1307	1297	1308	1304
Inclined plate	39.5	39.9	39.1	39.8	39.6	1345	1350	1340	1351	1347
SEd	1.209	0.974	1.350	2.018		73.767	75.455	58.384	64.956	
CD (P = 0.05)	2.541	2.047	2.837	4.240		155.714	158.527	122.663	136.470	

Conclusion

The study highlighted that, seed germination could be tested in Bael seeds using either sand media or germination paper, on using sand media, seeds should be sown at the depth of 2 cm and paper media modified RT methods has to be adopted for obtaining reproducible results.

REFERENCES

Abdul BA, Anderson JD (1973). Vigour determination in soybean seed by multiple criteria. Crop Sci. 13:630-633.

Anber MA (2010). Improving seed germination and seedling growth of some economically important trees by seed treatments and growing media. J. Hortic. Sci. Ornamental Plants 2(1):24-31.

Bahuguna VK, Rawat MMS, Joshi SR, Maithani GP (1987a). Studies on the viability, germination and longevity of *Terminalia myriocarpa* seed. J. Trop. Forest. 3(IV):318-323.

Bahuguna VK, Sood OP, Rawat MMS (1987b). Preliminary studies on the germination behaviour of *Adhatoda vasica* seeds. An important shrub for regeneration of Sub-Himalayan wastelands. Indian For. 113(6):256-261.

Docker BB, Hubble TCT (2008). Quantifying root reinforcement of river bank soils by four Australian tree species. Geomorphology 100(3-4):401-418.

Egharevba RK, Ikhatua MI, Kalu C (2005). The influence of seed treatments and growing media in seedling growth and development of African walnut, *Plukenetia conophorum*. Afr. J. Biotechnol. 4(8):808-811.

Goel RK, Maiti RN, Manickam M, Ray AB (1997). Antiulcer activity of naturally occurring pyranocoumarin and isocoumarins and their effect on prostanoid synthesis using human colonic mucosa. Ind. J. Expt. Biol. 35:1080-1083.

Islam R, Hossain M, Karim MR, Joarder OI (1995). Regeneration of *Aegle marmelos* (L.) Corr., plantlets *in vitro* from callus cultures of embryonic tissues. Cur. Sci. 69:494-495.

ISTA (1993). International Rules for Seed Testing. Seed Sci. and Technol. 21:1-288(suppl.).

Kala CP (2006). Ethnobotany and ethnoconservation of *Aegle marmelos* (L.) Correa. Indian J. Trad. Knowl. 5:541-550.

Mazumder R, Bhattacharya S, Mazumder A, Pattnaik AK, Tiwary PM, Chaudhary S (2006). Antidiarrhoeal evaluation of *Aegle marmelos* (Correa) Linn. root extract. Phytother. Res. 20:82-84.

Murugesan P, Bijimol G, Haseela H (2008). Effect of different substrates on growth of germinated oil palm hybrid seeds. Indian J. Hortic. 65(4):477-480.

Nadkarni KM (1954). Indian material Medica, 3rd edn. Popular Book Depot, Bombay, India. pp. 45-49.

Nawabahar (2008). Effect of media on seed germination of *Cupaniopsis anacardioides* (A. Rich.) Radlk. Indian J. For. 31(1):137-139.

Panse VS, Sukhatme PV (1995). Statistical Methods for Agricultural Workers. Indian Council for Agricultural Research, New Delhi, India.

Ponnachan PTC, Paulose CS, Panikar KR (1993). Effect of the leaf extract of *Aegle marmelos* (L.) Corr. In diabetic rats. Indian J. Exp. Biol. 31:345-347.

Purohit SS, Vyas SP (2005). Medicinal Plant Cultivation-A Scientific Approach. Agrobion. India. P. 282.

Rana BK, Sing UP, Taneja V (1997). Antifungal activity and kinetics of inhibition by essential oil isolated from leaves of *Aegle marmelos* (L.) Corr. J. Ethnopharmacol. 57:29-34.

Thapliyal RC, Rawat MMS (1991). Studies on the germination and viability of seed of two species of Himalayan Alders (*Alnus nitida* and *A. nepalensis*). Ind. For. 117(4):256-261.

Vilela AE, Ravetta DA (2001). The effect of seed scarification and soil media on germination, growth, storage and survival of seedlings of five species of Prosopis L. (Mimosaceae). J. Arid Environ. 48:171-184.

Willan RL (1985). A guide to forest seed handling with special reference to the tropics. FAO Forestry Paper, 20/2. FAO, Rome.

Zhe J, Paolo P, Robert F, Peter M, Paolo B (2009). An experimental comparison of silica gel and quartz sand grains as sediment media for growing vegetation at the laboratory scale. Aquat. Sci. 71:350-355.

6

Xylem vessel element structure explains why columnar apple trees flower early and have higher yield

Zhang Yugang and Dai Hongyi

College of Landscaping and Horticulture, Qingdao Agricultural University, Qingdao, Shandong Province 266109, China.

The vessel elements of secondary xylem in branches of columnar and normal apple trees were studied by isolation method and micrograph. The lengths and diameters were measured and the type of side wall (reticulate or pitted) was noted. Most of the perforation plates were simple. Normal apples had more abnormal vessel element cells than columnar apples. The average diameter of the xylem vessel elements of columnar apples was 43.27 μm, which was significantly greater than that of normal apples (32.64 μm). The lengths of the vessel elements did not differ significantly between columnar and normal apples. The results provided a theoretical basis for explaining the early flowering and fruiting of columnar apple trees and their higher fruit yield.

Key words: Columnar apple, isolation method, vessel element.

INTRODUCTION

The xylem vessel is a major component of plant vasculature, and the structure of vessel elements, the cells that make up xylem vessels, can affect growth and development of plants by affecting the rate of water and mineral transport. Columnar apple trees originated from the 'Weisai Ke Xu' mutant of the cultivar 'Asahi' and have many advantages over normal apple trees, such as very short internodes, high germination rates, weak branching, and generally short lateral branches from the trunk (Zhang et al., 2003). In particular, columnar apples flower and fruit earlier than normal apples, making columnar apples valuable resources for the genetic improvement of apples.

Previous research has investigated various aspects of columnar apples, including cultivar breeding (Dai et al., 2003), leaf anatomy (Liang et al., 2009), photosynthetic characteristics (Zhang et al., 2010), and vessel element biology based on element marker-based (Wang et al., 2002; Tian et al., 2005). Vessel element structure and the evolution of secondary xylem vessels has been reported in peach (Guo et al., 2008), mango (Chen and Tang,

2005), Grevillea (Chen and Tang, 2004a), Aquilaria (Chen and Tang, 2004b), Manilkara (Chen, 2007), but apple, especially columnar apple, has been little studied. In this study, the general internal structure of secondary xylem vessels of columnar apples was compared with that of normal apples using the segregation method and micro-photography to provide a theoretical basis for understanding the earlier flowering and fruiting phenology of columnar apples.

MATERIALS AND METHODS

Test material was collected in March 2011 from apple trees growing at the Qingdao Laixi Toyozane horticultural farm. South-facing branches were sampled from 2-year-old trees that were 1.5 m high and showed consistent growth.

The cultivars were as follows: Cultivar 1, Normal hybrid offspring of 'Fuji' × 'Telamon'; Cultivar 2, Normal hybrid offspring of 'Gala' × 'Telamon'; Cultivar 3, 'Fuji'; Cultivar 4, Columnar hybrid offspring of 'Fuji' × 'Telamon'; Cultivar 5, Columnar hybrid offspring of 'Gala' × 'Telamon'; Cultivar 6, 'Telamon'. Cultivars 1, 2 and 3 were normal

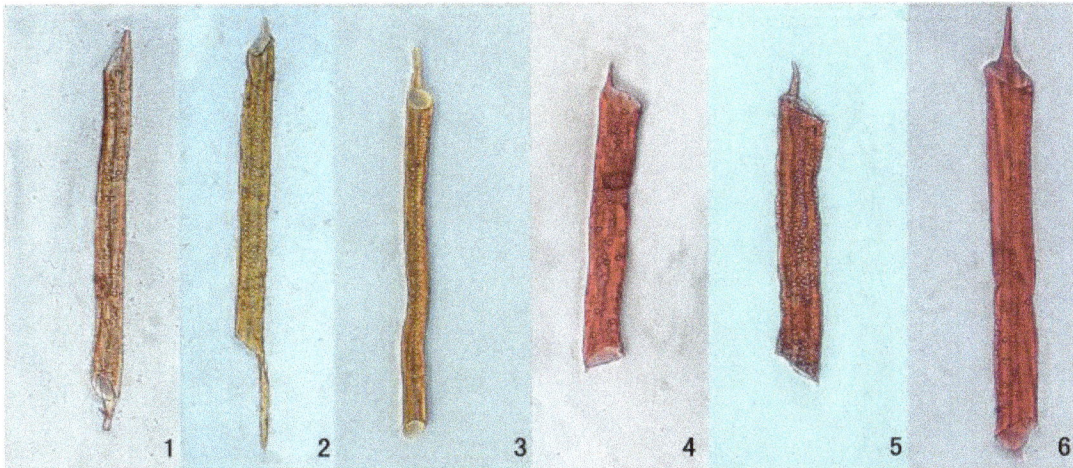

Figure 1. The pitted vessel element shape of xylem of branches from different cultivars:1, *'Fuji'*×*'Telamon'* normal apple; 2, *'Gala'*×*'Telamon'* normal apple; 3,*'Fuji'*. 4, *'Fuji'*×*'Telamon'* columnar apple; 5, *'Gala'*×*'Telamon'* columnar apple; 6, *'Telamon'*.

Figure 2. The reticulate vessel element shape of xylem of branches from different cultivars: 1, *'Fuji'*×*'Telamon'* normal apple; 2, *'Gala'*×*'Telamon'* normal apple; 3, *'Fuji'*. 4, *'Fuji'*×*'Telamon'* columnar apple; 5, *'Gala'*×*'Telamon'* columnar apple; 6, *'Telamon'*.

cultivars, and cultivars 4, 5 and 6 were columnar cultivars.

Experimental methods

Cut branches were washed and approximately 1 cm pieces cut from the same positions on each branch. These pieces were vertically sliced into cross-sectional strips 2 to 3 mm wide and dried for 3 to 4 days at 30 to 40°C in an incubator (Chen and Xie, 2003). Each piece was made into a temporary slide by staining with 0.5% Safranin dye, then observed and its vessel elements microscopically photographed with a Nikon E80i biological microscope. One hundred samples of each cultivar were randomly

observed. We measured vessel length (excluding tail) and diameter with Image-Pro Plus 6.0, then averaged and statistically analyzed the data using the DPSv7.05 software.

RESULTS AND ANALYSIS

Morphology of xylem vessel elements of apple branches

Xylem contains two types of vessel elements: pitted and reticulate. The type of xylem vessels found in shoots of columnar apple trees was not significantly different from those of normal apple trees under light microscopy (Figures 1 and 2). In normal apple trees,

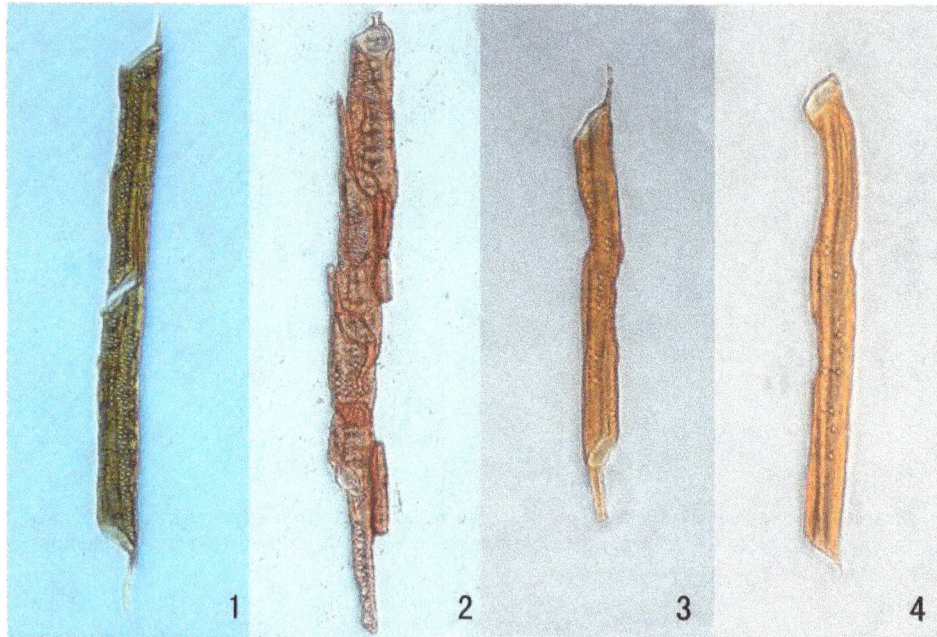

Figure 3. The connection of vessel element in vertical and abnormal cells: 1 and 2 are the connection of vessel element in vertical; 3 and 4 are the abnormal cells.

69.7% of the vessel elements were reticulate, compared with 75.7% in columnar apple trees. The perforation plates that serve as the vertical connections between vessel elements were also observed when possible. There were some abnormal vessel elements (vessel elements of uneven diameter) (Figure 3). The columnar apples had fewer deformities of their vessel elements (a 5.75% malformation rate) than normal apples (9.66%).

Vessel element length in two types of apple branches

The average lengths of the vessel elements were 243.359, 253.875, 260.776, 240.747, 218.674 and 293.383 µm for cultivars 1 to 6, respectively. There were no significant differences in average vessel element length between columnar and normal apple branches (Table 1).

Vessel element diameter in two types of apple branches

The average diameters of the vessel elements of cultivars 1 to 6 were 35.345, 32.255, 30.318, 43.830, 41.638 and 44.345 µm, respectively. The vessels of columnar apples were significantly wider than those of normal apples ($P < 0.01$), and the diameters of different cultivars in the same types had significant difference ($P < 0.05$) (Table 2).

DISCUSSION

The main function of xylem vessels is to transport water and minerals from the roots to other parts of the plant to support transpiration and photosynthesis. Columnar apples had more reticulate vessel elements, less secondary wall thickening, and fewer abnormal vessel

elements in their xylem than normal apples. These anatomical differences would enhance the transport rate of water and minerals in both horizontal and vertical directions. Vessel element structure explains why columnar apples have a higher transport efficiency of water and minerals and a photosynthetic efficiency than normal apples (Zhang et al., 2010), and also explains why columnar apples have the physiological characteristics of earlier flowering and fruiting and higher yield.

Physiological anatomist Zimmermann (1983) proved that the diameter of xylem vessels was directly related to the transport rate. That is, the larger the diameter of a xylem vessel, the higher its transport efficiency. The average diameter of xylem vessel elements of columnar apples was significantly greater than that of normal apples. The larger vessel diameter enhances the transport efficiency of water and minerals. Columnar apples thus have more efficient and rapid transport to the leaves, which benefits photosynthesis. The data presented here has provided an anatomical explanation for early flowering and fruiting in columnar apple trees (Zhang et al., 2010).

The relationship between the lengths of vessel elements and water transport is still debated. Bass et al. (1983) believed that the lengths of individual vessel elements were not directly related to water transport efficiency but that the total length of the xylem vessel was directly correlated to water transport efficiency. Carlquist (1975) pointed out that the longer the xylem vessel, the smaller the resistance of water transport. The relationship

Table 1. Comparison of the length of the vessel elements of xylem of branches between columnar and normal apple.

Number	Variety (Combination)	Type	Length of the vessel element(µm)		
			Average	Max	Min
1	*Fuji × Telamon*	Normal	243.359[bBC]	482.485	111.691
2	*Gala × Telamon*	Normal	253.875[bB]	518.067	105.897
3	*Fuji*	Normal	260.776[bB]	443.888	110.093
4	*Fuji × Telamon*	Columnar	240.747[bBC]	384.116	101.233
5	*Gala × Telamon*	Columnar	218.674[cC]	470.394	103.748
6	*Telamon*	Columnar	293.383[aA]	485.454	116.505

Note: Different letters of A, B, C and D represent significant difference at α = 0.01 level. Different letters of a, b, c and d represent significant difference at α = 0.05 level. Material 1 was '*Fuji*'×'*Telamon*' normal apple, Material 2 was '*Gala*'×'*Telamon*' normal apple, Material 3 was '*Fuji*', Material 4 was '*Fuji*'×'*Telamon*' columnar apple, Material 5 was '*Gala*'×'*Telamon*' columnar apple, Material 6 was '*Telamon*'.

Table 2. Comparison of the dimension of the vessel element of xylem of branches between columnar apple and normal apple.

Number	Variety (Combination)	Type	Dimension of the vessel element (µm)		
			Average	Max	Min
1	*Fuji × Telamon*	Normal	35.345[cC]	51.712	17.476
2	*Gala × Telamon*	Normal	32.255[dD]	53.886	11.989
3	*Fuji*	Normal	30.318[eD]	46.078	13.062
4	*Fuji × Telamon*	Columnar	43.830[aA]	57.226	22.230
5	*Gala × Telamon*	Columnar	41.638[bB]	59.653	26.096
6	*Telamon*	Columnar	44.345[aA]	66.021	22.862

Note: Different letters of A, B, C and D represent significant difference at α = 0.01 level. Different letters of a, b, c and d represent significant difference at α = 0.05 level. Material 1 was '*Fuji*'×'*Telamon*' normal apple, Material 2 was '*Gala*'×'*Telamon*' normal apple, Material 3 was '*Fuji*', Material 4 was '*Fuji*'×'*Telamon*' columnar apple, Material 5 was '*Gala*'×'*Telamon*' columnar apple, Material 6 was '*Telamon*'.

between vessel element length and apple type still needs further study.

ACKNOWLEDGEMENTS

We thank Zhou Wenzhe's help in this study. This work was financed by China Agriculture Research System Foundation (No. CARS-28-01-07), Shandong Provincial Young Scientist Foundation (No. BS2009NY023), Shandong Provincial Improved Variety Engineering System Foundation (No.620902) and Qingdao Agricultural University Doctoral Foundation (No. 630732).

REFERENCES

Baas P, Werker E, Fahn A (1983). Some ecological trends in vessel characters. IAWA Bull. 4:141-59.

Carlquist S (1975). Ecological strategies of xylem wvolution [M]. California:University of California Press.

Chen Q, Xie HX (2003). Improved methods of segregation of vessel and wood fiber. Bull. Biol. 38(5):55.

Chen SS (2007). Studies on the vessel elements of secondary xylem in *Manilkara zapota*. Acta Hort. Sin. 34(1):7-10.

Chen SS, Tang WP (2004b). Observation of vessel elements of secondary xylem in *Aquilaria agallocha*. J. Centr. China Normal Univ. 38(4):486-489.

Chen SS, Tang WP (2004a). Observation of vessel elements of secondary xylem in *Grevllea robusta*. Guihaia. 24(4):380-382.

Chen SS, Tang WP (2005). Observation of vessel elements of secondary xylem in *Mangifera indica*. Acta Botanica Yunnanica. 27(6):644-648.

Dai HY, Wang CH, Chi B, Zhu J, Wang R, Li GX, Zhuang LL (2003). Report on Breeding Columnar Apple Varieties. J. fruit sci. 20(2):79-83.

Guo XM, Xiao X, Xu XY, Dong FY, Zhang LB (2008). Observation on the vessel elements of secondary xylem in late-ripening peach trees. J. Fruit Sci. 25(1):22-26.

Liang MX, Dai HY, Ge HJ (2009). Comparison of Leaf Structure and Chloroplast Ultrastructure Between Columnar and Standard Apples. Acta Hort. Sin. 36(10):1504-1510.

Tian YK, Wang CH, Zhang JS, James C, Dai HY (2005). Mapping Co, a gene controlling the columnar phenotype of apple, with molecular markers. Euphytica. 145:181-188.

Wang CH, Wang Q, Dai HY, Tian YK, Jia JH, Shu HR, Wang B (2002). Development of a SCAR Marker Linked to Co Gene of Apple from an AFLP Marker. Acta. Hort. Sin. 29(2):100-104.

Zhang YG, Liang MX, Zhu J, Dai HY (2010). comparing photosynthetic characteristics of the columnar and normal hybrids from 'Gala'×' Telamon' Hybrids in common type of apple. J. fruit sci. 27(Special issue):35-37.

Zhang W, Zhu YT, Li GC (2003). Genetic analysis of several morphological characters of Columnar apple hybrids. China Fruit. 3:11-13.

Zimmermann MH (1983). Xylem structure and the ascent of sap. Springer Verlag, Berlin.

7

Laboratory studies on ovipositional preference of the peach fruit fly *Bactrocera zonata* (Saunders) (Diptera: Tephiritidae) for different host fruits

Imran Rauf, Nazir Ahmad, S. M. Masoom Shah Rashdi, Muhammad Ismail and
M. Hamayoon Khan

Nuclear Institute of Agriculture Tandojam - 70060, Pakistan.

The peach fruit fly *Bactrocera zonata* (Saunders) is a serious pest of different fruits and vegetables in Pakistan inflicting economical damages. A laboratory experiment was conducted to determine the oviposition preference of the peach fruit fly for different fruits including guava, banana, citrus, ber, chikoo and apple under free or no choice conditions. Results showed that the guava was the most preferred host with mean pupal recovery of 318.00 ± 4.61 pupa/fruit (p/f) under free choice and 434 ± 2.64 p/f under no choice conditions, followed by banana (266.00 ± 4.5 p/f) in free choice and ber (177.00 ± 2.08 p/f) in no choice experiment. Whereas, apple and citrus were least preferred hosts.

Key words: *Bactrocera zonata*, host preference, oviposition.

INTRODUCTION

Fruit flies (Diptera: Tephritidae) cause most of the damage to fruits and vegetables in the Indo-Pak subcontinent. The members of the sub-family Dacinae infest almost all kinds of fleshy fruits, including solanaceous and cucurbitaceous plants. Many species are specialized, and host specific in their feeding habits, while others are generalists and attack a wide range of fruits and vegetables (Kapoor et al., 1980).

The peach fruit fly, *Bactrocera zonata* (Saunders), is a serious polyphagous pest originated in the South and South-East Asia where it attacks more than 50 host plants, including guava, mango, peach, apricot, fig and citrus (White and Elson-Harris, 1992; Ghanim, 2009). About 11 species of fruit flies have been documented so far from Pakistan and the most prominent among them are *B. zonata*, *Bactrocera cucurbitae*, *Bactrocera* *dorsalis*, *Myiopardalis pardalina*, *Carpomiya incompleta*, *Carpomiya vesuviana*, *Dacus ferrugincus* and *Dacus diversus* (Abdullah and Latif, 2001; Abdullah et al., 2002; Stonehouse et al., 2002; Panhwar, 2005). The favorable hosts of these fruit flies species in Pakistan are guava (*Psidium guajava*), mango (*Mangifera indica*), apple (*Malus domestica*), ber (*Zizyphus jujube*), musk melon (*Cucumis melo*) and bitter gourd (*Momordica charantia*), (Khan and Musakhel, 1999; Sultan et al., 2000; Ahmad et al., 2005).

The scope of damage reported by the fruit flies species, *B. zonta* was 5 to 100% loss in Pakistan (Syed et al., 1970). Damage caused by fruit flies to fruit and vegetable growers in Pakistan is about 200 million US dollars annually at farm level with added losses to traders, retailers and exporters (Stonehouse et al., 1998).

The greatest threat caused by the fruit flies is the rejection of fruit commodities especially mangoes due to the presence of its maggots and fruits becoming unfit for human consumption (Stonehouse et al., 2002).

The oviposition behavior in insects has been a widely studied theme in the insect-plant interaction context. This behavior is connected with the insect's specificity to determine the host plants, changes in host and insect-plant co-evolution (Thompson and Pellmyr, 1991). In holometabolous insects, the oviposition behavior is decisive in the choice of selecting a proper host plant for development of their immature, once they start moving they look for the nutritional resources selected by their adult females for their nourishment (Singer, 1986; Renwick, 1989). How the females select the proper host for oviposition is a quite complex phenomenon. The physical and chemical factors associated with plants influences the choice and the balance between positive and negative stimuli that determine the final selection of the proper host (Eisemann and Rice, 1985; McInnis, 1989; Oi and Mau, 1989; Messina, 1990; Kostal, 1993). Keeping in view the mode of damage and host range of B. zonata, the present study was designed to evaluate host preference of B. zonata under laboratory conditions.

MATERIALS AND METHODS

The study was conducted in the insectary of the fruit fly management unit at the Nuclear Institute of Agriculture, Tandojam, Pakistan. Six fruit hosts, viz. Banana (Musa cavendish), guava (P. guajava), apple (M. domestica), chikoo (Manilkara zapota), citrus (Citrus reticulata) and ber (Z. jujube), were selected. Banana, guava, chikoo and ber were collected from the Institute Orchards, whereas, apple and citrus were purchased from a local fruit market. The experiment was conducted in two phases, that is, six hosts (collectively) were offered as a free choice for oviposition, and secondly, each fruit was offered as separate treatment (no choice test).

Adult rearing cage

Peach fruit fly (B. zonata) used in this study were obtained from the culture maintained in the laboratory with 27 ± 1ºC, 60 ± 5% relative humidity and 14:10 light and dark. Two hundred (200) pairs of fruit flies from this stock culture were sexed and released into a cage (1 × 1 × 1 m). Flies were maintained in the room at 27 ± 1℃ , 60 ± 5% relative humidity, under the natural light phase. A mixture of sugar, yeast and water was placed in a Petri dish as a food supplement in the cage. Furthermore, regular changing of cotton swabs with water, in Petri dish was essential to avoid microbial contamination, especially mould development.

Choice and no choice tests

In the first phase, each fruit was offered as separate treatment for oviposition after determining the weight of each fruit (500 g). In the second phase, all fruits were collectively offered as free choice for oviposition. The fruits were exposed to the females for 2 h. The sexually mature female B. zonata successfully laid eggs on these fruits and then each fruit was kept separately in plastic jars

(5" × 12") containing fine sawdust at the bottom (for pupation) covered with blotting paper (to absorb excess moisture). The maggots, which developed in the fruits, exited to pupate in the sawdust. Puparia were collected and the number of puparia that emerged from each fruit was counted. These treatments were replicated three times. Observations were recorded fruit-wise on incubation, pupal weight, pupal recovery, deformity, adult emergence and sex ratio for each fruit. The data were tabulated and analyzed using analysis of variance, with F-tests as criteria with STATIX software.

RESULTS

Data presented in Table 1 show results of host preference under no choice conditions that is, each fruit was exposed separately. Statistical analysis showed that guava had the highest pupal recovery (434 pupa/fruit) and citrus had the least pupal recovery (5 p/f). Guava, ber and banana showed highly significant results as compared to other fruits. In the case of pupal weight, apple showed the maximum (13.42 mg) and citrus showed the least (4.27 mg). Generally, pupal weight of different fruits varied significantly. Adult emergence percentage was significantly higher in guava (90.98%) and apple (90.30%) compared to all other fruits. Data regarding sex ratio also varied significantly. Maximum population percentage of males was observed in ber (48.55%), while that of females in apple (54.59%). Deformity of emerging flies was significantly high in chikoo, apple and ber, where it recorded 8.93, 7.47 and 6.98%, respectively.

Table 2 shows results of oppositional response and other biological characters in response to opposition like pupal weight, sex ratio percentage adult emergence and deformity, when offered as free choice. Of all the fruits, B. zonata prefers laying eggs in guava fruit which showed maximum pupal recovery (318 p/f) followed by banana (266 p/f). The least pupal recovery was recorded from apple (0.00). In case of pupal weight, weigh of pupa resulted from guava and chikoo are 11.03 and 10.50 g, respectively. They were the highest significant as compared to other treatments. There was no significant difference observed in sex ratio among all treatments. Guava, ber, chikoo and citrus showed a non-significant difference of the percentage of emergence to each other but showed significant results against banana and apple. Maximum deformity was observed in banana (5.38%) followed by guava (2.95%) which was significantly higher than all other fruits.

DISCUSSION

It has been well documented that the oviposition in fruit flies depends upon their decision to select the proper host which must support the activities of their offsprings (Fontellas-Brandalha and Zucoloto, 2004; Joachim-Bravo et al., 2001). Other factors that may affect the oviposition preference in fruit flies include odor, color and shape of

Table 1. Some biological parameters of *B. zonata* resulted from infestation of different fruits under no choice test.

Fruit	No. recovered pupae/fruit	Pupal weight (mg)	Sex ratio (%)		Deformity (%)	Adult emergence (%)
			Male	Female		
Citrus	5.00 ± 1.15[e]	4.27 ± 0.03[f]	9.52 ± 9.5[d]	11.43 ± 5.94[d]	0.00 ± 0.00[b]	20.95 ± 12.38[d]
Chikoo	8.33 ± 0.88[de]	8.77 ± 0.12[c]	23.09 ± 4.63[cd]	35.95 ± 5.53[bc]	8.93 ± 4.49[a]	67.97 ± 2.76[bc]
Banana	120.67 ± 3.17[c]	9.28 ± 0.01[b]	27.84 ± 1.20[bc]	20.62 ± 1.85[cd]	5.23 ± 0.44[ab]	53.69 ± 3.37[c]
Ber	177.00 ± 2.08[b]	7.77 ± 0.06[e]	48.55 ± 1.40[a]	21.82 ± 0.90[cd]	6.98 ± 0.57[a]	77.36 ± 1.73[ab]
Apple	13.00 ± 1.15[d]	13.42 ± 0.10[a]	28.23 ± 1.27[bc]	54.59 ± 8.43[a]	7.47 ± 3.93[a]	90.30 ± 5.78[a]
Guava	434.00 ± 2.64[a]	8.35 ± 0.27[d]	41.14 ± 2.00[ab]	46.54 ± 1.25[ab]	3.30 ± 0.40[ab]	90.98 ± 3.06[a]

Means followed by the same letters, within a column, do not significantly differ at the 5% level according to the LSD test.

Table 2. Some biological parameters of *B. zonata* resulted from infestation of different fruits under free choice test.

Fruit	No. recovered pupae/fruit	Pupal weight (mg)	Sex ratio (%)		Deformity (%)	Adult emergence (%)
			Male	Female		
Citrus	38.66 ± 2.9[c]	8.84 ± 0.30[b]	47.50 ± 1.33[a]	42.03 ± 1.77[a]	0.00 ± 0.00[c]	89.54 ± 0.77[a]
Chikoo	9.00 ± 1.15[d]	10.50 ± 0.98[a]	54.69 ± 6.87[a]	32.75 ± 2.26[b]	0.00 ± 0.00[c]	87.44 ± 8.42[a]
Banana	266.00 ± 4.5[b]	7.53 ± 0.03[b]	11.51 ± 0.70[b]	19.91 ± 0.64[c]	5.38 ± 0.67[a]	36.81 ± 1.27[b]
Ber	29.66 ± 1.76[c]	8.00 ± 0.21[b]	48.58 ± 3.68[a]	38.20 ± 1.90[a]	1.01 ± 1.01[c]	87.80 ± 1.49[a]
Apple	0.00 ± 0.00[d]	0.00 ± 0.00[c]	0.00 ± 0.00[c]	0.00 ± 0.00[d]	0.00 ± 0.00[c]	0.00 ± 0.00[c]
Guava	318.00 ± 4.61[a]	11.03 ± 0.04[a]	46.89 ± 1.55[a]	38.73 ± 1.69[a]	2.95 ± 0.5[b]	88.56 ± 3.34[a]

Means followed by the same letters, within a column, do not significantly differ at the 5% level according to the LSD test.

host fruits (Li-Li et al., 2008).

In our study, the fruit flies showed maximum infestation on guava at free or no choice test as compared to other fruits. Different studies have reported guava as a most preferred host for the fruit flies (Mohammad and Abdel-Galil, 2008). According to the studies of Alies Van Sauers-Muller (2005), fruit flies infestation depends upon the size and shape of the fruit specially the guava. He also found that large sized fruit showed more susceptibility. The pupal recovery on guava in both the case free and forced choices was similar, however significant differences was observed in the pupal weight, deformity and half emergence which were comparatively high in free choice while there was a slight difference in emergence percentage in both the cases (White, 2000; Kapoor, 1989).

Results of the present study revealed that the pupal recovery in banana was the second after guava with less pupal weight and less emergence percentage in both free and no choice tests but the deformity and emergence was maximum in free choice in case of banana. This is because of the reason that delicate skin and particular aroma of banana make it a suitable host (Li-Li et al., 2008) to attract the fruit flies. Our results are in agreement with Li-Li et al. (2008) and Jayanthi and Abraham (2002). In case of citrus, we observed that there was a least fruit fly activity on no choice test but the maximum emergence was observed from the recovered pupae in case of free choice. Lies Van Sauers-Muller

(2005) reported that citrus was not found as an important host for fruit flies.

The fruit fly activities especially the pupal recovery, pupal weight and emergence in free and no choice tests on chikoo was at normal par, however, a significant variation was observed in case of deformity. This may be because the egg laying was done on chikoo but rests of the activities were not supported. The results of more pupal recovery on ber suggested the preference of fruit flies to use it as a favorite host because of the nutrient contents provided to the offsprings, while the other parameters studied on ber were at normal par for both free and no choices. Alies Van Sauers-Muller (2005) also observed that the sweet varieties of the certain fruits showed more infestation, while the sour varieties of these fruits were free of fruit fly infestation. Among all the fruit choices, apple was the least preferred host in free choice, however, maximum pupal weight was observed in case of no choice. According to the studies of Prokopy and Roitberg (1984) and Oi and Mau (1989), the oviposition of the fruit flies depends upon smell and visual signs to trace and identify the oviposition sites for egg laying on fruits. Similarly, Prokopy and Duan (1998) observed that the females of fruit flies use their previous experience to select a proper fruit for laying the eggs. The females also showed a learning ability to select the most appropriate host (Cooley and Prokopy, 1986).

According to the findings of Phillips (1977), Cassidy (1978), Cooley and Prokopy (1986), Prokopy and Papaj

(1988), and Hoffmann (1988), when there is no marked choice among the hosts for fruit flies there is always a definite liking for those fruits which have been previously visited and contacted by the females. In case of the choice of a host over another, the preference for the original host remains dominant but can decrease in terms of percent. Certain fruit characters specially the nutritional status of fruit also plays a vital role in supporting the larval activities. However, the biochemical analysis can further exploit this relationship.

Conclusion

Our investigation established that guava is the most preferable fruit for oviposition by peach fruit fly *B. zonata*. Banana and ber were also attracted when offered as free choice and no choice, respectively, whereas, apple showed least infestation.

ACKNOWLEDGEMENT

Author thanks goes to Mr. Muhammad Sarwar and Mr. Shafkatullah Niazi for their scientific assistance during handling and rearing procedures of peach fruit fly in the laboratory.

REFERENCES

Abdullah K, Latif A (2001). Studies on baits and dust formulations of insecticides against fruit fly (Diptera: Tephritidae) on melon (*Cucumis melo*) under semi arid conditions of D. I. Khan. Pak. J. Biol. Sci. 4:334-335.

Abdullah K, Akram M, Alizai AA (2002). Nontraditional control of fruit flies in guava orchards in D. I. Khan. Pak. J. Agric. Res. 17:195-196.

Ahmad B, Anjum R, Ahmad A. Yousaf MM, Hussain M, Muhammad W (2005). Comparison of different methods to control fruit fly (*Carpomyia vesuviana*) on ber (*Zizyphus mauritiana*). Pak. Entomol. 27:1-2.

Alies Van Sauers-Muller (2005). Host Plants of the Carambola Fruit Fly, *Bactrocera carambolae* Drew & Hancock (Diptera: Tephritidae), in Suriname, South Am. Neotrop. Entomol. 34(2):203-214.

Cassidy MD (1978). Development of an induced food plant frequence in the Indian stick insect *Carausius morosus*. Entomologia Exp. Appl. Dordrecht 24:87-93.

Cooley SS, Prokopy RJ (1986). Learning in oviposition site selection by *Ceratitis capitata* flies. Entomologia Exp. Appl. Dordrecht 40:47-51.

Eisemann CH, Rice MJ (1985). Oviposition behaviour of *Dacus tryoni*: the effects of some sugars and salts. Entomologia Exp. Appl. Dordrecht 39:61-71.

Fontellas-Brandalha TML, Zucoloto FS (2004). Selection of oviposition sites by wild *Anastrepha obliqua* (Macquart) (Diptera: Tephritidae) based on the nutritional composition. Neotrop. Entomol. 33:557-562.

Ghanim NM (2009). Studies on the peach fruit fly, *Bactrocera zonata* (Saunders) (Tephritidae, Diptera). Ph. D. Thesis, Faculty of Agriculture, Mansoura University.

Hoffmann AA (1988). Early adult experience in *Drosophila melanogaster*. J. Insect. Physiol. Oxford. 34(3):197-204.

Jayanthi PDK, Abraham V (2002). A simple and cost-effective mass rearing technique for the tephritid fruit fly, *Bactrocera dorsalis* (Hendel). Curr. Sci. 82(3):10-13.

Joachim-Bravo IS, Fernandes OA, de Bortoli SA, Zucoloto FS (2001). Oviposition behavior of *Ceratitis capitata* Wiedemann (Diptera: Tephritidae): Association between oviposition preference and larval performance in individual females. Neotrop. Entomol. 30:559-564.

Kapoor VC (1989). Fruit flies in India sub- Continent. (C.F. World Crop pests. Fruit flies, their biology, natural enemies control. 3A:59-61. Ed. By Robinson and Hooper).

Kapoor VC, Hardy DE, Agarwal ML, Grewal JS (1980). Fruit fly *(Diptera: Tephritidae)* systematics of the Indian subcontinent. Export Indian Publisher Jullundur, India. 3:59-61.

Khan SM, Musakhel MK (1999). Resistance in musk-melon (Cucumis melo L.) against melon fruit fly (*Bactocera cucurbitae* Coq.) and its chemical control in Dera Ismail Khan. Pak. J. Biol. Sci. 2:1481-1483.

Kostal V (1993). Physical and chemical factors influencing landing and oviposition by the cabbage root fly on host-plant models. Entomologia Exp. Appl. Dordrecht 66:109-118.

Li-Li QI, Li-Yan, Jiang Q, Zhou S, DAI H (2008). Oviposition preference of oriental fruit fly, *Bactrocera dorsalis*. Chin. Bull. Entomol. P. 4.

McInnis DO (1989). Artificial oviposition sphere for Mediterranean fruit flies (Diptera: Teprhitidae) in field cages. J. Econ. Ent. Lanham, 82(5):1382-1385.

Messina FJ (1990). Components of host choice by two *Rhagoletis* species (Diptera: Tephritidae) in Utah. J. Kans. Soc. Lawrence 63(1):80-87.

Mohammad AA, Abdel-Galil FA (2008). Infestation predisposition and relative susceptibility of certain edible fruit crops to the native and invading fruit flies (DIPTERA: TEPHRITIDAE) in the new valley Oases-Eygpt. Assoc. Univ. Bull. Environ. Res. 11(1):89-96.

Oi DH, Mau RFL (1989). Relationship of fruit ripeness to infestation in "Sharwil" avocados by the mediterranean fruit fly and oriental fruit fly (Diptera: Tephritidae). J. Econ. Ent. Lanham 82(2):556-560.

Panhwar F (2005). Mediterranean fruit fly (*Ceratitis capitata*) attack on fruits and its control in Sindh, Pakistan. Publisher: Digital Verlag GmbH, Germany, www.chemlin.de.

Phillips WM (1977). Modification of feeding preference in the flea-beetle, *H. lythri*. Entomologia Exp. Appl. Dordrecht. 21:71-80.

Prokopy RJ, Duan JJ (1998). Socially facilitated egglaying behavior in Mediterranean fruit flies. Behav. Ecol. Sociobiol, Heidelberg, 42(2):117-122.

Prokopy RI, Roitberg BD (1984). Foraging Behavior of the fruit flies. Am. Sci. 72:41-49.

Prokopy RJ, Papaj DR (1988). Learning of apple fruit biotypes by apple maggot flies. J. Insect Behav. New York 1(1):67-74.

Renwick JAA (1989). Chemical ecology of oviposition in phytophagous insects. Experientia, Barel, 45:223-228.

Singer MC (1986). The definition and measurement of oviposition preference in plant-feeding insects. *In*: Miller JR, Miller TA, ed. Insect-plant interactions, New York, Springer-Verlag. pp. 66-94.

Stonehouse JM, Mumford JD, Mustafa G (1998). Economic losses to Tephritid fruit flies (Diptera: Tephritidae) in Pakistan. Crop Protect. 17:159-164.

Stonehouse JM, Mahmood R, Poswal A, Mumforda J, Baloch KN, Chaudhary ZM, Makhdum AH, Mustafa G, Huggett D (2002). Farm field assessments of fruit flies (Diptera: Tephritidae) in Pakistan: distribution, damage and control. Crop Protect. 21:661-669.

Sultan MJ, Sabri A, Tariq M (2000). Different control measures against the insect pests of bitter gourd (*Momordica charantia* L.). Pak. J. Biol. Sci. 3:1054-1055.

Syed RA, Ghani MA, Murtaza M (1970). Studies on the tephritids and their natural enemies in West Pakistan. III. *Dacus zonatus* (Saunders). Technical Bulletin, Commonwealth Institute of Biol. Control, pp. 1-16.

Thompson JN, Pellmyr O (1991). Evolution of oviposition behavior and host preference in Lepidoptera. A. Rev. Ent., Palo Alto, **36**:65-89.

White IM (2000). Identification of peach fruit fly, *Bactrocera zonata* (Saund.) in the eastern Mediterranean. The Natural history museum, London, UK. pp. 1-12.

White IM, Elson-Harris MM (1992). *Fruit flies of economic significance: their identification and bionomics*. C.A.B. International, U.K. P. 601.

8

Effect of salinity on the growth performance and macronutrient status of four citrus cultivars grafted on trifoliate orange

Qingjiang Wei[1], Yongzhong Liu[1], Ou Sheng[1,2], Jicui An[1], Gaofeng Zhou[1] and Shu-ang Peng[1]

[1]College of Horticulture and Forestry, Huazhong Agricultural University, Shizishan Street No.1, Wuhan 430070, China.
[2]Institute of Fruit Tree Research, Guangdong Academy of Agricultural Science, Tianhe District, Guangzhou 510640, China.

To test the scions whether or not affect salt tolerance of citrus at the whole-plant level, four common cultivars belonging to sweet oranges (Newhall and Lane late) and loose-skin mandarins (Egan#1 and Guoqing#1) were grafted on trifoliate orange, and exposed to 0, 30, 60 or 90 mM Sodium chloride (NaCl). Results showed that, the NaCl treatment induced less reduction in dry weight (DW), leaf area, photosynthetic rate (P_N) but greater reduction in leaf transpiration rate (Tr) in loose-skin mandarins than in sweet oranges. It was further found that the loose-skin mandarins accumulated less Na^+ and Cl^- in leaves while more of these ions in their roots. Additionally, the changes in main nutrient contents varied among the salt-treated cultivars, and the loose-skin mandarins showed less decreases of leaf Mg^{2+} and Ca^{2+}. Overall, our results revealed that the different cultivars grafted on the same rootstock showed differences in growth performance and ions distribution under salinity. Higher salt tolerance shown by loose-skin mandarins may depend on the ability of excluding Na^+ and Cl^- from leaves and also be associated with maintaining nutritional homeostasis especially the balance of Mg^+ and Ca^{2+} in their leaves.

Key words: Citrus, leaf gas exchange, nutrient concentrations, plant growth, salt tolerance.

INTRODUCTION

Plants are frequently threatened by salt stress that affects growth and development, productivity and quality. It is estimated that over 800 million hectares land in the world are salt-affected and the salinity problem continues increasing (Munns, 2005). Citrus is a salt-sensitive horticultural crop and cultivated in the production areas where some are salt-affected because of natural salty soils near around the coasts, arid climates and also cultivation practices. However, even in same citrus

orchards, variations in physiological responses to salinity are found among cultivars or scion-rootstock combinations. Thus, to investigate the exact details of different responses to salinity in various cultivars may help to know the possible salt-tolerance mechanism and obtain more insight into the roles of scion in salt-tolerance of citrus.

Plant response to salinity occurs in two phases: osmotic phase and ion-specific phase. The osmotic phase is a rapid response to the high salt concentration around root, while the ion-specific phase is a slow response and results from the accumulation of high concentrations of Na^+ and Cl^- in leaves, which lead to growth inhibition and nutritional imbalance in plant (Munns and Tester, 2008). To cope with the adverse effect of salinity, plants have evolved many physiological and biochemical mechanisms. As for citrus seedlings, one of the important strategies is to control Cl^- uptake by root and transport into leaves, a mechanism known as chloride exclusion (Zekri and Parsons, 1992; Garciaf-Legaz et al., 1993; Moya et al., 2003; Perez-Tornero et al., 2009). Previous studies on scion-rootstock combinations also show that use of the salt-tolerant rootstocks could reduce the accumulation of Na^+ and/or Cl^- in leaves of scions and improve this stress tolerance in citrus (Cerda et al., 1990; Al-Yassin, 2004). For example, Garcia-Sanchez et al. (2006) compared the salinity effect on 'Clemenules' mandarin grafted on either Cleopatra mandarin or Carrizo citrange rootstock found that trees on Cleopatra accumulate less Cl^- and more Na^+ than those grafted on Carrizo. In addition, other studies show that the rootstock not only regulate Na^+ and Cl^- concentration but also affect N, K^+ and Ca^{2+} accumulation in leaves of scions (Storey and Walker, 1998; Garcia-Sanchez et al., 2002). Although rootstock plays an important role in citrus tolerance to salinity, some evidences suggest that the scion itself limit ions accumulation in leaves, and that the effect of salinity on gas exchange and growth appear to be more associated with scions (Nieves et al., 1991; Garcia-Legaz et al., 1993).

Till date, the ability and possible mechanism of salt-tolerance in citrus have been extensively studied. However, it is noticed that most researches focused on evaluating the different tolerance of rootstock seedlings or the effect of rootstock on salt-tolerance of scions. Few studies were on the salt performance of cultivars (namely different scions on single rootstock), and cultivar variation in element distribution and its relationship to salt-tolerance of citrus have not been reported yet. Given that, in our experiment, four common citrus cultivars were used as scions and grafted onto trifoliate orange, and various Sodium chloride (NaCl) concentrations were supplied to these grafted trees. Their growth performance and macroelement (except for sulfur) accumulation were compared among different treatments of NaCl concentrations.

MATERIALS AND METHODS

Plant materials and growth conditions

Scion cultivars including Newhall (*Citrus sinensis* cv. Newhall), Lane late (*C. sinensis* cv. Lane late), Egan#1 (*C. reticulate* cv. Egan#1) and Guoqing#1 (*C. unshiu* cv. Guoqing#1) grafted on trifoliate orange (*Poncirus trifoliata* [L.] Raf.) were obtained from National Citrus Breeding Center at Huazhong Agricultural

University. Grafted plants were randomly chosen depending on uniform and homogeneous size, and the roots of all plants were severely pruned in order to stimulate new root development. Subsequently, all plants were washed with tap water to remove surface contaminants, followed by transplantation to 10 l black pots containing a washed mixture of quartz sand: perlite (1:1 by volume). Plants were grown from May to September 2009 in a greenhouse with daily temperatures between 21 and 32°C, and 50 to 80% of relative humidity, under natural photoperiod conditions. The plants were watered 4 weeks with a quarter strength modified Hoagland's Number 2 solution. Then salt treatments were applied when the scions were about 7 to 9 full-expanded leaves. The plants were watered every two days with half strength Hoagland's Number 2 solution plus 0 (control) 30, 60 or 90 mM NaCl, respectively. The modified full strength solution contained the following nutrients: 6 mM KNO_3, 4 mM $Ca(NO_3)_2$, 1 mM $NH_4H_2PO_4$, 2 mM $MgSO_4$, 9 μmol $MnCl_2$, 0.8 μmol $ZnSO_4$, 0.3 μmol $CuSO_4$, 0.01 μmol H_2MoO_4 and 50 μmol Fe-EDTA. To avoid building up toxic saline ions in culture media, the plants were irrigated with distilled water at 10 days intervals.

Measurements of leaf gas exchange

At the end of the experiment, the leaf gas exchange was measured using a portable photometer (LI-6400, USA). The measurement was performed on undamaged tissue of mature leaves chosen from the middle of the shoot of each plant (five replicates per treatment). All measurements were carried out between 08:00 and 11:00 a.m. under air CO_2 concentration of $385 ± 10$ μmol mol^{-1} and leaf temperature between 28 and 30°C. The values of net photosynthetic rates (P_N), transpiration rate (Tr) were recorded, and water use efficiency (WUE) was calculated as: WUE = P_N/Tr.

Plant harvest and measurement of mineral contents

After 2 months of salt treatments, the plants were harvested, and then leaves and roots were separated immediately. The leaf area was determined using a leaf area meter (LiCOR-3100). For mineral contents determination, the leaves and roots were rinsed twice with distilled water, oven-dried at 70°C to constant mass, weighed and ground to a fine powder to pass a 40-mesh screen.

Total nitrogen (N) of the plant tissue was determined by discrete auto analyzer (Smartchem200, Italy) after digested with concentrated sulfuric acid and hydrogen peroxide. For P, Na^+, K^+, Ca^{2+} and Mg^{2+} concentration, dried samples were digested in 0.1 M HCl and determined by inductively coupled plasma spectrometry (Iris Intrepid II, USA) according to the method of Sheng et al. (2009).

Determination of chloride concentration was performed as previously described (Moya et al., 2003). About 0.1 g dried powders were extracted in a 10 ml mixture of 0.1 M HNO_3 and 10% (v/v) CH_3COOH solution, and the samples were incubated at room temperature overnight. After filtering, 0.5 ml of the solution was used for determination in a silver ion titration chloridemeter (Model 926, Cambridge, UK).

Statistical analysis of the data

The experiment was set up as a completely randomized design including four NaCl levels and four scion cultivars with four replicates of each cultivar × NaCl combination. Data were subjected to analysis of variance (ANOVA), and the differences between means were separated following the Duncan's test at P<0.05. A two-factorial analysis was also performed to study the effects of cultivar, NaCl stress, and their interactions on the plants.

Table 1. Effect of external NaCl (mM) on dry weight of leaves(g), dry weight of roots(g), dry weight of total plants(g), leaf area(cm^2), photosynthetic rate (P$_N$, µmol m^{-2} s^{-1}) ,transpiration rate (Tr, mmol H$_2$0 m^{-2} s^{-1}) and water use efficiency (WUE, mmol mol^{-1}) of four cultivars grafted on trifoliate orange.

Cultivars	NaCl	DW leave	DW root	DW total	Leaf area	P$_N$	Tr	WUE
Newhall	0	10.26^a	8.77^a	26.16^a	1224^a	10.11^a	2.01^{ab}	5.05^{ef}
	30	8.9^b	5.87^b	20.58^b	1146.65^a	7.31^e	1.72^c	4.26^f
	60	5.49^c	5.01^c	14.72^c	715.3^b	4.81^{gh}	0.84^f	6.03^{def}
	90	0.88^d	2.03^h	4.14^h	122.28^g	3.01^j	0.44^g	6.87^{de}
Lane late	0	4.75^d	4.63^{cd}	13.51^c	569.15^c	10.52^a	1.89^b	5.57^{ef}
	30	4.56d^e	4.27^d	11.15^d	556.13^c	8.63^{bc}	1.43^d	6.28^{def}
	60	3.17^f	3.3^{ef}	8.4^e	287.88^{de}	5.28^{fgh}	0.75^f	7.06^{de}
	90	1.71^{hi}	2.88^{fg}	6.01f^g	165.73^{efg}	3.35^{ij}	0.49^g	6.89^{de}
Egan#1	0	4.23^{de}	3.93^{de}	11.53^d	558.23^c	7.82^{de}	1.40^d	5.62^{ef}
	30	4.24^{de}	4.15^d	11.22^d	568.85^c	7.16^e	1.09^e	6.66^{de}
	60	3.97^e	3.98^{de}	10.77^d	519.48^c	5.95^f	0.46^g	13.26^b
	90	2.11^{gh}	3.25^{ef}	7.37^{ef}	274.87^{def}	4.62^h	0.19^h	25.15^a
Guoqing#1	0	2.53^g	2.18^{gh}	6.84^f	326.75^d	9.34^b	2.11^a	4.43^f
	30	1.59^{hi}	2.36^{gh}	5.98^{fg}	269.1^{def}	8.15^{cd}	1.61^c	5.07^{ef}
	60	1.18^{ji}	2.23^{gh}	4.14^h	217.13^{defg}	5.54^{fg}	0.72^f	7.69^d
	90	0.64^j	1.85^h	3.72^h	140.75^{fg}	3.79^i	0.34^g	11.16^c
F$_{Cultivar}$		303.75***	113.66***	279.98***	98.88***	5.47**	58.21***	111.15***
F$_{NaCl}$		233.88***	61.55***	199.84***	83.55***	375.65***	600.02***	113.70***
F$_{Cultivar}$ × F$_{NaCl}$		47.53***	26.12***	47.89***	15.99***	10.94***	4.7***	37.28***

Mean values within a column followed by different letters are significant different (P = 0.05) according to Duncan's test (n = 4). F values for the cultivars, NaCl and their interaction are also shown. Ns,*, ** Indicate non-significant differences, significant differences at P<0.05 and 0.01, respectively.

Correlation analysis among leaf ion concentrations and some other determined parameters was conducted using Pearson's method.

RESULTS

Plant dry weight, leaf area, net photosynthetic (P$_N$), transpiration rate (Tr) and water use efficiency (WUE)

Plant dry weigh and leaf area were significantly affected by cultivar, salt and their interaction. Although the dry weigh was generally reduced by salinity, the degree of reduction varied among cultivars (Table 1). For example, Leaf dry weight showed the least reduction in Egan#1. For the most reduction, it was found in Guoqing#1 at 30 and 60 mM NaCl, while in Newhall when 90 mM NaCl were supplemented. Root dry weight decreased in sweet oranges whereas increased slightly in loose-skin mandarins at relatively low NaCl concentrations (30 and 60 mM). At 90 mM NaCl, the root biomass decreased in all treated cultivars, and the reduction was more in sweet oranges especially in Newhall, as compared with loose-skin mandarins (Table 1). With respect to total plant dry weight, it decreased as NaCl level increased, and the

reduction was the least in Egan#1, followed by Guoqing#1, Lane late and Newhall. As with plant dry weight, leaf area sharply decreased under salinity, and showed similar reduction pattern with leaf biomass among cultivars. It was further found that the leaf area of Egan#1 and Guoqing#1 exhibited no significant difference till 90 mM NaCl supplemented, but significant decreases were observed on Newhall and Lane late when the higher than 60 mM NaCl supplemented (Table 1).

In addition, the P$_N$, Tr, and WUE were also significantly affected by cultivar, salt and cultivar × NaCl interaction. As shown in Table 1, both P$_N$ and Tr were significantly decreased by NaCl stress, but they showed different reductions among cultivars. Compared with the controls, the reduction in P$_N$ was the least in Egan#1 (8.44 to 40.92%), followed by Guoqing#1(12.74 to 59.42%), Lane late (17.97 to 68.16%) and Newhall (27.7 to 70.23%) when external NaCl concentrations increased from 30 to 90 mM. In contrast, the reduction in Tr was greater in loose-skin mandarin than that in sweet oranges. Based on the changes in P$_N$ and Tr, it is clear the loose-skin mandarin possessed higher WUE under 60 and 90 mM NaCl stress.

Table 2. Effect of external NaCl (mM) on leaf mineral concentration (mg g^{-1} DW) of four cultivars grafted on trifoliate orange.

Cultivars	NaCl	Na$^+$	Cl$^-$	N	P	K$^+$	Ca^{2+}	Mg^{2+}
Newhall	0	0.99^h	0.81^f	34.8^a	2.19^a	26.27^a	35.75^a	4.14^d
	30	7.23^e	8.57^e	33.22^b	2.02^b	24.82^{ab}	31.24^{bcd}	3.46^{ef}
	60	14.91^c	27.51^c	29.58^{cde}	1.89^b	24.01^{ab}	30.52^{cde}	3.53^e
	90	21.81^a	42.06^a	28.02^{ef}	1.71^c	23.06^{bc}	28.8^{def}	2.71^g
Lane late	0	1.14^h	0.67^f	33.07^b	1.7^c	25.2^{ab}	33.37^b	5.62^a
	30	5.82^f	9.39^e	30.65^c	1.63^{cd}	23.73^b	31.58^{bcd}	4.91^b
	60	12.74^d	24.82^d	28.47^{de}	1.53^{def}	23.14^{bc}	27.83^{fg}	4.69^{bc}
	90	20.67^a	41.8^a	26.6^f	1.4f^g	20.96^c	26.02^g	3.4^{ef}
Egan#1	0	0.75^h	0.91^f	33.02^b	1.59cde	24.4ab	32.36^{bc}	4.67^{bc}
	30	3.59^g	11.12^e	30.85^c	1.44^{efg}	24.65^{ab}	31.01^{bcd}	4.3^{cd}
	60	8.31^e	22.72^d	29.69^{cde}	1.45^{efg}	23.03^{bc}	29.76^{edf}	4.09^d
	90	15.53^c	33.94^b	29.87^{cd}	1.4^{fg}	23.48^b	30.43^{cde}	3.67^e
Guoqing#1	0	0.77^h	1.11^f	31.13^c	1.49^{defg}	24.44^{ab}	31.06^{bcd}	4.27^{cd}
	30	3.39^g	10.66^e	29.57^{cde}	1.38^{fg}	23.48^b	29.45^{edf}	3.57^e
	60	12.05^d	24.86^d	28.14^{def}	1.32^g	24.1^{ab}	30.38^{cde}	3.44^{ef}
	90	17.87^b	35.61^b	28.43^{de}	1.11^h	22.85^{bc}	28.38^{ef}	3.05^{fg}
FCultivar		16.62***	3.19N^s	1.14^{Ns}	29.53***	2.18^{Ns}	5.71*	64.39***
FNaCl		175.64***	170.01***	7.46***	10.91***	8.12***	26.49***	73.57***
FCultivar × FNaCl		3.16^{Ns}	2.85^{Ns}	0.69^{Ns}	0.28^{Ns}	1.02^{Ns}	3.36*	2.89^{Ns}

Mean values within a column followed by different letters are significant different (P = 0.05) according to Duncan's test (n = 4). F values for the cultivars, NaCl and their interaction are also shown. Ns,*, ** Indicate non-significant differences, significant differences at P<0.05 and 0.01, respectively.

Mineral elements accumulation in the leaves

All the leaf mineral elements determined in this study were significantly influenced by salinity. In addition, Na$^+$, P, Ca^{2+} and Mg^{2+} contents were significantly influenced by cultivars, while only the Ca^{2+} content was significantly influenced by cultivars × NaCl interaction (Table 2).

Our results showed that both Na$^+$ and Cl$^-$ contents sharply increased in the leaves of the salt-treated plants. Accumulation of Na$^+$ was significant lower in loose-skin mandarins than that in sweet oranges under the same NaCl concentrations. For leaf Cl$^-$, the loose-skin mandarins generally accumulated higher contents at 30 mM NaCl whereas lower contents at 60 and 90 mM NaCl. However, no significant differences in the leaf Cl$^-$ accumulation was found among cultivars at 30 and 60 mM NaCl. At 90 mM NaCl, the Cl$^-$ content was significant lower in loose-skin mandarins than that in sweet oranges (Table 2).

Total nitrogen (N) reductions in loose-skin mandarins were almost parallel among different salt levels. By contrast, the leaf N in sweet oranges decreased proportionally with increasing salinity and showed greater decreases than that in loose-skin mandarins at 90 mM NaCl. Leaf P was also reduced by salinity in all cultivars

(Table 2). Even in the control treatment, leaf P varied among cultivars with the highest value occurring in Newhall and the lowest values occurring in Egan#1 and Guoqing#1. The K$^+$ decreased in the salt-treated plants, but it showed no significant change in all cultivars expect the sweet oranges at 90 mM NaCl, as compared with the controls. Additionally, leaf Ca^{2+} was also decreased in all cultivars by salinity especially higher NaCl concentrations, and the severity of reduction depended on the cultivars. Greater decreases of the Ca^{2+} were obtained in Newhall and Lane late (Table 2). For leaf Mg^{2+}, it was significantly reduced in all cultivars and showed the least reduction in Egan#1 under salinity condition. Other three cultivars possessed similar reductions in the leaf Mg^{2+} at 30 or 60 mM NaCl, but Newhall and Lane late showed much more reductions when 90 mM NaCl supplemented (Table 2).

Mineral elements accumulation in the roots

In the roots, all the mineral elements contents were significantly affected by cultivars and NaCl. One exception is the Mg^{2+} content, which was not significantly changed by NaCl. Moreover, the Na$^+$, Cl$^-$ and P content

Table 3. Effect of external NaCl (mM) on root mineral concentration (mg g^{-1} DW) of four cultivars grafted on trifoliate orange.

Cultivars	NaCl	Na$^+$	Cl$^-$	N	P	K$^+$	Ca^{2+}	Mg^{2+}
Newhall	0	0.68^g	1.03^g	27.2^b	2.04^{cd}	19.68^a	13.88^{ab}	3.23^{bcde}
	30	4.4^f	10.38^f	25.64^{bc}	1.83^{def}	14.93^{cd}	11.47^d	3.18^{cde}
	60	6.43^e	13.34^d	25.24^{bc}	1.78^{efg}	12.92^{efg}	11.69^d	3.05^e
	90	6.7^e	13.21^{de}	19.84^{ef}	1.43^{hij}	12.43^{fg}	11.06^d	3.14^{de}
Lane late	0	0.45^g	0.86^g	30.67^a	2.62^a	19.51^a	14.17^{ab}	3.72^a
	30	6.32^e	11.18d^e	27.28^b	2.17^{bc}	14.31^{de}	12.55^{bcd}	3.63^{abc}
	60	8.89^d	15.2^d	23.13^{cd}	1.51^{hij}	12.58^{fg}	11.66^d	3.45^{abcde}
	90	8.51^d	13.94^d	19.02^{fg}	1.28^{jk}	12.43^{fg}	11.16^d	3.47^{abcde}
Egan#1	0	0.58^g	0.39^g	30.2^a	2.3^b	20.82^a	12.77^{bcd}	3.69a^b
	30	9.29^d	13.93^d	27.44^b	1.9^{de}	15.99^{bc}	12.42^{bcd}	3.72^a
	60	11.09^c	19.23^{bc}	24.39^c	1.54^{ghi}	13.74^{def}	12.61^{bcd}	3.57^{abcd}
	90	13.94^a	20.8^{ab}	21.84^{de}	1.32^{ijk}	14.2^{de}	12.05^{cd}	3.53^{abcd}
Guoqing#1	0	0.58^g	0.74^g	24.28^c	1.97^{cde}	17.21^b	15.06^a	3.92^a
	30	4.4^f	13.66^d	20.89^{def}	1.62^{fgh}	12.64^{fg}	13.74^{abc}	3.86^a
	60	10.98^c	18.21^c	20.52^{ef}	1.55^{ghi}	10.84^h	11.16^d	3.14^{de}
	90	12.95^b	21.5^a	16.95^g	1.16^k	11.53^{gh}	11.06^d	3.05^e
F$_{Cultivar}$		115.55***	35.21***	33.31***	10.41***	28.98***	8.96***	15.87***
F$_{NaCl}$		731.55***	448.61***	81.56***	95.25***	175***	12.27***	1.76^{Ns}
F$_{Cultivar}$ × F$_{NaCl}$		29.74***	7.38***	2.42^{Ns}	4.73***	0.48^{Ns}	1.02^{Ns}	0.1^{Ns}

Mean values within a column followed by different letters are significant different (P = 0.05) according to Duncan's test (n = 4). F values for the cultivars, NaCl and their interaction are also shown. Ns,*, ** Indicate non-significant differences, significant differences at P<0.05 and 0.01, respectively.

was significantly affected by cultivars × NaCl interaction (Table 3).

The accumulation of Na$^+$ and Cl$^-$ in root exhibited an opposite trend with that in leaves, and they always accumulated more in Egan#1 and Guoqing#1 than in Lane late and Newhall. It was further found that, compared with 60 mM NaCl treatment, the 90 mM NaCl enhanced root Na$^+$ and Cl$^-$ accumulation in loose-skin mandarins, but it had almost no significant effect on the saline ions contents in the roots of sweet oranges (Table 3).

Root N concentration was decreased by every NaCl treatment and higher NaCl-treated concentration caused higher reduction in the N accumulation (Table 3). Despite this, Newhall showed significant decrease only when 90 mM NaCl supplemented. The influence of salinity on the P accumulations in the roots was general greater than that on leaves, and the root P of all treated plants were significantly decreased by various salt concentrations. Root K$^+$, in contrast with leaf K$^+$, was significantly decreased in all cultivars by various salinity levels, but no great difference in reduction were found among cultivars under the same NaCl treatment. Salinity also reduced root Ca^{2+} concentration in all cultivars except for Egan#1, which showed no significant decrease in the Ca^{2+} under

all NaCl conditions. Although the leaf Mg^{2+} was greatly reduced by salinity, root Mg^{2+} showed no significant decrease in any of the cultivars except Guoqing#1 at higher salt levels.

DISCUSSION

Growth response to Sodium chloride (NaCl) stress

The ANOVA for cultivars, NaCl levels and their interaction was found to be significant for growth and photosynthetic parameters, indicating that all cultivars responded differently to salt stress with respect to these parameters. Salt stress greatly inhibits plant growth, and biomass production is always used as a direct index of salt tolerance in plants (Misra et al., 1997; Rahnama et al., 2011). Here, Pearson correlation analysis showed that the leaf area was significantly correlated with the plant dry weigh (Table 4). Thus, it might also be a good indicator of plant growth under salinity. Under normal condition, the sweet oranges always had more vigorous growth. However, they showed greater reductions in biomass and leaf area than loose-skin mandarins under salinity (Table 1). This suggests that Newhall and Lane

Table 4. Pearson correlation coefficients between leaf ion concentrations and some other determined traits.

Parameter	DW leaf	DW root	DW total	Leaf area	P_N	Tr	WUE	Na+	Cl-	N	P	K+	Ca2+	Mg2+
DW leaf	1													
DW root	0.936**	1												
DW total	0.985**	0.967**	1											
Leaf area	0.955**	0.886**	0.942**	1										
PN	0.492**	0.439**	0.490**	0.471**	1									
Tr	0.549**	0.437**	0.532**	0.538**	0.815**	1								
WUE	-0.494**	-0.264*	-0.307*	-0.3*	-0.287*	-0.467**	1							
Na+	-0.494**	-0.407**	-0.469**	-0.501**	-0.844**	-0.828**	0.255Ns	1						
Cl-	-0.577**	-0.474**	-0.551**	-0.583**	-0.847**	-0.898**	0.33*	0.965**	1					
N	0.725**	0.676**	0.732**	0.72**	0.676**	0.701**	-0.25Ns	-0.732**	-0.764**	1				
P	0.81**	0.822**	0.842**	0.781**	0.341**	0.482**	-0.303*	-0.23Ns	-0.355**	0.589**	1			
K+	0.471**	0.449**	0.458**	0.505**	0.482**	0.478**	-0.227Ns	-0.476**	-0.505**	0.564**	0.437**	1		
Ca2+	0.572**	0.573**	0.61**	0.606**	0.624**	0.55**	-0.116Ns	-0.613**	-0.639**	0.745**	0.496**	0.481**	1	
Mg2+	0.279*	0.239Ns	0.242Ns	0.234Ns	0.532**	0.465**	-0.131Ns	-0.663**	-0.658**	0.481**	0.108Ns	0.28*	0.40**	1

Ns, *, ** Indicate non-significant differences, significant differences at $P<0.05$ and 0.01, respectively.

late are relative sensitive to salt stress, as compared with Egan#1 and Guoqing#1.

Distribution of Na+ and Cl-

In citrus, the relative salt-tolerant rootstocks always accumulate lower Cl- concentration in their leaves, and it is proposed that the physiological disturbances produced by salinity are more related to leaf Cl- build-up (White and Broadley, 2001; Moya et al., 2003; Brumos et al., 2010). In this study, we found that the leaf Cl- content generally showed greater correlation coefficients with the plant dry weight, leaf area, P_N and Tr, than that of leaf Na+ content (Table 4). In spite of this, both leaf Cl- and Na+ contents were generally lower in Egan#1 and Guoqing#1 than in Newhall and Lane late under various NaCl levels. Thus, it is difficult to differentiate between the effects of Na+ and Cl-. Syvertsen et al. (2010) reported that, high WUE are related to low leaf Cl-. Our results showed that the WUE was weak related to other determined traits, but it was still more related to leaf Cl- than other elements (Table 4). Therefore, we conclude that that the salinity induced less decrease of leaf water status in loose-skin mandarins, contributing to the less Na+ and/or Cl- accumulation in the leaves.

Furthermore, we found that Egan#1 and Guoqing#1 accumulated more Na+ and Cl- in their roots than that in Newhall and Lane late (Table 3). The result combined with the saline ions distributions in leaves indicated that the Egan#1 and Guoqing#1 had superior ability to exclude Na+ and Cl- from their leaves than other two tested cultivars, which are beneficial to keep a high photosynthetic activity for more biomass production. It was further found that the root Na+ and Cl- continued to increase in loose-skin mandarins but not in sweet oranges at 90 mM NaCl. This could be explained by that the 90 mM NaCl treatment lead to an excess of the upper limit of Na+ and Cl- that the roots of sweet oranges can loaded. And this high NaCl concentration break down the ions exclusion ability in sweet oranges, but it has no obvious effect on that of loose-skin mandarins. This is in line with the earlier observation that the ions exclusion ability in citrus could be lost at high salinity (White and Broadley, 2001; Brumos et al., 2010).

Variations in main nutrient contents

Plants acquire essential nutrients from their root system environment, but salinity may disturb the nutrients absorption and translocation in plant by ways of competitive interactions or affecting the ion selectivity of membranes (Grattan and Grieve,

1992; Munns and Tester, 2008; Wang et al., 2012). To explore the nutrient status of citrus under NaCl stress, we also investigated major nutrients in leaves and roots of the tested cultivars. Results showed that, the salinity changed N, P, K^+, Ca^{2+} and Mg^{2+} concentrations in all treated plants, but the changes were varied depended on cultivars and plant parts.

Leaf N accumulation showed less decreases in Egan#1 and Guoqing#1 under salinity. This could be partly attributed to the lower leaf Cl^- concentration in leaves of the two cultivars due to the competitive interactions between Cl^- and N, which showed an obvious negative correlation (r = -0.764). Although P is significant correlated with plant dry weight and leaf area, it seems that there is no obvious difference in P reductions among cultivars differing in salt tolerance in this study. Similarly, Chatzissavvidis et al. (2008) also reported that, the interactions between salinity and P depended on complex interactions of many factor.

Under NaCl stress, high concentration of Na^+ in plants always inhibits the accumulations of other cations. Our data showed that the NaCl treatment decreased K^+ content in roots but not in leaves of all cultivars. Previous studies reported that accumulation of K^+ in leaves could keep the balance of Na/K within plants and thus regulate osmotic balance against large effluxes of saline ions (Zhu, 2003; Al-Yassin, 2004). However, a strong relationship between leaf K^+ and salt tolerance did not find in this study. Different with the leaf K^+, both leaf Ca^{2+} and Mg^{2+} showed less reductions in the relative salt-tolerance cultivars, Egan#1 and Guoqing#1, indicating the ions play positive roles in these cultivars response to salt stress. Indeed, the Ca^{2+} was reported to be important in maintaining cell membrane integrity and regulating selective absorption of K/Na (Parida and Das, 2005). Furthermore, salinity reduced the root Ca^{2+} concentration in all cultivars except for Egan#1. This is consistent with earlier findings that salinity reduces Ca^{2+} accumulation in more salt-sensitive cultivars (Ruiz et al., 1997; Gimeno et al., 2010). The magnesium is the center of the chlorophyll molecule. Thus, the less reduction of leaf Mg^{2+} concentrations in Egan#1 and Guoqing#1 may help them to keep a higher P_N. This was further supported by the observation that the leaf Mg^{2+} had a positive correlation with P_N (Table 4). However, root Mg^{2+} showed no significant changes in most of the salt-treated cultivars. This indicates that salinity inhibited Mg^{2+} translocation from roots to leaves, but it had no effects on Mg^{2+} uptake by roots.

Conclusion

One familiar strategy of improving the salt-tolerance of citrus is to graft the scions onto salt-tolerant rootstocks. Our results revealed that the Egan#1 and Guoqing#1

showed superior salt-tolerance than the Newhall and Lane late under salinity, although they were grafted on the same rootstocks. This suggests that the toxic action of NaCl could be partly mitigated by use of suitable scion. Results presented here indicated that, the high leaf WUE were related to control of Cl^- accumulation in the leaves of Egan#1 and Guoqing#1. In addition, the homeostasis of leaf Mg^{2+} and Ca^{2+} might also be involved in the salt adaptation of the loose-skin mandarins.

ACKNOWLEDGEMENTS

This work was financially supported by National Modern Agriculture (Citrus) Industry System of Special Funds (No. CARS-27) and Special Fund for Agro-scientific Research in the Public Interest (No. 201203075).

REFERENCES

Al-Yassin A (2004). Influence of salinity on citrus: a review paper. J. Cent. Eur. Agric. 5:263-272.

Brumos J, Talon M, Bouhlal R, Colmenero-Flores JM (2010). Cl homeostasis in includer and excluder citrus rootstocks: Transport mechanisms and identification of candidate genes. Plant Cell Environ. 33:2012-2027.

Cerda A, Nieves M, Guillen MG (1990). Salt tolerance of lemon trees as affected by rootstock. Irrigation Sci. 11:245-249.

Garciaf-Legaz MF, Ortiz JM, Garci-Lidon A, Cerda A (1993). Effect of salinity on growth, ion content and CO_2 assimilation rate in lemon varieties on different rootstocks. Physiol. Plant. 89:427-432.

Garcia-Sanchez F, Jifon JL, Carvajal M, Syvertsen JP (2002). Gas exchange, chlorophyll and nutrient contents in relation to Na^+ and Cl^- accumulation in 'Sunburst' mandarin grafted on different rootstocks. Plant Sci. 162:705-712.

Garcia-Sanchez F, Perez-Perez JG, Botia P, Martinez V (2006). The response of young mandarin trees grown under saline conditions depends on the rootstock. Eur. J. Agron. 24:129-139.

Gimeno V, Syvertsen J, Rubio F, Martinez V, Garcia-Sanchez F (2010). Growth and mineral nutrition are affected by substrate type and salt stress in seedlings of two contrasting citrus rootstocks. J. Plant Nutr. 33:1435-1447.

Grattan SR, Grieve CM (1992). Mineral element acquisition and growth response of plants grown in saline environments. Agric. ecosyst. Environ. 38:275-300.

Misra AN, Sahu SM, Misra M, Singh P, Meera I, Das N, Kar M, Sahu P (1997). Sodium chloride induced changes in leaf growth, and pigment and protein contents in two rice cultivars. Biol. Plant 39:257-262.

Moya JL, Gomez-Cadenas A, Primo-Millo E, Talon M (2003). Chloride absorption in salt-sensitive Carrizo citrange and salt-tolerant Cleopatra mandarin citrus rootstocks is linked to water use. J. Exp. Bot. 54:825-833.

Munns R (2005). Genes and salt tolerance: bringing them together. New Phytol. 167:645-663.

Munns R, Tester M (2008). Mechanisms of salinity tolerance. Annu. Rev. Plant Biol. 59:651-681.

Nieves M, Cerda A, Botella M (1991). Salt tolerance of two lemon scions measured by leaf chloride and sodium accumulation. J. Plant Nutr. 14:623-636.

Parida AK, Das AB (2005). Salt tolerance and salinity effects on plants: a review. Ecotox. Environ. Safe. 60:324-349.

Perez-Tornero O, Tallon CI, Porras I, Navarro JM (2009). Physiological and growth changes in micropropagated citrus macrophylla explants due to salinity. J. Plant Physiol. 166:1923-1933.

Rahnama A, Poustini K, Tavakkol-Afshari R, Ahmadi A, Alizadeh H (2011). Growth properties and ion distribution in different tissues of bread wheat genotypes (*Triticum aestivum* L.) differing in salt tolerance. J. Agron. Crop Sci. 197:21-30.

Ruiz D, Martínez V, Cerda A (1997). Citrus response to salinity: Growth and nutrient uptake. Tree Physiol. 17:141-150.

Sheng O, Song SW, Peng SA, Deng XX (2009). The effects of low boron on growth, gas exchange, boron concentration and distribution of 'Newhall' navel orange (*Citrus sinensis* Osb.) plants grafted on two rootstocks. Sci. Hortic. 121:278-283.

Storey R, Walker RR (1998). Citrus and salinity. Sci. Hortic. 78:39-81.

Syvertsen JP, Melgar JC, Garcia-Sanchez F (2010). Salinity tolerance and leaf water use efficiency in citrus. J. Am. Soc. Hortic. Sci. 135:33-39.

Wang H, Wu Z, Zhou Y, Han J, Shi D (2012). Effects of salt stress on ion balance and nitrogen metabolism in rice. Plant Soil Environ.58:62-67.

White PJ, Broadley MR (2001). Chloride in soils and its uptake and movement within the plant: A review. Ann. Bot-London. 88:967-988.

Zekri M, Parsons LR (1992). Salinity tolerance of citrus rootstocks: Effects of salt on root and leaf mineral concentrations. Plant Soil. 147:171-181.

Zhu JK (2003). Regulation of ion homeostasis under salt stress. Curr. Opin. Plant Biol. 6:441-445.

9

Identification of culturable endophytes in 'Champaka' pineapple grown in an organic system

Olmar Baller Weber[1], Sandy Sampaio Videira[2] and Jean Luiz Simões de Araújo[2]

[1]Embrapa Tropical Agroindustry, Dra. Sara Mesquita, 2270, Pici, 60511-110, Fortaleza, Ceará, Brazil.
[2]Embrapa Agrobiology, BR 465, km07, 23890-000, Seropédica, Rio de Janeiro, Brazil.

The objective of the study was to characterize culturable diazotrophic endophytic bacteria from 'Champaka' pineapple plants during the fruiting period in an organic system. Micropropagated plants were inoculated with the diazotrophic endophytic bacterium, strain AB 219a, during acclimatization phase. In the field, the plants were fertilized with different dosages and sources of organic compost. After 17 months of field cultivation, roots and stems were collected for the isolation of endophytes, DNA extraction, 16S rDNA amplification and molecular analysis. A total of nine bacterial groups were identified, and species *Burkholderia silvatlantica*, *Azorhizobium caulinodans*, *Pantoea eucrina*, *Erwinia* sp. and unculturable bacterium occurred in inoculated plants. In contrast, non-inoculated plants were associated endophytes related to *Burkholderia cenocepacia*, *Azospirillum brasilense*, *Enterobacter oryzae*, *Erwinia* sp. and *Sphingobium yanoikuyae*. Segments of 360 base pair from nifH gene were amplified from representative endophytes within identified species, except from the *Pantoea eucrina* strain AB 295. 16S rDNA phylogenetic analysis of endophytes isolated from inoculated plants revealed distinct families: Xanthobacteraceae, Burkholderiaceae and Enterobacteriaceae; and from non-inoculated plants, Rodospirillaceae and Sphingomonadaceae families were also identified. The 'Champaka' pineapple plants at fruiting stage associate with endophytes related to α, β and γ *Proteobacteria*, after plantlets inoculation or not with the diazotrophic bacterium *Burkholderia silvatlantica* (AB 219a). Most of those endophytes present nifH gene, a characteristic for nitrogen-fixing bacteria.

Key words: Biodiversity, nitrogen-fixing bacteria, *Ananas comosus,* microbial ecology.

INTRODUCTION

The association of diazotrophic bacteria with fruit plants has been described by many different researchers. In 1980, bacteria from the *Azospirillum* genus were initially identified in the rhizosphere of tropical fruits (Subba-Rao, 1983; Ghai and Thomas, 1989). Later, several bacteria belonging to the genera *Azospirillum, Herbaspirillum* and *Burkholderia* have been isolated from the roots and aerial parts of banana and pineapple plants (Weber et al., 1999). *Gluconoacetobacterdiazotrophicus* (Tapia-Hernández et al., 2000), and bacteria from the genera *Klebsiella, Bradyrhizobium* and *Serratia* (Ando et al., 2005) have been isolated from pineapple plants.

In pineapple, the benefits of plant-diazotrophic bacteria associations have been documented. The growth of seedlings micropropagated from the cultivar 'Champaka' was enhanced in the presence of strain AB 219, which was isolated from the 'Smooth Cayenne' pineapple and related to *Asaia bogorensis* (Weber et al., 2003). The induction of plant growth was reported for 'Victoria' cultivar, after the seedlings inoculation with several endophytic diazotrophic and epiphytic bacteria (Baldotto et al., 2010).

The better performance of pineapple plants in the presence of diazotrophic endophytic bacteria may be due primarily to the production of phytoregulators, but the biological nitrogen fixation associated with diazotrophic bacteria in the fruit crop may also be a contributing factor. Baldotto et al. (2010) observed an increase in leaf nitrogen (up to 193%) in 'Vitoria' cultivar seedlings inoculated with diazotrophic bacteria. In addition, Ando et al. (2000) described a lower abundance of natural $\delta^{15}N$ in the leaves of the 'Pattavia' cultivar when compared to other pineapple plants in the same growing area. Later, Ando et al. (2005) reported sequences of the nifH gene similar to those of the genera Klesiella and Serratia in leaves of 'Pattavia' and 'Smooth Cayenne' cultivars.

Considering the economic and social importance of the pineapple cultivation in Brazil, one of the leading producers (FAO, 2012), particularly in the Northeast region, where small farmers account up to 40% of the national fruit production (IBGE, 2012), the development of agro-technology to reduce production costs and to improve fruit quality is needed. Some problems are associated with cultivation of 'Pérola', 'Smooth Cayenne' and 'Champaka' (Cayenne Champac, Golden or MD2) suckers, when directly obtainned form fields. These cultivars are susceptible to fusarium wilt, a disease caused by Fusarium subglutinans, which is devastating (Reinhardt et al., 2002) and can leaded to impracticability of some production areas.

Biofertilizers containing plant growth promoting microorganisms increase the production of fruit plants (Mia et al., 2010). Besides, seedling inoculation with growth-promoting diazotrophic bacteria is environmental friendly. Inoculants containing Azospirillum spp. (Hungria et al., 2010) and other N_2-fixing bacteria (Moreira et al., 2010) have been tested in graminous plants. According to Abreu-Tarazi et al. (2010), micropropagated 'Gomo de Mel' pineapple seedlings contain bacteria related to Actinobacteria, α and β Proteobacteria. Also, in fruit crops it could become a profitable technology; so more detailed studies are needed on the identification and selection of fruit diazotrophic bacteria and their interactions with other endophytes. Knowledge of structures of endophytic bacterial populations in plants can help us to understand the plant-bacterium relationships. In this study we characterize culturable endophytes from the 'Champaka' pineapple plants at fruiting stage after seedlings inoculation with diazotrophic bacterium and cultivation in an organic irrigated orchard intercropped with sapota.

MATERIALS AND METHODS

Isolation and quantification of culturable endophytes

The 'Champaka' pineapple plants were inoculated or not with a a diazotrophic bacterium, 10^8 cells of the strain AB 219a, during acclimatization phase, and evaluated for root and stem endophytic colonization at fruiting stage. In the field, both plant groups were fertilized with three doses (40.17, 80.35 and 120.5 m^3 ha^{-1}) of three sources of composts; a) bovine manure, shredded leaves of wax palm (Copernicia cerifera) and sugar cane bagasse, b) plant debris and phosphate rock powder and c) sugar cane bagasse, coconut fiber, bovine manure, phosphate rock powder, rock powder (MB-4) and fruit residues of West Indian cherry (Malpighia emarginata DC), in an irrigated orchard intercropped with Sapota. Representative plants (54) from field treatments in three replicates were selected and fresh roots and stems were collected (taken after fruit harvest at 17 months), and these samples were taken to the Embrapa Tropical Agroindustry in Fortaleza, Brazil, for processing and isolating the culturable endophytes.

Root fragments (< 2 mm in diameter) and stems (1 g of fresh mass) without spots or necrosis were superficially sterilized in a solution of 1% Chloramine-T (monochloramine), washed three times in sterile water, macerated and serially diluted with sterile saline (up to 10^{-7}). The diluted suspensions were inoculated into penicillin-type vessels containing 5 ml of semi-solid JNFb N-free medium, according to Döbereiner et al. (1995). The vessels were incubated for five days in a BDO chamber regulated for 30°C, and the most probable number (MPN) of endophytes was determined for all treatments.

The MPN of endophytes from roots and aerial parts were submitted to analysis of variance, using the GLM procedure (General Linear Model) of the SAS® System (SAS Institute, 2000). The comparison of average values obtained from field treatments was achieved by using the F-test for contrasts. After bacterial growth in flasks was observed, the cultures were transferred to new vessels containing JNFb semi-solid medium. Typical bacterial growth, forming subsurface pellicles in the semi-solid medium were streaked onto Petri dishes containing solid Dygs medium (g liter^{-1}) 10 glucose, 2 malic acid, 1.5 peptone, 2 yeast extract, 0.5 K_2HPO_4, 0.5 $MgSO_4$ $7H_2O$, 1.5 glutamic acid, 15 agar) and the pH was adjusted to 6.5 (Weber et al., 1999). Cells from isolated colonies on solid medium were again inoculated into semi-solid JNFb medium, and the process was repeated until pure cultures were obtained and stoked in freezer (50% glycerol, at -18°C). Culturable endophytes from roots and stems of the 'Champaka' pineapple plants, and the strain AB 219a, used for plantlets inoculation can be found in the culture collections at Embrapa Tropical Agroindustry http://www.cnpat.embrapa.br) and/or at Embrapa Agrobiology (http://www.cnpab.embrapa.br).

Physiological characterization of culturable endophytes

Endophytes from different field treatments, and the strain AB 219a were activated in liquid Dygs medium and after reaching exponential growth; they were evaluated for their ability to use different carbon sources: D-fructose, sucrose, mannitol, inositol, D-galactose, succinate, oxalate, citrate, (D+) rhamnose, (L+) tartrate, and D, L-malate in semi-solid media containing salts of JNFb (pH 5.8), according to Weber et al. (1999). Also, the strain AB219a, used for plantlets inoculation, was evaluated for its ability to use different organic substrates under nitrogen fixation conditions.

DNA extractions, 16S rDNA amplification, sequencing and phylogenetic analysis

Representative culturable endophytes from field treatments, and the strain AB 219a were grown on solid Dygs medium and then used to total DNA extraction. The 16S rDNA gene was amplified by polymerase chain reaction (PCR) with primers Y1 and Y3 (Young et al., 1991). For sequencing, the PCR product was precipitated with 5M NaCl and 70% ethanol and sequenced using the MegaBACE1000 automated DNA sequencer (Amersham Bioscience)

Plant	Compost	Root[1]			Stem		
		Dosage (m³ ha⁻¹)					
		40.17	80.35	120.5	40.17	80.35	120.5
Control	A	6.089 ± 0.053	6.082 ± 0,055	5.930 ± 0.219	2.424 ± 0.717	2.608 ± 0.399	2.561 ± 0.721
	B	5.941 ± 0.666	5.945 ± 0.793	6.342 ± 0.416	2.223 ± 0.482	2.305 ± 0.506	3.460 ± 0.529
	C	6.170 ± 0.744	6.149 ± 0.509	6.721 ± 0.407	2.603 ± 0.510	2.414 ± 0.085	2.616 ± 0.472
Average		6.067	6.059	6.331	2.417	2.442	2.879
Inoculated	A	6.089 ± 0.014	6.005 ± 0.030	5.849 ± 0.362	2.221 ± 0.194	3.522 ± 0.105	2.750 ± 0.328
	B	6.408 ± 0.898	6.102 ± 0.368	6.717 ± 0.296	2.270 ± 0.314	2.912 ± 0.707	2.919 ± 0.215
	C	6.759 ± 0.434	6.873 ± 0.225	6.148 ± 0.664	3.290 ± 0.937	2.085 ± 0.238	3.086 ± 0.234
Average		6.419	6.327	6.238	2.594	2.840	2.918

with the Y1 and Y3 primers and five others (16S362f, 16S786f, 16S1203f, 16S110r, and 16S850r) (Soares-Ramos et al., 2003). After 16S sequence assembly with approximately 1,500 base pairs, each sequence was subjected to similarity analysis using the BLAST algorithm (Altschul et al., 1997) and the non-redundant NCBI database (http://www.ncbi.nlm.nih.gov/). The sequences with the greatest similarity to each isolate were selected for alignment using Clustal W (Thompson et al., 1994). A phylogenetic tree was constructed using the neighbor-joining method (Saitou and Nei, 1987) and the program MEGA4 (Tamura et al., 2007) based on distance calculations according to Kimura (1980). Bootstrap analysis was conducted with 1,000 replicates (Felsenstein, 1985).

Polymerase chain reaction amplification of nifH gene

The nifH gene (360 base pair) from representative culturable endophytes from filed treatments, and from the strain AB 219a were amplified using primers PolF (5'-TGC GAY CCS AAR GCB GAC TC-3') and PolR (5'-ATS GCC ATC ATY TCR CCG GA-3') (Poly et al., 2001). The PCR mix reactions were performed in a final volume of 25 µl containing: 0.2 mM of each dNTP; buffer 1x; 3 mM MgCl₂; 0.5 µM of each primer; DMSO 5%; 1 U GoTaq DNA polymerase and 25-50 ng genomic DNA. The touchdown PCR protocol was used for increased specificity and sensitivity in PCR amplification. The cycling conditions were as follow: denaturation step at 95°C for 10 min followed by 16 touchdown cycles involving denaturation at 95°C for 45 s, annealing for 45 s, and primer extension at 72°C for 45 s, with a 0.5°C decrease in annealing temperature per cycle starting at 72°C. Additional 5 cycles were performed at annealing temperature of 57°C for 45 s and one cycle of final extension at 72°C for 10 min. The reactions were carried out in an Eppendorf Mastercycler® thermocycler and amplified fragments were analyzed by horizontal electrophoresis on a 1.5% agarose gel at 100 V for 2 h. Gels were stained with ethidium bromide, visualized under UV light and photographed using Kodak® Gel Logic 100 Imaging System.

RESULTS

The endophyte AB 219a used for seedlings inoculation was a stocked subculture of diazotrophic bacteria. It was previously named as AB 219 (Weber et al., 2010), and 16S rDNA gene had been partially sequenced and had been related to Asaia bogorensis (Weber et al., 2003). Here the strain AB 219a was subjected to complete sequencing of the ribosomal gene, these sequences were submitted to NCBI database (accession HQ 706106), and could be now identified as Burkholderia silvatlantica.

The 'Champaka' pineapple plants presented endophytic bacterial colonization at fruiting stage, regardless of seedlings inoculation and field treatments, and higher population densities was detected in roots when compared to plant stems (Table 1). Field compost treatments influenced the MPN of endophytes in roots, as could be observed with contrasts of organic sources A and C ($p > 0.0112$); and in stems by contrasts with the compost dosages 40.17 and 120.5 m³ ha⁻¹ ($p > 0.0353$); as so with the seedlings inoculation ($p > 0.0001$). The strong evidence of beneficial endophytic association in inoculated pineapple stem may be due to the preferential site colonization of the strain AB 219a.

Culturable endophytes were obtained from inoculated (12 strains) and non-inoculated plants (13 strains) (Table 2), and they are able to grow within different organic substrates under nitrogen-fixing conditions (Table 3). The consequent bacterial growth in semi-solid N-free media allows establishing nine groups: first three bacterial groups (I to III), from control plants; other three

Table 2. Endophytes isolated from the 'Champaka' pineapple plants, as a function of the absence or presence of inoculation with the strain AB 219a and the sources of compost.

Plant	Source of compost	Strains (n°)	Root endophytes	Stem endophytes
Control	A	4	AB 280 (*a)	AB 294 (a), AB 291 (a), AB 301 (c)
	B	5	AB 287 (a), AB 288 (b)	AB 300 (b), AB 296 (c), AB 297 (c)
	C	4	AB 285 (b)	AB 305 (A), AB 292 (a), AB 293 (c)
Inoculated	A	3	AB 281 (a), AB 282 (a)	AB 290 (b)
	B	3	AB 286 (b), AB 289 (c)	-
	C	4	AB 284 (a)	AB 295 (a), AB 304 (a), AB 299 (c)

Dosage of compost used (*a = 40.17 m^3 ha^{-1}, b = 80.35 m^3 ha^{-1} and c = 120.5 m^3 ha^{-1}). Selected strains: AB 286 (=BR12271), AB 287 (=BR12272), AB 299 (=BR12217), AB 292 (=12274), AB 280 (=BR12268), AB 281 (=BR12269), AB 284 (=BR12270), AB 285 (=BR12273), AB 294 (=BR12275), AB 295 (=BR12276), AB 301 (=BR12278) and the strain used for the plantlet inoculation (AB 219a = BR12266) were submitted to the culture collection of Embrapa Agrobiology (BR) in Seropédica, Brazil; and all other strains were maintained in the culture collection of Embrapa Tropical Agroindustry in Fortaleza.

Table 3. Ability of grouped endophytes from the 'Champaka' pineapple inoculated with the strain AB 219a, non-inoculated (Control) and both (inoculated or not inoculated), to use carbon sources in semi-solid media containing salts of JNFb (pH 5.8).

Group	Plant treatment	Carbon sources										
		(D+) Fructose	Sucrose	Mannitol	Inositol	(D+) Galactose	Succinate	Oxalate	Citrate	(L+) Rhamnose	(L+) Tartrate
I	Control	+	++	+	+	++	+	-	++	-	++
II	Control	+	-	+	+	+	+	++	++	+	++
III	Control	+++	++	+	+	++	+++	++	+	++	++
IV	Inoculated	++	++	+++	++	++	++	++	++	++	+
V	Inoculated	+	+	+	+	+	+	++	++	+	+
VI	Inoculated	-/+	-/+	+	+	+	+	+	+	+	-
VII	Both	-/+	-/+	-/+	+	+	++	-	+	+	-/+
VIII	Both	+	-/+	+	+	+	++	-/+	+	+	++
IX	Both	+	++	++	+	+	+	-/+	++	++	+

*Strains from groups: I (AB 280, AB 288), II (AB 291, AB 292), III (AB 293, AB 297, AB 301, AB 305), IV (AB 284, AB 304), V (AB 282), VI (AB 281, AB 286), VII (AB295, AB 300), VIII (AB 294, AB 299), and IX (AB 285, AB 287, AB 290, AB 296). **Good (+++), medium (++), poor (+), or null (-) growth after five days of incubation at 30 °C. All isolates grew within JNFb medium.

Bacterial groups (IV to VI), from inoculated plants; and last three endophyte groups (VII to IX), from both plant treatments, with or without bacterial inoculation. We should mention that strains AB 284 and AB 304 (group IV) isolated from root and from stem of inoculated plants that received compost C (Table 2), exhibited growth in semi-solid media (Table 3) similar to growth we observed for the bacterium AB 219a. Also, a higher

Table 4. Analyses of sequence similarity of the 16S rDNA gene in selected endophytes from the 'Champaka' pineapple plants inoculated or not with the strain AB 219a and grown in an organic system.

Strain	Bacterial group	Treatment	Plant part	16S base pairs	Proximity of bacterial group BLASTn	E-value	Coverage (%)	ID (%)
AB 280	I	Control	Root	1507	CP000959.1 *Burkholderia cenocepacia* (MC0-3)	0.0	100	99
AB 292	II	Control	Steam	1471	AB120764.1 *Sphingobium yanoikuyae*	0.0	98	99
AB 301	III	Control	Steam	1447	DQ288687.1 *Azospirillum brasilense* (MTCC4036)	0.0	99	99
AB 294	VIII	Control	Steam	1522	EF522135.1 *Erwinia* sp. (CU208)	0.0	100	99
AB 285	IX	Control	Root	1504	EF488760.1 *Enterobacter oryzae* (Ola 01)	0.0	97	99
AB 287	IX	Control	Root	1500	EF488758.1 *Enterobacter oryzae* (Ola 50)	0.0	98	99
AB 284	IV	Inoculated	Root	1522	AY965240.1 *Burkholderia silvatlantica* (SRMrh-20)	0.0	98	99
AB 281	VI	Inoculated	Root	1485	AP009384.1 *Azorhizobium caulinodans* (ORS 571)	0.0	95	99
AB 286	VI	Inoculated	Root	1362	AP009384.1 *Azorhizobium caulinodans* (ORS 571)	0.0	100	99
AB 295	VII	Inoculated	Steam	1519	HQ455824.1 *Pantoea eucrina* (CT194)	0.0	99	99
AB 299	VIII	Inoculated	Steam	1503	EF522135.1 *Erwinia* sp. (CU208)	0.0	100	99
AB 290	IX	Inoculated	Steam	1505	FM872505.1 Uncultured bacterium (clone FA01C07)	0.0	99	98

* Strain AB 219a presented 1492 16S base pairs, E value 0.0 and coverage 100%, and ID 99%. **GenBank accession sequences of selected endophytes: AB 299 (=BR12217) HQ706104, AB 292 (=12274) HQ706105, AB 280 (=BR12268) HQ706107, AB 281 (=BR12269) HQ706108, AB 284 (=BR12279) HQ706109, AB 285 (=BR12270) HQ706110, AB 290 (=BR12273) HQ706111, AB 294 (=BR12275) HQ706112, AB 295 (=BR12276) HQ706113, AB 301 (=BR12278) HQ706114; and the AB 219a (=BR12266) HQ706106used for plantlet inoculation.

endophytes MPN was detected in roots of plants from plant treatment receiving bacterial incoculum and compost C in comparison to the control plants receiving just that organic fertilizer ($p > 0.012$), and the differences between those treatments (Table 1) may be related to organic constituents of the compost C and soil nutrient balance.

Representative culturable endophytes from all bacterial groups (Table 3) were submitted to DNA extraction and the 16S rRNA gene was sequenced (Table 4), with identification of different bacterial species (Figure 1). From stems of inoculated pineapple plants, endophytes closely related to *Erwinia* sp. (strain AB 299), *Pantoea eucrina* (strain AB 295) and uncultured bacterium (strain AB 290), also, species belonging to Enterobacteriaceae (γ *Proteobacteria*) were obtained; from roots of the inoculated plants we

isolated endophytes related to *Azorhizobium caulinodans* (AB 281 and AB 286), Xanthobacteraceae (α *Proteobacteria*), to *Burkholderia silvatlantica* (strain AB 284), Burkholderiaceae (β *Proteobacteria*). Plants association with endophytic bacteria may persist in an organic cropping system, so the diazotrophic bacterium (strain AB 219a) used for plantlets inoculation (Weber et al., 2010), also was closely related to *B. silvatlantica* (Table 4). From stems of the control plants we obtained endophytes closely related to *Azospirillum brasilense* (strain AB 301), belonging to Rodospirillaceae, to *Sphingobium yanoikuyae* (strain AB 292), belonging to Sphingomonadaceae, these both species are from α*Proteobacteria*, and to *Erwinia* sp. (strain AB 294), γ *Proteobacteria*; from roots of the control plants were obtained endophytes closely

related to *B. cenocepacia* (strain 280) and *Enterobacter oryzae* (strains AB 285 and AB 287), positioned into Burkholderiaceae and Enterobacteriaceae families, respectively.

Endophytes from different groups (Table 3) and closely related to different bacterial species (Table 4) were evaluated for *nifH* gene, using the PolF/PolR primers (Poly et al., 2001) for *nifH* fragment amplification. Also, the AB 219a, used for plantlets inoculation (Weber et al., 2010), was tested. Employed primers amplified a single and correct sized band in eleven of twelve DNA strains tested (Figure 2). The gene presence was a strong evidence of bacterial nitrogen-fixing capacity, so only for the strain AB 295, related to *Pantoea eucrina*, the *nifH* gene fragment was not detected by using that primer combination. Also, we should observe that AB 295 and AB 300 grew

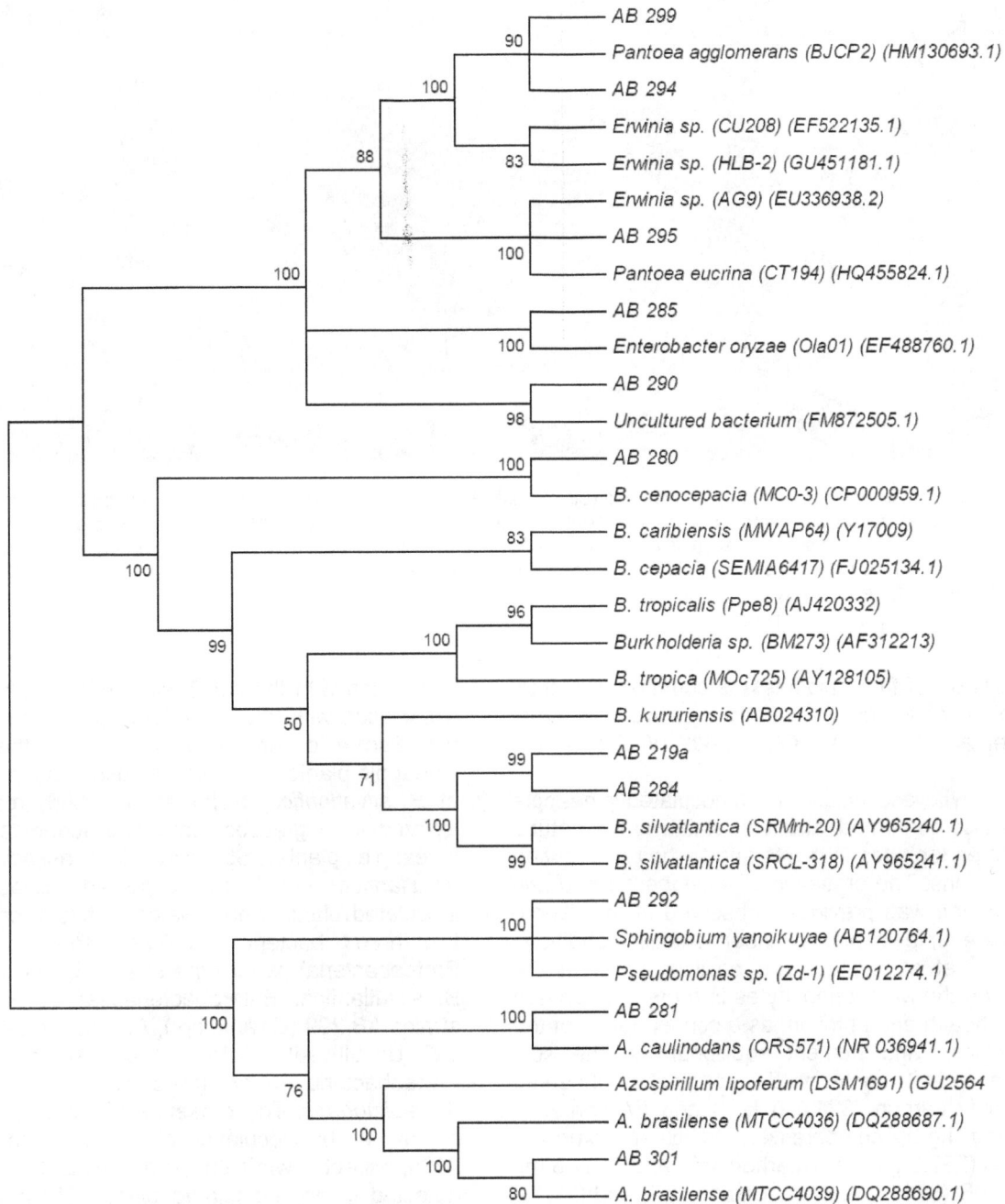

Figure 1. Phylogenetic tree based on 16S sequencing of selected culturable endophytes isolated from the 'Champaka' pineapple plants inoculated or not with the strain AB 219a and grown in an organic system. The program MEGA 4 and the neighbor-joining method were used to construct the tree.

poorly in semi-solid media (Table 3), under nitrogen-fixing conditions.

DISCUSSION

The 'Champaka' pineapple plants at fruiting stage, regardless of the seedlings inoculation with *B. silvatlantica* (AB 219a) and the organic fertilization in field (Weber et al., 2010), presented endophytic bacterial colonization in roots and stems. The density of endophytes was higher in plants that received the bacterial inoculum, suggesting that association with the AB219a may persist during a plant cycle. In the presence of that bacterial strain and the treatment with low dosage of organic compost (40.17 m^3ha^{-1}) the MPN of endophytesreached10^6 cells g^{-1} of fresh root. The higher population density in roots (1.7 10^6 cells g^{-1} of fresh material) in comparison with stems

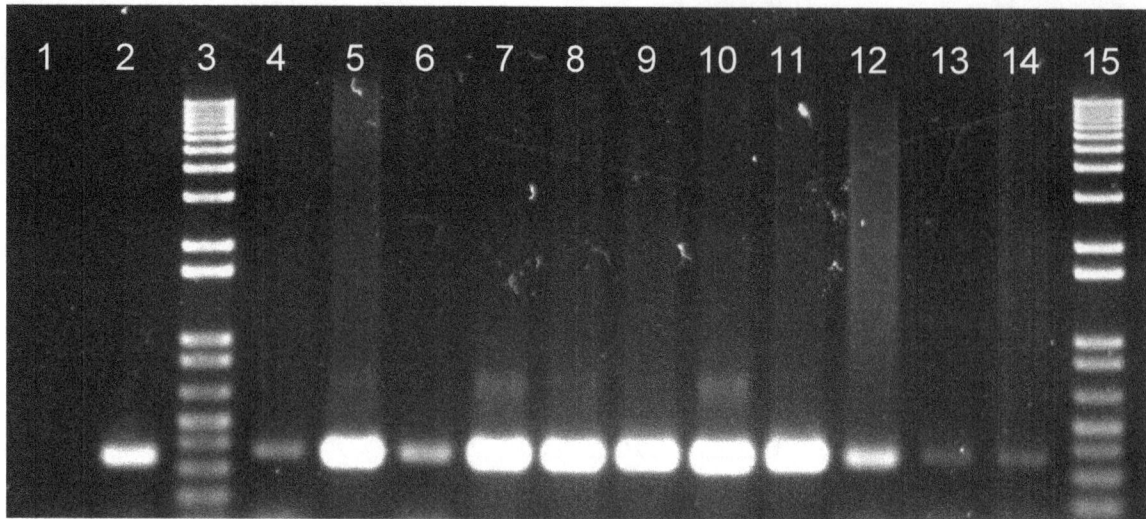

Figure 2. Agarose gel electrophoresis analysis of *nif*H gene(360 base pair) of selected endophytes from the 'Champaka' pineapple plants, and the fragment amplification by using primers PolF/PolR. Numbers: 1 negative control; 2 positive control (*Azospirillum brasilense*, strain SP245); 3 and 15 molecular marker; and 4 to 14 representing the following strains: AB 299, AB 219a, AB 280, AB 281, AB 284, AB 285, AB 286, AB 287, AB 290, AB 292, AB 294, respectively.

(4.8 10^2 cells g^{-1} of fresh biomass) is common in plants, and has been observed in pineapple cultivars: 'Vitória' (Baldotto et al., 2010) and "Champaka" (Weber et al., 2010).

The endophytic association in uninoculated pineapple plants was also confirmed, and it is related to the natural presence of endophytes in plants and to their distribution in field conditions. The presence of *Actinobacteria*, *α and β Proteobacteria* was previously observed in the 'Gomo de Mel' pineapple explants under axenic conditions (Abreu-Tarazi et al., 2010). The positive effect of the compost C on the MPN endophytes in roots may be due to the plant health and nutrition, as a consequence of the organic fertilizer and nutrient equilibrium in the soil. Weber et al. (2010) analyzing the compost C observed higher organic carbon (321.0 g kg^{-1}) and Fe contends (12431.3 mg kg^{-1}), comparatively to other composts employed: A (292.6 g organic carbon kg^{-1}, and 2431.3 mg Fe kg^{-1}) and B (321.0 g organic carbon kg^{-1}, and 2431.3 mg Fe kg^{-1}); and these nutrients may affect the plant symbiotic association. Organic fertilizers benefits on population of the diazotrophic endophytic bacteria were demonstrated in sugarcane, in comparison to conventional fertilizer management (Pariona-Llanos et al., 2010). Also, Shu et al. (2012), investigatingthe structureofnitrogen-fixing bacteriain rhizosphere andpadysoil andanalyzing relative abundanceofnifH, observed greaterabundance and diversity ofnitrogen-fixing bacteriain the soils underorganic management in comparison to conventional cropping system.

Our results demonstrate that pineapple plants associate with culturable endophytes (strains AB 284 and AB 304) closely related to *B. silvatlantica*, after plantlets

inoculation with the AB 219a.Persistent plant endophytic association with this bacterial specie can be considered, but also wild strains may colonize the 'Champaka' pineapple plants. It should be observed that in description of *B. silvatlantica*, Perin et al. (2006) reported strains isolated from grasses (corn and sugar cane) and from pineapple plants. So, that plant association with *B. silvatlantica* could be expected. Endophytes from inoculated plants were related to five species belonging to three bacteria families: Burkholderiaceae (β Proteobacteria), where the strain AB 284 was related to *B. silvatlantica*; Enterobacteriaceae (γ Proteobacteria), strains AB 299 (*Erwinia* sp.), AB 295 (*P. eucrina*) and AB 290 (Uncultivable bacterium); and Xanthobacteraceae (α Proteobacteria), where the strain AB 281 was related to *A. caulinodans*. The presence of these endophytes may be favored by inoculation of plantlets with the strain AB 219a, except *Erwinia* sp. (strain AB 294), which was also detected in non-inoculated plants. These control plants were associated with endophytes related to four other species: *E. oryzae* (strains AB 285 and AB 287), belonging to Enterobacteriaceae; *A. brasilense* (strains AB 301), Rodospirillaceae; *Sphingobium yanoikuyae* (strain AB 292), Sphingomonadaceae; and *B. cenocepacia* (strain AB 280), Burkholderiaceae. The 'Champaka' pineapple plant colonization by those different endophytes is an evidence of a non-specific plant-bacterial relation, and the microorganisms effectively contributing to plant-growth are difficult to determine, as reported by Moreira et al. (2010).

The *nifH* gene of grouped endophytes from the'Champaka' pineapple plants was detected, except for the strain AB 295 (*Pantoea eucrina*). This evidence is characteristic of nitrogen-fixing bacteria (Marin et al., 2003; Ando et al., 2005; Luvizotto et al., 2010; Shu et al., 2012). Also, our results indicate that Poly PCR primers (Poly et al., 2001) were able to amplify *nifH* sequences of pineapple endophytes DNA; however, the amplification efficiency differed between them (Figure 2). Some information had been previously reported about the contribution of endophytes we identified in this work. Bacteria related to *A. brasilense*, are able to colonize bananas (Weber et al., 1999; Mia et al., 2010) and especially graminous plants, leading to an increase of 16 to 30% in the production of wheat and corn grains (Hungria et al., 2010). *A. caulinodans* strains promote growth in rice seedlings (Senthilkumar et al., 2008), and *Sphingobium yanoikuyae* strains have activity of indol acetic acid production (Poonguzhaly et al., 2006). Other endophytes related to *Enterobacter oryzae* (Peng et al., 2009), *Pantoea eucrina* and *Erwinia* sp. may also influence the association of selected strains in pineapple cultivation. In further work we should consider the interaction of those endophytes with pineapple plants and the genetic similarity of strains related to *B. silvatlantica*. Based on the results obtained in the present study we conclude that pineapple 'Champaka' plants at fruiting stage associate with endophytes related to α, β and γ Proteobacteria, after plantlets inoculation or not with the diazotrophic bacterium AB 219a. Most of those endophytes present *nifH* gene, a characteristic of nitrogen-fixing bacteria.

ACKNOWLEDGEMENT

The work was partially supported by Bonafrux Organics, Banco do Nordeste do Brasil (BNB).

REFERENCES

Abreu-Tarazi MF, Navarrete AA, Andreote FD, Almeida CV, Tsai SM, Almeida M (2010). Endophytic bacteria in long-term in vitro cultivated "axenic" pineapple microplants revealed by PCR–DGGE. World J. Microbiol. Biotechnol. 26:555-560.

Altschul SF, Madden TL, Schäffer AA, Zhang J, Miller DJ (1997). Gapped BLAST and PSI-BLAST: a new generation of protein database search programs. Nucleic. Acids. Res. 25(17):3389-3402.

Ando S, Goto M, Meunchang S, Thongra-ar P, Fugiwara T, Hayashi H, Yoneyama T (2005). Detection of *nifH* sequences in sugarcane (*Sacharum officinarum* L.) and pineapple (*Ananas comosus* (L.) Merr.). Soil Sci. Plant. Nutr. 25(2):303-308.

Ando S, Meunchang S, Vadisirisuk P, Yoneyama T (2000). Estimation of nitrogen imput by N2 fixation to field-grown pineapples. Acta Hort. 529:203-210.

Baldotto LEB, Baldotto MA, Olivares FO, Vianna AP, Bressan-Smith R (2010). Seleção de bactérias promotoras de crescimento no abacaxizeiro cultivar Vitoria durante a aclimatização. Rev Bras Ci Solo 34(2):349-360.

Döbereiner J, Baldani VLD, Baldani JI (1995). Como isolar e identificar bactérias diazotróficas de plantas não leguminosas. Brasília: Embrapa- SPI/Embrapa-CNPAB. P. 60.

Felsenstein J (1985). Confidence limits on phylogenies: an approach using the bootstrap. Evolution 39(4):783-791.

FAO (Food and Agriculture Organization of the United Nations) (2012).

FAOSTAT database. Available at http://faostat.fao.org/site/567/DesktopDefault.aspx?PageID=567#anc or Accessed 3 May 2012.

Ghai SK, Thomas GV (1989). Occurrence of *Azospirillum* spp. in coconut-based farming systems. Plant Soil 114:235-241.

Hungria M, Campo RJ, Souza EM, Pedrosa F (2010). Inoculation with selected strains of *Azospirillum brasilense* and *A. lipoferum* improves yields of maize and wheat in Brazil. Plant Soil 331:413-425.

IBGE (Instituto brasileiro de Geografia e Estatística) (2012). Sistema IBGE de Recuperação Automática. Banco de dados agregados, agricultura, produção. Available at http://www.sidra.ibge.gov.br/bda/agric/default.asp?t=2&z=t&o=11&u1=1&u2=1&u3=1&u4=1&u5=1&u6=1Accessed 3 May 2012.

Luvizotto DM, Marcon J, Andreote FD, Dini-Andreote F, Neves AAC, Araújo WL, Pizzirani-Kleiner AA (2010) . Genetic diversity and plant-growth related features of *Burkholderia* spp. from sugarcane roots. W J. Microbiol. Biotechnol. 26(10):1829-1836.

Kimura M (1980). A simple method for estimating evolutionary rate of base substitutions through comparative studies of nucleotide sequences. J. Mol. Evol. 16:111-120.

Marin VA, Teixeira KRS, Baldani JI (2003) Characterization of amplified polymerase chain reaction *glnB* and *nifH* gene fragments of nitrogen-fixing *Burkholderia* species Lett. Appl. Microbiol. 36"77-82.

Mia MAB, Shamsuddin ZH, Mahmood M (2010). Use of plant growth promoting bacteria in banana: a new insight for sustainable banana production. Int. J. Agric. Biol. 12(3):1814-9596 09-279/SBC/2010/12-3-459-467.

Moreira FMS, Silva K, Nóbrega RSA, Carvalho F (2010). Bactérias diazotróficas associativas: diversidade, ecologia e potencial de aplicações. Comunicata Sci. 2:74-99.

Pariona-Llanos R, Ferrara FIS, Gonzales HHS, Barbosa HR (2010). Influence of organic fertilization on the number of culturable diazotrophic endophytic bacteria isolated from sugarcane. Eur. J. Soil. Biol. 46:387-393.

Peng G, Zhang W, Luo H, Xie H, Lai W, Tan Z (2009). *Enterobacter oryzae* sp. nov., a nitrogen-fixing bacterium isolated from the wild rice species *Oryza latifólia*. Int. J. Syst. Evol. Microbiol. 59(7):1650-1655.

Perin L, Martinez-Aguilar L, Paredes-Valdez G, Baldani JI (2006). *Burkholderia silvatlantica* sp. nov., a diazotrophic bacterium associated with sugar cane and maize. Int. J. Syst. Evol. Microbiol. 56:1931-1937.

Poly F, Monrozier LJ, Bally R (2001). Improvement in the RFLP procedure for studying the diversity of *nifH* genes in communities of nitrogen fixers in soil. Res. Microbiol. 152:95-103.

Poonguzhaly S, Madhaiyan M, Sa T (2006). Cultivation-dependent characterization of rhizobacterial communities from field grown Chinese cabbage *Brassica campestris* spp *pekinensis* and screening of traits for potential plant growth promotion. Pl Soil. 286:167-180.

Reinhardt DH, Cabral JR, Souza LFS, Sanches NF (2002). Pérola and Smooth Cayenne pineapple cultivars in the state of Bahia, Brazil: growth, flowering, pests, diseases, yield and fruit quality aspects. Fruits 57:43-53.

SAS Institute (2000). SAS/STAT user's guide. Release 9.1 SAS Institute, Inc., Cary.

Saitou N, Nei M (1987). The neighbor-joining method: a new method for reconstructing phylogenetic trees. Mol. Biol. Evol. 4(4):406-425.

Senthilkumar M, Madhaiyan M, Sundaram SP, Sangeetha H, Kannaiyan S (2008). Induction of endophytic colonization in rice (*Oryza sativa* L.) tissue culture plants by *Azorhizobium caulinodans*. Biotechnol. Lett. 30(8):1477-1487.

Shu W, Pablo GP, Jun Y, Danfeng H (2012). Abundance and diversity of nitrogen-fixing bacteria in rhizosphere and bulk paddy soil under different duration of organic management. W M Microbiol. Biotechnol. 28:493-503.

Soares-Ramos JRL Ramos HJO, Cruz LM, Chubatsu LS, Pedrosa FO Rigo LU, Souza EM (2003). Comparative molecular analysis of *Herbaspirillum* strains by RAPD, RFLP, and 16S rDNA sequencing. Gen. Mol. Biol. 26(4):537-543.

Subba-Rao NS (1983) Nitrogen-fixing bacteria associated with plantation and orchard plants. Can. J. Microbiol. 29(8):863-866.

Tamura K, Dudley J, Nei M, Kumar S (2007). MEGA4: molecular evolutionary genetics analysis (MEGA) program version 4.0. Mol.

Biol. Evol. 24(8):1596-1599.

Tapia-Hernández A, Bustillos-Cirstales MR, Jiménez-Salgado T, Caballero-Melado J, Fuentes-Ramírez LE (2000). Natural endophytic occurrence of *Acetobacter diazotrophicus* in pineapple plants. Microbial. Ecol. 39 (1):49-55.

Thompson JD, Higgins DG, Gibson TJ (1994). Clustal W: improving the sensitivity of progressive multiple sequence alignment through sequence weighting, positions-specific gag penalties and weight matrix choice. Nucleic Acids. Res. 22:4673-4680.

Weber OB, Baldani VLD, Teixeira KRS, Kirchhof G, Bldani JI, Döbereiner J (1999). Isolation and characterization of diazotrópfica bacteria from banana and pineapple plants. Plant Soil. 210:103-113.

Weber OB, Correia D, Silveira mrs,Crisóstomo LA, Oliveira EM,Sá EG (2003). Efeito da bactéria diazotrófica em mudas micropropagadas de abacaxizeiros cayenne champac em diferentes substratos. pesq. agropec. bras., 38:689-696.

Weber OB, Lima RN, Crisóstomo LA, Freitas JAD, Carvalho ACP, Maia AH (2010). Effect of diazotrophic bacterium inoculation and organic fertilization on yield of Champaka pineapple intercropped with irrigated sapota. Plant Soil 327:355-364.

Young JPW, Downer HL, Eardly BD (1991). Phylogeny of the phototrophic rhizobium strain BTAi1 by polymerase chain reaction-based sequencing of a 16S rRNA gene segment. J. Bacteriol. 173(7):2271-2277.

10

Optimization of ethanol production from apple pomace through solid-state fermentation using enzymes and yeasts combination through response surface methodology

Manoj Kumar Mahawar[1], Anupama Singh[2], B. K. Kumbhar[2] and Manvika Sehgal[3]

[1]Department of Agricultural Engineering, IARI New Delhi, India.
[2]Department of PHPFE, GBPUAT Pantnagar, Uttarakhand, India.
[3]Department of Microbiology, GBPUAT Pantnagar, Uttarakhand, India.

Apple pomace (AP) which accounts for 25% of original fruit mass is a by-product from the apple processing industry. Solid-state fermentation of AP was conducted on laboratory scale in 250 ml flask at 30°C, agitation speed of 55 rpm at different pH levels of 4, 4.5 and 5.0 using the Y_{51} strain, *Saccharomyces cerevisiae* (SC) ATCC 9673 and their combination. The sample was treated with both α-amylase and cellulase enzyme collectively for higher reducing sugar content. Ethanol yield of 5.23% (v/v) was obtained in the case wherein strain Y_{51} was inoculum at pH of 4.5 and the fermentation period was 72 h. Response surface methodology was used to design the experiments as well as for the data analysis. Optimization of various process conditions was done using software Design-Expert 7.1.6.

Key words: Apple pomace, *saccharomyces cerevisiae*, Y_{51}, cellulase, optimization.

INTRODUCTION

Sustainable food production and waste valorisation have become important issues in modern life and are becoming important issues in the food industry. Food producers generate high amounts of biological by-products and waste that could be used for other purposes as well. The use of agro-industrial wastes in solid state fermentation is economically important and can minimize various environmental problems. The direct disposal of agro-industrial residues as a waste on the environment represents an important loss of biomass, which could be bioconverted into different metabolites, with a higher commercial value (Vendruscolo et al., 2007).

AP is the solid residue that remains after the extraction of juice from apple and its disposal as such causes considerable economic (Miller et al., 1982) and environmental (Hang and Woodams, 1986) problems. Conventional process of juice recovery removes 75% of fresh weight as juice and 25% as pomace (Vendruscolo et al., 2008). More than 500 food processing plants in India produces about 1.3 million tones of AP annually which involves annual disposal expenditure of 0.5 million US dollars (Jewell and Cummings, 1984). Hence, there is a strong need to have an integrated approach for AP waste utilization and its treatment.

Globally, several million tones of AP are generated. Owing to the high carbohydrate content, it is used as a substrate in a number of microbial processes for the production of organic acids, enzymes, single cell protein, ethanol, low alcoholic drinks and pigments (Bhushan et al., 2008). AP also serves as the potential source of

Table 1. Selected process variables and their assigned levels.

Independent variable		Coded levels		
		-1	0	1
Name	Code	Actual levels		
Yeast strains	X_1	Y_{51}	SC	Combination
pH	X_2	4.0	4.5	5.0
Time	X_3	24	48	72

ethanol. Jain and Singh (2006) has reported the ethanol yield of 4.074% (v/v) in inoculated fermentation using Y_{51} strain at pH of 4.5 and the sample was kept for 72 h of incubation at 30 °C. Kumar and Sahgal (2008) reported the ethanol yield of 5.02% (v/v) when Y_{51} was inoculated to the substrate combination of 75% AP plus 25% molasses at 72 h of fermentation.

Enzymatic hydrolysis of various feedstocks for ethanol production has been widely attempted with considerable amount of success (Aswathy et al., 2010). From the preliminary experiments it was found that AP sample when treated with the enzyme combination of α- amylase and cellulase having higher amount of reducing sugar, that is, 10.85% as compared to the samples which are treated separately. The α-amylase and cellulase concentration were calculated as 25 and 2 mg/g of dry matter in AP sample, respectively (Kumar and Wyman, 2009).

Optimisation of different parameters, by the traditional 'one-factor-at-a-time' method requires a considerable amount of time and effort. An alternative potential approach is a statistical approach, such as response surface methodology (RSM), one of the most widely used statistical techniques for bioprocess optimisation (Liu and Tzeng, 1998). Optimization includes finding "best available" values of some objective function given a defined domain including a variety of different types of objective functions and different types of domains. RSM can be used to evaluate the relationship between a set of controllable experimental factors and outcomes. The interactions among the possible influencing factors can be evaluated with a restricted number of experiments.

This study comprised of improvement of ethanol yield from apple pomace by using the various combinations of enzymes and yeast isolates. The combined effect of independent variables on the responses is also being investigated using second order model. The study also reveals the optimum condition for the maximum ethanol yield by using RSM.

MATERIALS AND METHODS

Substrate procurement and pretreatment

Apples of *Red delicious* variety were procured from the local market of Pantnagar as per requirement and stored in the refrigerator at

4 °C until needed for experiments. AP was prepared using hydraulic press in the bioconversion laboratory of the Department of Post Harvest Process and Food Engineering (PHPFE), GBPUAT Pantnagar. Sterilization of AP was done by using autoclave unit at 15 psi (121 °C) for 15 to 20 min. The initial moisture content of AP was calculated by using hot-air oven method and was observed to be 80% (wb).

Characteristics of apple pomace

Initially apple pomace has 5.80 to 7.20% of reducing sugar, after the treatment of α- amylase and cellulase collectively it increased to10.65 to 13.10%. AP is a poor source of nitrogen. Therefore, from the micronutrient analysis salt of the essential nutrient Nitrogen, that is ammonium sulphate 0.02 g/150 ml of sample was added before fermentation.

Microorganisms and enzymes used

Saccharomyces cerevisiae ATCC 9673 was procured from IMTC Chandigarh while Y_{51} strain was procured by natural fermentation of AP in the bioconversion laboratory of the department. Both strains (*SC* 9673 and Y_{51}) were tested for the ethanol producing quality and were found to have independent growth which ensures that the growth of one does not affect the growth of another. Hence, combination of both yeast strains was considered as an independent variable in the experimental design. Yeast isolates (Y_{51}, *SC* 9673) were grown on Yeast- Peptone- Dextrose medium for 48 h at 30 °C kept at 120 rpm in incubator shaker.

Experimental design

Design Expert 7.1.6 (Stat-Ease Inc., Minneapolis, USA) was used for experimental design and also for statistical and regression analysis of the data. A Box- Behnken design with three independent variables was used: Type of yeast (X1), pH (X2) and fermentation time (X3). The coded and actual range of the selected variables is given in Table 1, which resulted in 17 experimental runs, including five central points.

The variables which were kept constant during the experimental run are fermentation temperature (30 °C), agitation speed (55 rpm), dilution level (1:10), sample size (150 ml), amount of amylase (0.068 g/ 150 ml), amount of cellulase (0.0055 g/ 150 ml), enzyme treatment (α-amylase + cellulase) with the incubation period (1 h) and inoculum rate 10% (v/v) of the sample size as standardized by Jain and Singh (2006).

Fermentation for production of ethanol

Fermentation experiments were carried out at the process

Table 2. Response surface design and corresponding response values for ethanol production.

Exp. No.	Variable			Responses			
	Yeast strains	pH	Time (h)	Utilized sugar conc. (%)	Change in pH	Cell count 10^6 cfu/ml	Ethanol conc. (%v/v)
1	SC	5.0	24	1.24*	1.32**	0.09*	0.65
2	SC	4.0	72	7.65	0.53	150	0.94
3	Y$_{51}$	4.0	48	4.29	0.36*	57	3.95
4	SC	4.0	24	1.57	1.04	0.109	0.75
5	COM	4.5	48	3.94	0.64	110	1.26
6	SC	4.5	48	4.78	0.55	63	0.98
7	COM	4.0	48	4.26	0.58	8.9	1.09
8	SC	4.5	48	4.94	0.52	69	1.05
9	SC	4.5	48	5.22	0.48	73	1.12
10	COM	4.5	24	1.35	0.66	9.5	0.58*
11	COM	4.5	72	8.86**	1.05	158**	1.43
12	Y$_{51}$	4.5	72	8.61	0.76	156	5.23**
13	Y$_{51}$	5.0	48	4.70	1.13	82	4.86
14	SC	4.5	48	5.08	0.54	59	1.17
15	Y$_{51}$	4.5	24	1.52	1.31	0.103	2.42
16	SC	4.5	48	4.64	0.45	67	0.89
17	SC	5.0	72	7.85	1.14	146	1.28

*Minimum value; ** Maximum value.

conditions as mentioned in Table 1. Samples were withdrawn at the specified intervals and were tested for their sugar, ethanol, pH and viable count individually. The change of pH was monitored by digital pH meter. Colony forming unit were determined by serial dilution pour plating method (Seeley et al., 1991). The amount of sugar was estimated by Dinitrosalicylic acid method of Miller (1972). The fermentation worth was distilled and then amount of ethanol was estimated using GC according to the method of Lancas and de Moreas (2007). A calibration curve was constructed using ethanol standards in water. *n* Propanol at a concentration of 5% (v/v) was used as an internal standard to correct for unequal injection volumes in gas chromatography. The distilled samples and the standards all contain the same concentration of the internal standard. The calibration curve is constructed by dividing the peak area of the ethanol by the peak area of the internal standard and plotting the ratio against the concentration of the ethanol. The peak area ratio is independent of injection volume. The amount of ethanol was determined by the formula:

$$\frac{X \% \text{ Ethanol}}{\text{Peak area ratio for sample}} = \frac{5\% \text{ Ethanol}}{\text{Peak area ratio for sample}}$$

Data analysis

A full second order mathematical model was fitted into each response. The adequacy of the model was tested using coefficient of determination (R^2) and Fisher's F-test. The effects of variables on responses were then interpreted. If the model was found adequate, the best fit equations were developed in order to draw contour plots for showing the effect of independent variables on those responses and to select the optimum range of variables for an acceptable product.

RESULTS AND DISCUSSION

The experimental data given in Table 2 was analysed employing multiple regression technique to develop response functions and variable parameters optimized for best outputs.

Sugar utilization

The maximum sugar (8.86%) was utilized in the case of combination of yeast strains with the initial fermentation pH of 4.5 and the fermentation period of 72 h, it was due to the fact that CFU's of yeast cells was maximum in this case. Sugar was metabolized by yeast cells for its growth and was subsequently converted into ethanol. Yeast favours an optimum value of pH (4.3-4.7) for their growth and hence utilize maximum amount of sugar during this range (Neuberg, 1958).

The coefficient of determination (R^2) for the regression model for utilized sugar was 99.20%. Model was highly significant ($p<0.05$) with F as 95.92. Effect of independent variables was highly significant ($p< 0.01$) at linear level, while the level of significance at quadratic level was 5%.

Change in pH

Coefficient of determination (R^2) was 97.43%, model was

highly significant (p<0.05) with F value of 29.48. It was observed that all the 3 parameters viz. yeast strain, pH and fermentation time affected the change in pH at 1% level of significance.

Cell count

During the fermentation process throughout, an followed by a decrease. Yeast, during the log phase of increasing pattern was observed for the cell count their growth cycle utilized the nutrients and hence leads to ethanol production. Higher value indicates the conditions where sample of pomace was treated with combination of yeast strain at pH level of 4.5 and fermentation time was 72 h. Fermentation time affected the cell concentration at 1% level of significance.

Ethanol yield

Maximum conversion of sugar into ethanol 5.23% (v/v) was observed when the sample was treated with Y_{51} strain at the pH level of 4.5 and for 72 h of fermentation. The reason behind this is that strain Y_{51} converts the sugar into ethanol under the optimum initial pH of 4.5 which is favourable for yeast growth. Similar results for Y_{51} strain were obtained by Jain and Singh (2006) and Kumar and Sahgal (2008). Effect of independent variables on ethanol yield was highly significant (p< 0.01) at linear and quadratic level. Yeast strain affected the ethanol yield at 1% level of significance and fermentation time affected at 5% level of significance.

Effect of yeast strains, pH and fermentation time on utilized sugar

Utilizedsugar = 4.93 - 0.089X1 - 0.005X2 + 3.41X3 - 0.18X1X2 + 0.11X1X3 + 0.13X2X3 - 0.064X12 -0.57X22 + 0.22X32   (1)

Full second order model, Equation (1) was fitted into utilized sugar and experimental conditions using multiple regression analysis.

Effect of yeast strains, pH and fermentation time on change in pH

Model was highly significant (p<0.05) with F value of 29.48 and hence found to be satisfactory in describing change in pH content.

Change in pH = 0.51 - $0.079X_1$ + $0.21X_2$ - $0.11X_3$ - $0.18X_1X_2$ + $0.23X_1X_3$ + $0.085X_2X_3$ + $0.055X_1^2$ + $0.11X_2^2$ + $0.38X_3^2$ (2)

Effect of yeast strains, pH and fermentation time on cell concentration

The coefficient of determination (R^2) for the regression model for this parameter was 95.40%. Model was highly significant (p<0.05) with F value of 16.13 and therefore was sufficient in describing cell concentration.

CFU = (6.620E+007) - (1.088E + $006X_1$) + (1.526E + $007X_2$) + (7.502E + $007X_3$) + (1.902E + $007X_1X_2$) - (1.849E + $006X_1X_3$) (9.952E + $005X_2X_3$) + (2.563E + $006X_1^2$) + (4.288E + $006X_2^2$) + (1.214E + $007X_3^2$)   (3)

Effect of yeast strains, pH and fermentation time on ethanol yield

The coefficient of determination (R^2) for the regression model for ethanol yield was 96.47%, which implies that the model could account for 96.47% data. Lack of fit was significant but model can be considered adequate as it had a high R^2 value.

Ethanol yield = 1.04-$1.51X_1$ + $0.17X_2$ + $0.56X_3$ - $0.19X_1X_2$ - $0.49X_1X_3$ + $0.11X_2X_3$ + $1.63X_1^2$ + $0.12X_2^2$-$0.26X_3^2$ (4)

The result of regression analysis of all the dependent parameters is given in Table 3.

Process optimization

The objective of the study was to get the optimized condition where the best product can be obtained among the experiments performed. The optimized condition could be a single point or a range of points in which all the possible combinations would yield good results.

Optimization of independent variables

Optimization is a process of making compromises between responses, to achieve a common target. The responses namely ethanol yield, utilized sugar, change in pH and cell count were considered for optimization. The goal setup for optimization is given in the Table 4.

The validity of the model was proved by fitting different values of the variables into the model equation and by carrying out the experiment at those values of the variables. During optimization 10 solutions were obtained, out of which the one that suited the criteria most was selected. The most suitable optimum point is given in the Table 5. The model F- value was found to be highly significant at 1% level of significance in case of all the responses observed. Hence second order model was fitted to predict all the dependent parameters. The contours were drawn using the best fit model equations

Table 3. Result of regression analysis for dependent parameters.

Source	Ethanol conc.		Utilized sugar		Change in pH		CFU	
	Coeff.	P value	Coeff.	P value	Coeff.	P value	Coeff.	P value
Cons	1.04	0.03	4.93	0.01	0.51	0.01	6.620E+007	0.07
X_1	-1.51	0.01***	-0.089	47.38	-0.079	2.40**	-1.088E+006	87.18
X_2	0.17	30.23	-5E-003	96.72	0.21	0.01***	1.526E+007	5.13*
X_3	0.56	0.69***	3.41	0.01***	-0.11	0.65***	7.502E+007	0.01***
X_1X_2	-0.19	40.67	-0.18	30.75	-0.18	0.26***	1.903E+007	7.73*
X_1X_3	-0.49	5.20*	0.11	54.67	0.23	0.05***	-1.849E+006	84.63
X_2X_3	0.11	61.59	0.13	45.05	0.085	6.47**	-9.952E+005	91.68
X_1^2	1.63	0.01***	-0.064	70.61	0.055	19.12	2.563E+006	78.31
X_2^2	0.12	57.85	-0.57	0.96*	0.11	1.90**	-4.288E+006	64.68
X_3^2	-0.26	25.04	0.22	22.23	0.38	0.01***	1.214E+007	21.76
R^2%	96.47		99.20		97.43		95.40	
F value	21.27		95.92		29.48		16.13	
LOF	S		NS		S		S	

***, **, * Significant at 1, 5 and 10% level of significance respectively; ns = Non significant, s = significant, cons = constant; X_1 = yeast strains, X_2 = pH , X_3 = Fermentation time (min).

Table 4. Constraints for optimization.

Variable	Goal	Lower limit	Upper limit
Yeast strains	None	-1	+1
pH	Is in range	-1	+1
Fermentation time	Maximum	-1	+1
Ethanol yield	Maximum	0.58	5.23
Utilized sugar	Maximum	1.24	8.86
Change in pH	Maximum	0.36	1.32
Cell count	Maximum	90000	1.58E+008

Table 5. Optimum levels of variables.

Independent variable	Coded levels	Actual levels
Yeast strains (X_1)	-1	Y_{51}
pH (X_2)	1	5
Fermentation time (X_3)	1	72 h

for the centre point as well as for optimum point as shown in the Figures 1 to 4.

Conclusion

The present study has shown a promising potential for utilising apple pomace as a novel substrate for the production of ethanol. However, the yield 5.23% (v/v) was low, but it can be further improved with different possible combinations of sugar reducing enzymes and also by blending AP with other potential sources for production of ethanol, that is, sugar cane, sugar beet, molasses, corn, grains (wheat, maize, and barley), tubers, biomass etc. The statistically based optimisation procedure, using response surface methodology was proved to be an effective technique in optimising fermentation conditions.

ACKNOWLEDGEMENT

Authors express their thanks to the Department of PHPFE and Microbiology, GBPUAT Pantnagar (Uttarakhand), India.

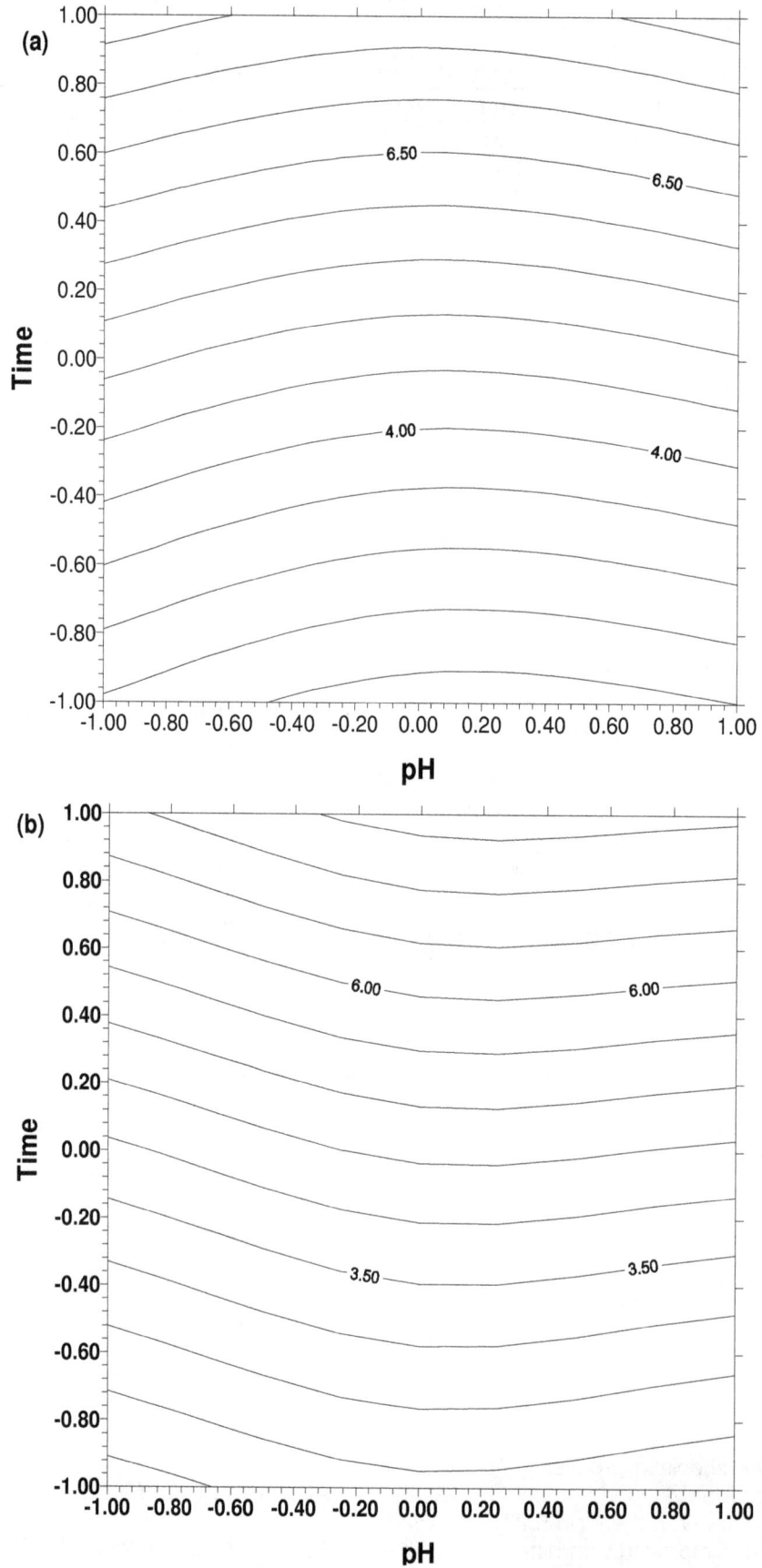

Figure 1. Contour plots for utilized sugar. (a) At centre point; (b) At optimum point.

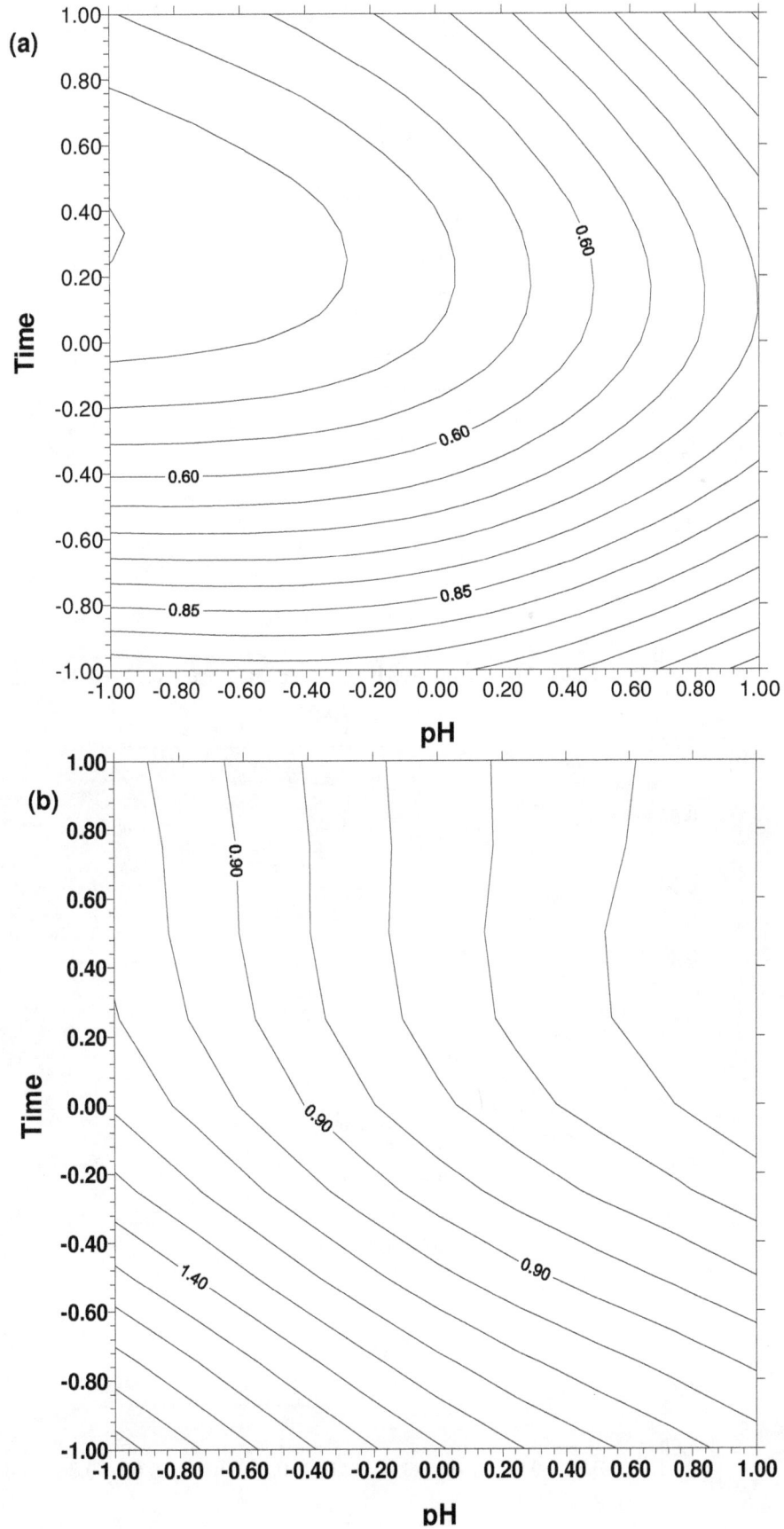

Figure 2. Contour plots for change in pH. (a) At centre point; (b) At optimum point.

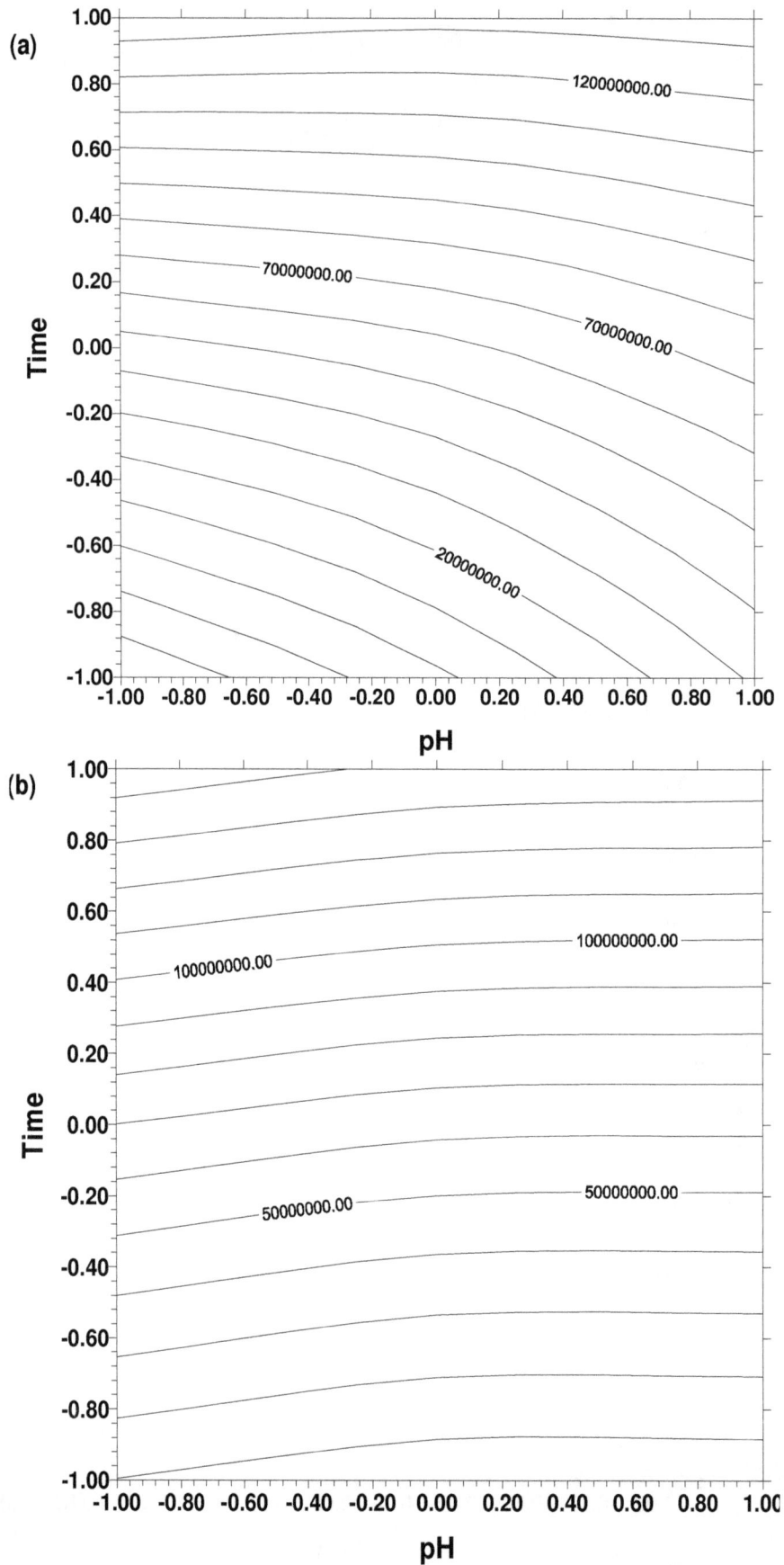

Figure 3. Contour plots for cell count. (a) At centre point; (b) At optimum point.

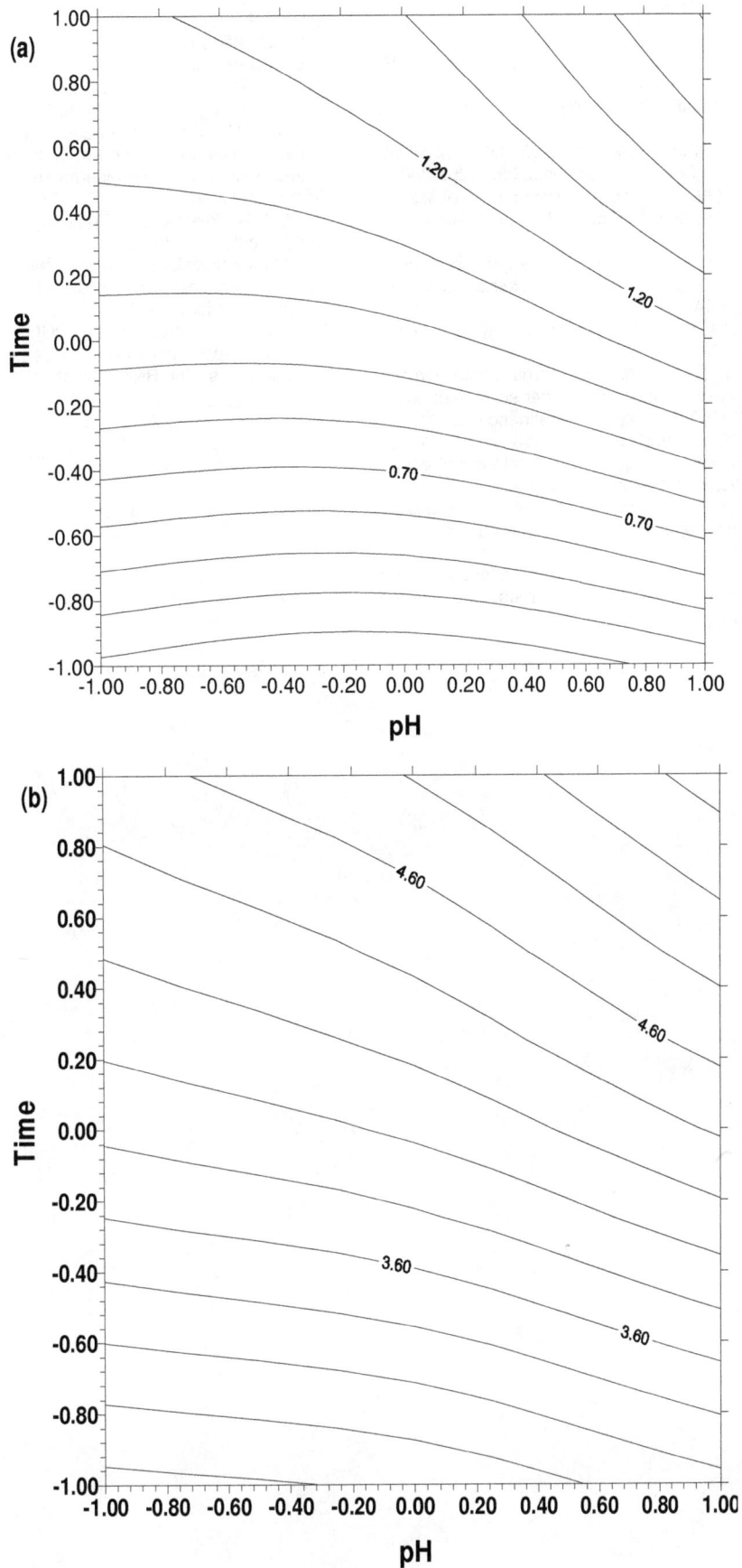

Figure 4. Contour plots for ethanol yield. (a) At centre point; (b) At optimum point.

REFERENCES

Aswathy US, Sukumaran RK, Devi GL, Rajasree KP, Singhania RR, Pandey A (2010), Bio-ethanol from water hyacinth biomass: An evaluation of enzymatic saccharification strategy. Bioresour. Technol. 101(3):925–930.

Bhushan S, Kalia K, Sharma M, Singh B, Ahuja PS (2008), Processing of AP for bioactive molecules. Crit. Rev. Biotechnol. 28(4):285-296.

Hang YD, Woodams EE (1986). Solid state fermentation of apple pomace for citric acid production. Mircen. J. Appl. Microbial. Biotechnol. 2:283-287.

Jain A, Singh A (2006). Ethanol production from apple pomace in natural and inoculated fermentation. M. Tech Thesis. GBPUAT Pantnagar, Uttarakhand (INDIA).

Jewell WJ, Cummings KJ (1984). Apple pomace energy and solids recovery. J. Food Sci. 49:407-410.

Kumar G, Sahgal M (2008). Production of ethanol from apple pomace using yeast strain with improved fermentation under solid state. M.Sc Thesis (Microbiology). GBPUAT Pantnagar, Uttarakhand (INDIA).

Kumar R, Wyman CE (2009). Effects of cellulase and xylanase enzymes on the deconstruction of solids from pretreatment of poplar by leading technologies. Biotechnol. Prog. 25:302-314.

Lancas FM, de Moreas M (2007). Analysis of alcoholic beverages by gas chromatography. In Encyclopaedia of chromatograph. New York: Taylor and Francis.

Liu BL, Tzeng YM (1998). Optimization of growth medium for production of spores from Bacillus thuringiensis using response surface methodology. Bioprocess Eng. 18:413-418.

Miller JE, Weathers PJ, McConville FX, Goldberg M (1982). Saccharification and ethanol fermentation of apple pomace. Biotechnol. Bioeng. Symp. 12:183-191.

Miller GL (1972). Use of dinitrosalicylic acid reagent for determination of reducing sugars. Anal. Chem. 31:426-428.

Neuberg C (1958). Mechanism of ethanol formation in alcoholic fermentation by yeast cells. In: Industrial Microbiology. Prescott, S.C. and Dunn, C.G. (ed.). Mcgraw Hill, New York.

Seeley HWJ, VanDemark PJ, Lee JJ, (1991).Microbes in Action, 4th ed, W.H. Freeman and Company, New York, USA.

Vendruscolo F, Pitol LO, Koch F, Ninow JL (2007). Produ¸c˜ao de prote´ına unicelular a partir do baga¸co de ma¸c˜a utilizando fermenta ¸c˜ao em estado s´olido. Revista Brasileira de Tecnologia Agroindustrial 1:53-57.

Vendruscolo F, Albuquerque PCM, Streit F, Esposito E, Ninow JL (2008). Apple pomace: A versatile substrate for biotechnological applications. Crit. Rev. Biotechnol. 28:1-12.

11

Surface color based prediction of oil content in oil palm (*Elaeis guineensis* Jacq.) fresh fruit bunch

K. Sunilkumar* and D. S. Sparjan Babu

Indian Council of Agricultural Research, Pedavegi–534 450 West Godavari District, Andhra Pradesh, India.

The oil palm, a perennial oil yielding crop, is the richest source of vegetable oils which can produce 4 to 6 MT of palm oil (mesocarp oil) and 0.4 to 0.6 t of palm kernel oil per hectare per annum. The extent of oil available/extractable depends on the ripeness stage of the fruits. The present study was undertaken to evaluate different maturity stages of oil palm fruits in terms of color and oil content, establish their inter relationship and to develop prediction models based on color values so that non destructive ripeness evaluation could be achieved. Models developed with Red, Green and Blue (RGB) values showed that the oil content on fresh fruit bunch (FFB) could be predicted with 57 to 66% efficiency. Models developed using L*a*b* values could predict oil content in fresh fruit bunches up to 79% accuracy. Validations of the models were done with different data sets. The RGB based model showed 64% accuracy in prediction. The L*a*b* model upon validation could predict oil percent up to 89% accuracy. The L*a*b* based model would be ideal for incorporating in gadgets like colorimeters for the purpose of color based grading of FFB and prediction of oil content. Further, it will be useful in automation of harvesting through a machine vision system. This will finally help in harvesting at correct stage of ripeness and objective grading as well as price fixation of oil palm FFB.

Key words: Oil palm, fresh fruit bunch, color grading, L*a*b* color space, mesocarp oil.

INTRODUCTION

The oil palm, a perennial oil yielding crop, is the richest source of vegetable oils which can produce 4 to 6 MT of palm oil (mesocarp oil) and 0.4 to 0.6 t of palm kernel oil per hectare per annum. The maturity of oil palm fresh fruit bunches (FFB) at harvest is an important factor affecting quantity and quality of oil recovered. In case of immature fruits, the oil content is less which in turn results in low oil extraction ratio (OER). It was revealed (Oo et al., 1986) that 41.4% of mesocarp oil is accumulated between 16 and 20 weeks after pollination and another 18.9% later. Over mature fruits will drop from the bunch which incurs more labour for collection of the same. Otherwise loss of the equivalent quantity of oil would be the result. Further, the free fatty acid (FFA) content of over mature fruit tends

to increase rapidly after harvest. Hence, harvesting at correct maturity stage and grading plays a key role in achieving maximum OER. Manual grading can be biased and the farmer may not be convinced of the price of the produce. Development of suitable grading system can help to separate the FFB into distinct groups based on maturity and price fixing could be made on that basis. Hence, a reliable system of maturity evaluation and grading forms the basis for optimizing the oil yield and quality as well as for price fixation of FFB (Alfatni et al., 2008; Sunilkumar and Sparjan, 2010).

Colour is an ideal maturity index which can be measured easily, non-destructively and is distinct at different levels of ripeness for many agricultural commodities. Although, many colour scales were developed for the purpose, the predominant one used for fruit and vegetable grading is Hunter 'Lab' or CIE L*a*b*. L* a* b* is the set of standards adopted by the International Commission on Illumination (CIE: *Commission*

*Corresponding author. E-mail: sunilk.icar@gmail.com.

International de l'Eclairage) in order to define the color in absolute terms and is a widely used and accepted colour measurement system. It is a simplified cube root version of the Adams-Niclearson space produced by plotting the quantities of L* a* b* in rectangular coordinates. The L* refers to lightness or darkness of the colour; a* indicates the change from Red to Green colour and b* indicates the change between Yellow and Blue. Standard colour chips are available for assessment of maturity of including peaches (Delwiche and Baumgardner, 1985) and colour charts are used to assess tomatoes (McGlasson et al., 1986; U.S.D.A, 1975). Colorimeter showed potential for maturity assessment of tomatoes (Yang et al., 1987). Reports suggested that that there is positive association between oil content and colour development in oil palm (Choong et al., 2006; Razali et al., 2008). The Malaysian Palm Oil Board indicated the possibility of using a colour meter for objective grading of oil palm FFB (Omar et al., 2003; Razali et al., 2011). A Lab color space based prediction model for maturity of Granny Smith apple was developed (Tijskens et al., 2010).

Machine vision system of grading could be developed based on the standard color values that reflect oil content in the mesocarp. The prediction model would form the basis of grading different categories of FFB depending on color and the current understanding is that total oil content is a measure of true ripeness (Rajanaidu et al., 1988). A machine vision system was developed (Xu and Zhao, 2010) for strawberry grading which used a* channel as dominant color criteria with 88.85 accuracy in grading. The CIE LAB color space was employed for evaluating egg plant quality (Chong et al., 2008). A machine vision system based on RGB color values was developed for automated sorting of pomegranate arils (Choong et al., 2006). Extraction of features like calyx color, size and shape for grading of egg plant through machine vision was developed (Chong et al., 2008).

Alfatni et al. (2008) analyzed color density to determine the ripeness of bunch based on RGB color model to distinguish between ripeness categories of oil palm FFB; and he found overall relationship between content based on Pearson co relations r^2 is 0.84.

The present study was aimed at analyzing the changes in color and oil content of fruits at varying stages of ripeness and establishing their relationship. The association between oil content and color of fruit bunch could be employed in developing models for non destructive estimation of oil content in the oil palm fresh fruit bunch. This would help avoid laborious and time consuming laboratory method of oil estimation and can help mechanise the grading procedure.

MATERIALS AND METHODS

There are mainly two fruit types in oil palm based on color: *Nigrescence* as well as the *Virescence* type. During ripening the

fruit color of *Nigrescence*, changes from black /dark purple to dark red while in case of *virescence* the color changes from dark green to deep orange and hence they were studied separately. The present paper deals with the measurement of color and development of prediction models for *virescence* type which is having distinct color changes during ripening. The ripeness stages were decided according to the description provided by Malaysian Palm Oil Board (MPOB) with required modification.

The experiment was conducted by harvesting bunches at varying ripeness stages viz. unripe, under ripe, ripe and over ripe of Tenera variety grown under irrigated conditions. Color parameters were measured in two scales viz. in CIE L*a* b* color space as well as in RGB values. Different bunch samples were used for measurement of color with respect to different scales. Hunter colorimeter D25LT was used for obtaining color values on Lab scale. The bunches were cleaned for dirt and placed on the sample holder position in such way that the middle portion of the bunch was focused and then the bunch was rotated to collect images/values from three positions along the same circumference. The instrument was standardized by using standards (perfect reflecting difusser) given by the Hunter lab Associates. Standardization is done by placing the black glass at the sample port and then setting the scale which is reflected to a calibrated standard white tile (TS-102030).

For RGB values, the images of bunch with varying ripeness stages from unripe to over ripe stages were captured (avoiding irregular colored portions) using Sony DSC-F828 digital camera. The distance between camera and bunch sample was kept constant and images were captured following the method suggested by Choong et al. (2006). The pixel size was 1280 × 960. The images were then converted into three basic colors (RGB) using Adobe Photo shop 7.0. The intensity of each color was recorded by performing histogram analysis which counts the total number of pixels in each grayscale value and graphs it. Then, mean of three images taken from three positions along the same circumference was taken as value of a particular sample.

Laboratory analysis were carried out simultaneously using the mesocarp samples to determine the change in oil content (Soxhlet method) on wet weight basis at each ripeness (color) stage under study. Statistical analysis of the data was performed by descriptive analysis, correlation analysis and linear regression models were developed through Step wise as well as Enter method using SPSS version 17.0

RESULTS AND DISCUSSION

L* a* b* values and oil content

The total mesocarp oil content varied from 6.46 to 68.13% with a standard deviation of 16.09 (Table 1). The high variance in oil content could be attributed to the presence of significant number of unripe bunches having low oil content. With respect to L*, values ranged from 34.24 to 52.03 with standard deviation of 4.16. The a* values, which stands for the change from green to red, had second highest variance ranging from -7.56 to 33.92 with a standard deviation of 9.87. The b* values varied from 3.09 to 49.54 with a standard deviation of 6.774. The variance was maximum for oil content which was followed by a* and b*. Correlation with oil content was positive and significant for a* (0.85) followed by, L*/b* (0.53) and a*/b* (0.36), whereas the correlation was significantly negative for L* (0.64) and b* (0.30).

Table 1. Descriptive statistics for L*a*b* values.

Variable	N	Minimum	Maximum	Mean	Std. Deviation	Variance
Oil (%)	104	6.46	68.13	42.50	16.09	259.04
L*	104	34.24	52.03	42.78	4.16	17.31
a*	104	-7.56	33.92	23.00	9.87	97.44
b*	104	3.09	49.54	37.84	6.77	45.88
L*2a*	104	1.01	12.97	1.25	1.17	1.36
L*/b*	104	-19.60	4.60	0.42	4.68	21.86
a*/b*	104	-.24	8.67	0.69	0.85	0.73

Table 2. Descriptive statistics for RGB values.

Variable	N	Minimum	Maximum	Mean	Std. Deviation	Variance
Oil (%)	50	0.35	60.50	39.96	17.47	305.26
Red	50	57.69	230.06	134.72	43.02	1851.023
Green	50	39.41	114.74	77.29	21.58	465.77
Blue	50	14.56	81.95	42.72	14.59	213.14
Red/Green	50	0.91	2.90	1.79	0.48	0.23

Table 3. Linear regression models with L*a*b* and their diagnostic attributes.

Model no.	Equation	R^2	Adjusted R^2	Std error of the estimate	R^2 change	P value
1	Oil (%)=10.537+1.39a*	0.726	0.724	8.46	0.73	**
2	Oil (%)=64.076+1.159a*-1.127L*	0.791	0.787	7.43	0.07	**
3	Oil (%)=77.9-1.576L*+0.566a*+0.529b*-10.727 L*/a*-0.086 L*/b*+18 a*/b*	0.801	0.789	7.39	0.80	**

RGB values and oil content

The mean oil content varied from 0.35 to 60.50% with an average of 39.96 and standard deviation of 17.47 (Table 2). The red color values ranged from 57.69 to 230.06 with an average of 134.72 and standard deviation of 43.024. Correlation analysis indicated that positive and significant association exist between oil content and Red/Green ratio (0.76) followed by Red (0.56). Whereas Blue values showed negative association (0.316) with oil content. Further, Green and Red/Green values showed indirect positive association with oil content through Red. Similar correlation of oil content with most of the color values was reported earlier (Alfatni et al., 2008) using individual fresh fruit bunch images in oil palm.

Model development

L* a* b* values

Linear regression was performed by Enter as well as Stepwise methods for developing suitable prediction models for oil percent based on the fruit surface color and the models are presented in Table 3. The regression standardized normal P-P plot is presented in Figure 1.

RGB values

Different models could explain 57 to 66% variation in fruit surface color. Of these (Table 4), model 2 and 3 could be considered promising where adjusted R^2 was high and R^2 change was low. In case of model 4, though adjusted R^2 was equally good, the R^2 change was also more and hence excluded from validation stage. The model 1 was having comparatively less efficacy of prediction. The regression standardized histogram and normal P-P plot are presented in Figure 2.

Model validation

Model based on L* a* b* values

For validation of the model, different data sets of 26 values were used (Table 5). The variance component was more or less equal to that of the original data set for

Table 4. Linear regression models with RGB and their diagnostic attributes.

Model no.	Equation	R^2	Adjusted R^2	Std error of the estimate	R^2 change	P value
1	Oil (%)=-9.233+27.49R/G	0.579	0.571	11.45	0.579	**
2	Oil (%)=-32.73+30.991R/G+0.233G	0.646	0.631	10.62	0.066	**
3	Oil (%)=-57.812+37.054G+0.318B	0.689	0.669	10.05	0.043	**
4	Oil (%)= - 6.571+.012R+.210G+.320B+36.29R/G	0.689	0.662	10.16	0.689	**

Table 5. Descriptive statistics for L*a*b* values of validation data.

Variable	N	Minimum	Maximum	Mean	Std. Deviation	Variance
Oil (%)	26	2.69	60.50	42.21	16.56	274.14
L*	26	34.60	50.48	42.60	3.95	15.56
a*	26	-6.56	33.17	22.06	11.51	132.50
b*	26	24.28	48.68	37.29	6.17	38.11
L*/a*	26	-17.00	4.21	0.41	4.39	19.31
L*/b*	26	1.02	1.65	1.16	0.149	0.02
a*/b*	26	-0.21	1.09	0.61	0.36	0.13

Table 6. Descriptive statistics for RGB values of validation data.

Variable	N	Minimum	Maximum	Mean	Std. Deviation	Variance
Oil (%)	69	23.48	59.41	43.69	9.89	97.89
Red	69	86.81	189.15	134.28	29.19	852.45
Green	69	38.09	107.14	74.99	19.81	392.38
Blue	69	27.47	61.47	39.35	9.02	81.33
Red/Green	69	1.15	2.50	1.85	0.37	0.14

Dependent Variable: WetOil

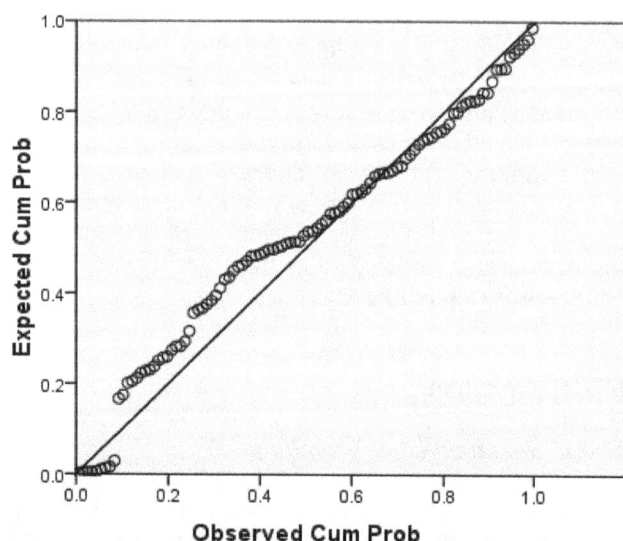

Figure 1. Normal P-P plot of regression standardized residual.

most of the parameters. The fit between observed and predicted oil percent was 0.88 for model 1 (Figure 3) and 0.89 for model 2 (Figure 4), which indicated good accuracy of prediction. This is very high for a heterogeneous biological material like oil palm FFB. The result was in agreement with the findings of Xu and Zhao (2010) in case of strawberry grading.

Model based on RGB values

Validation of the Model 2 was carried out with a different data set of 69 values and the descriptive statistics are presented in Table 6. The oil content varied from 23.48 to 59.41% with a mean of 43.68 and standard deviation of 9.89. The variance components are comparatively lower for the validation data set than that used for model development. The variance for red, green, Blue color values as well as red to green ratio was lesser for validation data set. The variance of oil content was reduced to 32% of the original data set. The relation between observed oil and predcted oil as per model 2 is given in

Dependent Variable: Oil percent

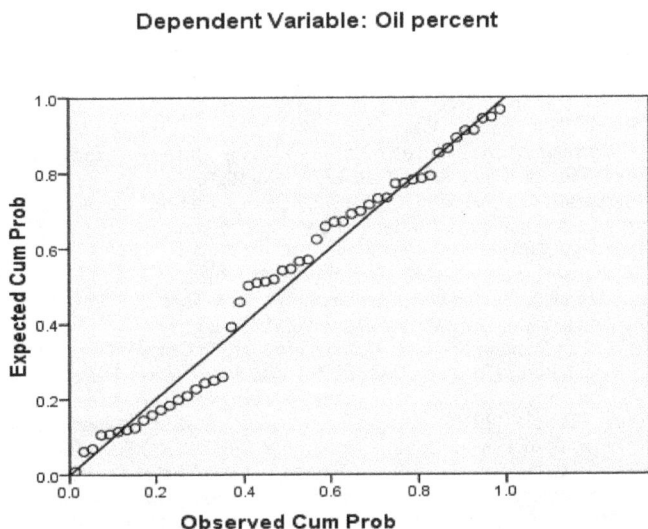

Figure 2. Normal P-P plot of regression standardized residual.

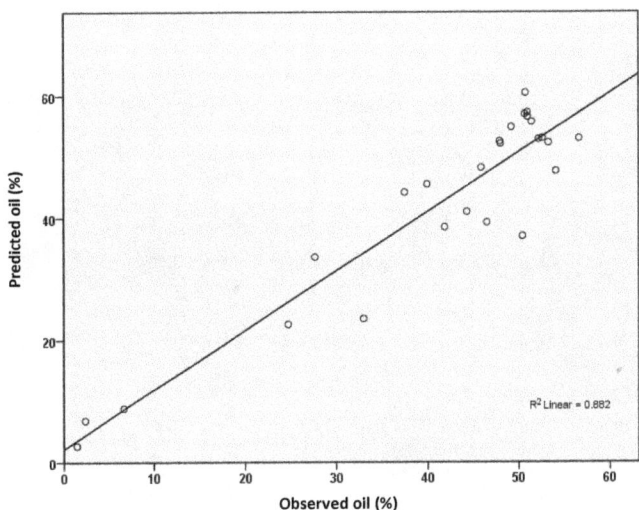

Figure 3. Validation of model 1.

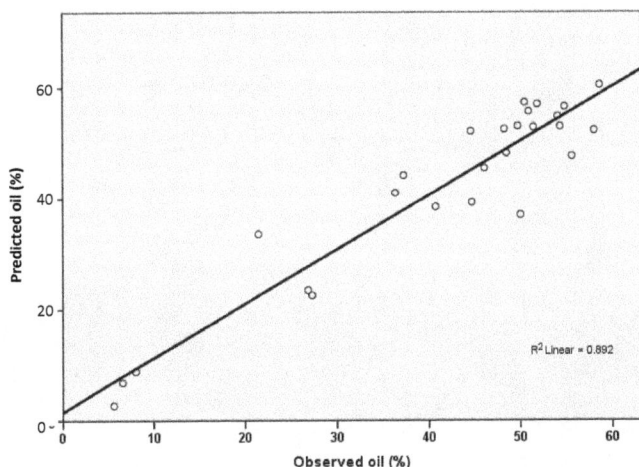

Figure 4. Validation of model 2.

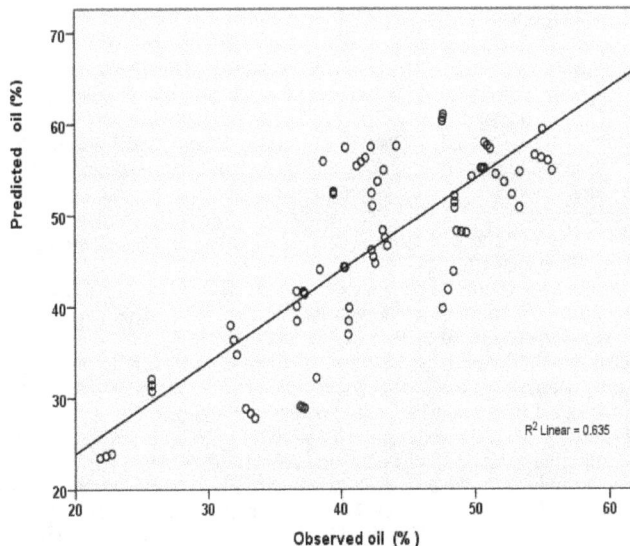

Figure 5. Validation of RGB model 3.

Figure 5. The fit for the two values was 0.63 using the model. Oil (%)=-57.812+37*Green+0.31*Blue, indicating reasonable accuracy for prediction purpose. The oil content of fruit in mesocarp and digital value was increased with increasing the stages of maturity which was also found by Abdullah et al. (2002), Omar et al.(2003), Rashid et al. (2004), Choong et al.(2006) and Balasundram et al.(2006). Thus it could be established that from L*a*b* values of the FFB images has strong relationships with oil content of mesocarp.

Conclusion

The oil palm is a very labour intensive crop and harvesting operations account for 76 to 85% of total labour cost (Balasundram et al., 2006). Present maturity assessment is based on the number of fruits fallen and normally five fruits is the standard. However, while five fruits are visible on the ground, many more must have fallen and got trapped in the frond (leaf) bases in the crown. Moreover, fruits from the outer layer of the bunch which are having maximum oil in the mesocarp are lost in this way, aggravating oil loss. Harvesting of under ripe bunches lead to reduction in extractable oil (OER). As a result of present study, prediction models were developed to estimate the oil content through non destructive means. The model viz., Oil (%)=64.076+1.159a*-1.127L* was the most promising and the same could be validated with accuracy of 89%. This will help in proper assessment of maturity by non destructive means and hence enable timely harvesting. Moreover the models could be incorporated in gadget like colorimeter for rapid and reliable maturity assessment in field. This will also be useful for mechanization of oil palm FFB harvesting through computer vision system.

REFERENCES

Abdullah ZM, Guan CL, Abdul MDM, Mohid AMN (2002).Color vision system for ripeness Inspection of Oil palm (*Elaeis guineensis*). J. Food Process. Preserv. 26(3):213-235.

Alfatni MSM, Shariff ARM, Shafri HZM, Saaed OMB, Eshanta OM (2008). Oil palm fruit bunch grading system using red, green and blue digital number. J. Appl. Sci. 8(8):1444-1452.

Balasundram SK, Robert, PC, Mulla DJ (2006). Relationship between oil content and fruit surface color in oil palm (*Elaeis guineensis* Jacq) J. Plant Sci. 1(3):217-227.

Chong VK, Kondo N, Ninomiya K, Nishi T, Monta M, Namba K, Zhang Q (2008). Features extraction for eggplant fruit grading system using machine vision, Appl. Eng. Agric. 24(5):675-684.

Choong TSY, Abbas S, Shariff AR, Halim R, Ismail MHS, Yunus R, Ali S, Ahmadun FR (2006). Digital image processing of palm oil fruits 2(2):1-4 http://www.bepress.com/ijfe/vol2/iss2/art7

Delwiche MJ, Baumgardner RA (1985). Ground colour as peach maturity index. J. Am. Soc. Hort. Sci. 110:53-57.

McGlasson WB, Beattie BB, Kavanagh EE (1986). Tomato ripening guide. Department of New South Wales, Ag Fact. H84.85

Omar I, Khalid AM, Haniff HM, Wahid, BM (2003) Color meter for measuring fruit Ripeness. MPOB Information series, TT No.182.

Oo KC, Lee KB, Ong SH (1986). Changes in fatty acid composition of the lipid classes I developing oil palm mesocarp. Phytochemistry 25:405-407.

Rajanaidu N, Ariffin AA, Wood BJ, Sarjit S, Pushparajah E (1988). Ripeness standards and harvesting criteria for oil palm bunches., In H.A.Halim et al (Eds), Proc. International Oil palm-Agriculture. PORIM incorporated Soc. Of planters (ISP), KualaLumpur, Malaysia, pp. 224-230.

Rashid S, Nor A, Adnam RM, Shattri M, Rohaya H, Roop G (2004). correlation between oil content and DN values, Department of Biological and Agriculture, Universiti Putra Malaysia, GISdevelopment.net.

Razali MH, Wan Ismail WI, Ramli AR, Sulaiman MN (2008), "Modeling of Oil Palm Fruit Maturity for the Development of an Outdoor Vision System," Int. J. Food Eng. 4(3):1396-1396.

Razali HM, Ishak WIW, Ramli RA, Nasir S, Harun, MH (2011). Technique on Simulation for Real Time Oil Palm Fruits Maturity Prediction. Afr. J. Agric. Res. 6(7):1823-1830.

Sunilkumar K, Sparjan DS (2010).studies on fruit maturity standards of oil palm in Andhra Pradesh under irrigated conditions. Int. J. Oilpalm 7(1&2):5-8

Tijskens L, Schouten RE, Konopacki PJ, Hribar J, Simcic M (2010). Modelling the biological variance of yellow aspect of Granny Smith Apple colour. J. Sci. Food Agric. 90(5):798-805.

U.S. Department of Agriculture, 1975.Colouer classification requirements in tomatoes. *USDA Visual aid TM-L-1*.The John Herry Company. P.O. Box 17099, Lansing Michigan 48901.

Xu L, Zhao Y (2010). Automated strawberry grading system based on image processing. Comput. Electron. Agric. 71 (SY):S32-S39.

Yang CC, Brennan P, Chinnan MS, Shewfelt RL (1987). Characterization tomato ripening process as influenced by individual seal packaging and temperature. J. Food Qual. 10:21-33.

12

Quality improvement in lemon (*Citrus limon* (L.) Burm.) through integrated management of fruit cracking

Savreet Sandhu[1] and J. S. Bal[2]

[1]Punjab Agricultural University, Regional Research Station, Bathinda-151001, Punjab, India.
[2]Department of Agriculture, Khalsa College, Amritsar-143001, Punjab, India.

The studies on management of fruit cracking were carried out for three consecutive years (2006 to 2008) using integrated approach. The plant material was selected from "Punjab Government Progeny Orchard and Nursery, Attari, Amritsar". The investigation comprised three sets of experiments during the fruiting years 2006 and 2007. The first experiment comprised irrigation and mulching practices; the second consisted of graded doses of farmyard manure (FYM), inorganic fertilizer and biofertilizer and in the third experiment foliar spray of 1-Naphthaleneacetic acid (NAA), K_2SO_4 and Borax were applied. The statistically best treatments accrued from three different experiments during 2007 and 2008 were combined and tested in 2008. It was revealed that the optimum utilization of different orchard cultural practices viz. proper water management, appropriate fertilizer programme and good preventive spray schedule brought profound changes in fruit cracking intensity. Hence, the consortium of intelligent management practices such as irrigation at 20% available soil moisture depletion (ASMD), mulching with black polythene, application of FYM (75 kg/tree), inorganic fertilizer (Nitrogen 350 g/tree), azotobacter (18 g/tree) and foliar spray of NAA at 40 ppm in lemon cv. Baramasi substantially reduced the cracking losses by 94.5% and resulted in impressive impact on fruit quality.

Key words: Irrigation, mulching, organic fertilizer, inorganic fertilizer, biofertilizer, growth regulator, nutrients.

INTRODUCTION

In the changing global scenario, success of citrus cultivation depends largely on the ideal quality attributes to ensure better marketability. Lemon, a leading acid citrus fruit is highly lucrative because it bears fruit in many flushes making it available throughout the year. However, the summer crop is beset with severe fruit cracking which is one of the most exasperating problems causing heavy financial losses. What is more, the summer crop has been observed to be prone to severe fruit cracking. Fruit cracking is a worldwide problem which affects a number of fruits and losses are sometimes high. Cracking is manifested as a meridian fissure of the peel, usually developing from the stylar end and reaching the equatorial zone or even extending beyond that. Irrespective

to its origin, crack develops as a consequence of disruption between peel and pulp growth. It was assessed that during the phase of cell enlargement, if the peel does not restart its growth, when the pulp expansion takes place, the fruit splits.

Hoffmann (2007) explained citrus fruit splitting as one of the most serious problems experienced by citrus fruit growers. Although many studies have dealt with this complex phenomenon, the basic mechanism involved in fruit cracking remains unclear. According to Hoffmann, the split usually starts at the blossom end of the fruit, which is the weakest point in the rind. The split may be short and shallow or it may be deep and wide, exposing the segments of the juice vesicles. Splitting appears to be

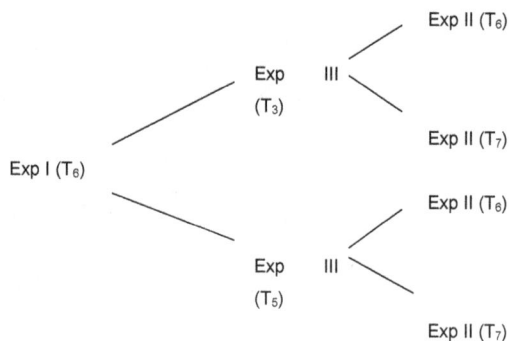

Figure 1. Treatment combinations

most closely related to extreme fluctuations in temperature, humidity, soil moisture and fertilizer levels. It is thought that the problem is caused by a combination of these factors rather than a single one. Splitting is usually observed when growing conditions become erratic such as water stress and sudden rainfall with uneven fertilizer supply. The optimal growing conditions including reasonable cultural practices, sufficient water supply with mineral nutrition and mulching can significantly reduce the occurrence of splitting. Hoffmann advocated frequent irrigation along with the use of compost and slow release fertilizers to feed the tree and foliage spray of trace elements at the most receptive time to replenish the nutrients. Therefore, for lemon cultivation to be successful, efficient management of water and nutrients is of utmost importance to produce high quality of lemon fruits free from fruit cracking. The present study is an endeavour in this direction.

MATERIALS AND METHODS

The present studies were conducted at "Punjab Government Progeny Orchard and Nursery, Attari, Amritsar" during the years 2006 to 2008. In the trial, 8 year old lemon trees, uniform in size and vigor, free from attack of diseases and pests were selected on which given treatments were applied. The investigations were planned in three sets of experiments during 2006 and 2007 and a separate one during 2008. The statistical analysis was done using RBD having four replications. The percentage data was analysed using arc sine transformation. The total number of fruits on the tree was counted on 11th June each year when fruit cracking was first observed and recorded. Cracked fruits were counted regularly at weekly interval. These were picked out and then removed. The percentage of cracked fruits was calculated on the basis of the total number of fruits initially found on the tree.

Experiment I: Effect of various irrigation and mulching treatments on fruit cracking in Baramasi lemon

Every year in April, the irrigation and mulching treatments were started, a week after fruit set and continued until the harvest. The plants received the standard fertilizer dose as recommended by PAU, Ludhiana. The treatment details were:

T_1: Irrigation at 10-15 days interval (control)

T_2: Irrigation at 40% ASMD
T_3: Irrigation at 20% ASMD
T_4: Control and mulching with black polythene sheet
T_5: Irrigation at 40% ASMD and mulching with black polythene sheet
T_6: Irrigation at 20% ASMD and mulching with black polythene sheet

Experiment II: Effect of organic manure, inorganic fertilizer and biofertilizer on fruit cracking in Baramasi lemon

The standard fertilizer dose used was as recommended by PAU, Ludhiana for 8 years old citrus trees (75 kg/tree FYM and 350 g/tree nitrogen). The whole quantity of farm yard manure was applied in December. Nitrogen dose was given in two split doses, the first part in February and the second in April after fruit set. The biofertilizer used in this experiment was Azotobacter and obtained from PAU, Ludhiana. The dose of biofertilizer used was 2 kg/acre or 18 g/tree and the applied method of application was to mix 2 kg of biofertilizer
with 200 L of water and drenching near the root zone of the plants (Indiamart, 2007). The treatment details were:

T_1: Control (Standard dose viz. 75 kg/tree FYM and 350 g/tree N).
T_2: FYM (standard dose *viz.* 75 kg/tree) + Azotobacter (18 g/tree)
T_3: Inorganic fertilizer (standard dose *viz.* 350 g/tree N) + Azotobacter (18 g/tree)
T_4: FYM (standard dose *viz.* 75 kg/tree) + inorganic fertilizer (half the standard dose *viz.* 175 g/tree N) + Azotobacter (18 g/tree)
T_5: FYM (half the standard dose *viz.* 38 kg/tree) + inorganic fertilizer (standard dose *viz.* 350 g/tree N) + Azotobacter (18 g/tree)
T_6: FYM (standard dose *viz.* 75 kg/tree) + inorganic fertilizer (standard dose *viz.* 350 g/tree N) + Azotobacter (18 g/tree)
T_7: FYM (1.25 standard dose *viz.* 94 kg/tree) + inorganic fertilizer (1.25 times standard dose *viz.* 438 g/tree N) + Azotobacter (18 g/tree)

Experiment III: Effect of foliar spray of growth regulator (NAA) and micronutrients on fruit cracking in Baramasi lemon

This experiment consisted of foliar sprays of NAA (20 ppm), NAA (40 ppm), K_2SO_4 (4%), K_2SO_4 (8%), Borax (0.5%) and Borax (1%). The whole plant spray was given with the help of knapsack sprayer during forenoon. Each year during May, 2 sprays were administered at an interval of 15 days. The first spray was given on 10th May and the second on 25th May. The treatment details were:

T_1: Control (untreated)
T_2: NAA (20 ppm)
T_3: NAA (40 ppm)
T_4: K_2SO_4 (4%)
T_5: K_2SO_4 (8%)
T_6: Borax (0.5%)
T_7: Borax (1%)

Experiment IV: Effect of treatment combinations on fruit cracking in Baramasi lemon

The best statistic results of treatments during 2006 and 2007 were combined and researched in a separate experiment during 2008. The treatment combinations are presented diagrammatically as Figure 1.

The treatment details were:

T_1: (Irrigation at 10-15 days interval + FYM (75 kg/tree) + inorganic fertilizer (350 g/tree N) + spray of plain water) Control.

Table 1. Effect of irrigation and mulching on the fruit cracking in lemon.

Treatments	Fruit cracking (%)	
	2006	2007
T$_1$ (Control viz. Irrigation at 10-15 days interval)	35.29 (36.40)*	36.30 (37.01)*
T$_2$ (Irrigation at 40% ASMD)	20.46 (26.82)	21.41 (27.49)
T$_3$ (Irrigation at 20% ASMD)	13.14 (21.17)	15.21 (22.90)
T$_4$ (Irrigation at 10-15 days interval and mulching)	29.20 (32.68)	30.09 (33.24)
T$_5$ (Irrigation at 40% ASMD and mulching)	15.65 (23.28)	17.53 (24.74)
T$_6$ (Irrigation at 20% ASMD and mulching)	7.21 (15.50)	8.28 (16.66)
CD at 5%	2.73	2.52
CV%	6.98	6.21

*Transformed values.

T$_2$: (Irrigation at 20% ASMD and mulching + FYM (75 kg/tree) + inorganic fertilizer (350 g/tree N) + Azotobacter (18 g/tree) + foliar spray of NAA at 40 ppm)
T$_3$: (Irrigation at 20% ASMD and mulching + FYM (113 kg/tree) + inorganic fertilizer (525 g/tree N) + Azotobacter (18 g/tree) + foliar spray of NAA at 40 ppm)
T$_4$: (Irrigation at 20% ASMD and mulching + FYM (75 kg/tree) + inorganic fertilizer (350 g/tree N) + Azotobacter (18 g/tree) + foliar spray of K$_2$SO$_4$ at 8%)
T$_5$: (Irrigation at 20% ASMD and mulching + FYM (113 kg/tree) + inorganic fertilizer (525 g/tree N) + Azotobacter (18 g/tree) + foliar spray of K$_2$SO$_4$ at 8%)

RESULTS AND DISCUSSION

Experiment I: Effect of irrigation and mulching on fruit cracking in Baramasi lemon

A glance over the data in Table 1 shows that during 2006, higher fruit cracking (35.29%) increased in the control treatment. T$_6$ proved to be the most effective treatment, significantly lower over control, by registering minimum fruit cracking (7.21%). The treatment T$_6$ maintained its superiority in the next trial (year 2007), also by cutting short the percentage of fruit cracking significantly over all other treatments including control. The fruit cracking in this treatment was recorded to be 8.28%, compared to 36.30% in control. Heavy losses due to fruit cracking have also been reported in Kagzi Kalan (Sharma and Shukla, 2002). The soil moisture seemed to be a major contributing factor in fruit cracking as its incidence was lowered with enhanced moisture supply. As a result of sudden increase in water content of soil and atmospheric humidity after long dry spell, the tissues of fruit skin did not cope with the rapid increase of the fruit internal tissues (Chandra, 1988), resulting in the bursting of the skin (Lu and Lin, 2011) because of internal turgor pressure of the fruit (Measham et al., 2010). The use of black polythene mulch also attributed to minimize the extent of fruit cracking in lemon. Mulching might have played an important role in plant establishment, growth and fruiting of kinnow as reported by Lal et al. (2003),

which could be due to moisture conservation of soils. The frequent irrigation and mulching with black polythene would have changed the micro-climate of the trees in comparison to trees receiving irrigation at longer interval and unmulched. Hence, a reduction in fruit cracking could be attributed to better moisture regulation together with its conservation through mulching. High moisture content certainly have reduced the temperature of tree canopy, leaf, fruit and soil and increased the atmospheric humidity and created favourable conditions for continued growth of the peel for a longer period. Therefore, the texture of peel would have attained ability to resist the pressure of the expanding juice vesicles. Subsequently, it would have helped to reduce the splitting of fruit. It can be further concluded that high temperature and low humidity during the period of fruit growth rendered the peel inelastic, affecting the

Experiment II: Effect of organic manure, inorganic fertilizer and biofertilizer on fruit cracking in Baramasi lemon

Minimum percentage of fruit cracking was recorded in treatment T$_7$ during two years of research study giving values 18.89 and 19.93%, respectively while maximum extent of fruit cracking was evidenced in T$_1$ to the tune of 35.29% during first trial year and 36.30% in second year (Table 2). The extent of fruit cracking varied significantly in T$_6$ and T$_7$ as compared to control. Taking into consideration the effect of different treatments on fruit cracking, T$_7$ was found to be the best practice. However, the effect of treatment T$_6$ was found to be statistically at par with the best treatment. Thus, the treatment T$_6$ proved to be most judicious fertilizer application practice from an economic point of view. When all three nutrient sources viz. FYM, inorganic fertilizer and biofertilizer (Azotobacter) were applied, it reduced the fruit cracking percentage. This can be attributed to improved nutrient and water availability as result of application of required nutrient sources, leading to vital plant and fruit growth,

Table 2. Effect of organic and inorganic nutrient sources on the fruit cracking in lemon.

Treatments	Fruit cracking (%)	
	2006	2007
T₁ {Control (75 kg/tree FYM + 350 g/tree N)}	35.29 (36.40)*	36.30 (37.01)*
T₂ {FYM (75 kg/tree) + Azotobacter (18 g/tree)}	33.15 (35.11)	34.26 (35.78)
T₃ {Inorganic fertilizer (350 g/tree N) + Azotobacter (18 g/tree)	31.55 (34.15)	32.82 (34.93)
T₄ {FYM (75 kg/tree) + inorganic fertilizer (175 g/tree N) + Azotobacter (18 g/tree)}	28.75 (32.40)	30.00 (33.19)
T₅ {FYM (38 kg/tree) + inorganic fertilizer (350 g/tree N) + Azotobacter (18 g/tree)	26.60 (31.03)	27.77 (31.78)
T₆ {FYM (75 kg/tree) + inorganic fertilizer (350 g/tree N) + Azotobacter (18 g/tree)}	20.28 (26.72)	21.37 (27.49)
T₇ {FYM (94 kg/tree) + inorganic fertilizer (438 g/tree N) + Azotobacter (18 g/tree)}	18.89 (25.72)	19.93 (26.47)
CD at 5%	2.32	2.61
CV%	4.94	5.43

*Transformed values.

Table 3. Effect of foliar spray of growth regulator (NAA) and nutrients on the fruit cracking in lemon.

Treatments	Fruit cracking (%)	
	2006	2007
T₁ (control) untreated	35.29 (36.40)*	36.30 (37.01)*
T₂ {NAA (20 ppm)}	17.06 (24.34)	17.57 (24.72)
T₃ {NAA (40 ppm)}	11.86 (20.04)	12.65 (20.75)
T₄ {K₂SO₄ (4%)}	17.60 (24.69)	18.40 (25.30)
T₅ {K₂SO₄ (8%)}	12.39 (20.49)	13.09 (21.11)
T₆ {Borax (0.5%)}	18.03 (25.08)	18.80 (25.67)
T₇ {Borax (1%)}	19.68 (29.30)	19.88 (26.44)
CD at 5%	3.38	3.41
CV%	9.00	8.90

*Transformed values.

because of the development of better root system along with increase in number of rootlets. This is corroborated with the findings of Prahraj et al. (2002). Bio-fertilization helps a better proliferation of roots, which ultimately results in sturdy and healthy plants showing resistance to biotic and abiotic stresses. These results conform with that of Sharma and Thakur (2001). The biofertilizers produces nitrate substances along with auxins. The presence of both promoted the deposition of the exogenous calcium in the cell walls of pericarp. Higher concentration of structural calcium in cell wall of pericarp provided cracking resistance. The higher capacity in binding exogenous calcium in the cell wall of pericarp suggests higher concentration of negatively charged structural component, that is, glacturonic acid residues which can be one of the material bases for cracking resistance (Zhong et al., 2006). It was further suggested that availability of such nutrients in the early stage of fruit ontogeny is important for cracking resistance. Azotobacter is capable of elaborating small qualities of growth promoting substances like B-vitamin and phyto-hormones like IAA, and with inorganic N and FYM might have improved the physiology of the plants (Nair and

Najachandra, 1995), thereby, reducing fruit cracking. The increase in auxin status in plants could have increased peel thickness as the auxins have the tendency of faster and prolonged cell division in peel (Amiri et al., 2012). The optimized standards of fertilizer application might also have played role in keeping pace between the growth of the cell wall and the cortex leading in increase in elasticity of peel which in turn have helped to cut short fruit cracking.

Experiment III: Effect of foliar spray of growth regulator (NAA) and nutrients (K₂SO₄ and Borax) on fruit cracking in Baramasi lemon

The data presented in Table 3 revealed that maximum fruit cracking was increased in control (untreated) trees, with a rate of 35.29% in 2006 and 36.30% in 2007. The treatment T₃ (NAA at 40 ppm) proved to be most effective in minimizing the fruit cracking in lemon in two consecutive years of research trial, with records of 11.86% in first year and 12.65% in second year. The treatment T₅ was statistically at par with T₃. All the spray treatments had a profound effect on the fruit cracking percentage and the elastic and plastic properties of the citrus rind are thought to be involved in resistance to puncture. Application of auxins caused enlargement of cells by increasing the elasticity or permeability of cell wall (Cline and Trought, 2007). Thus, the peripheral tissues of the fruit would keep pace with the growth of cortex resulting in the control of fruit cracking, given that one of the main reasons for fruit cracking is attributed to the differential growth rates of the peripheral and cortex tissues. Low level of potassium was thought to be responsible for splitting of Hamlin orange (Morgan et al., 2005). Earlier findings of Bar-Akiva (1975) in Valencia orange also lend support to the present results, who further reported that reduction of splitting may be a potassium mediated effect, via strengthening of the fruit rind as seen from the increasing rind thickness of fruits in potassium treated trees. The decline in fruit cracking

Table 4. Effect of various treatment combinations on the fruit cracking in lemon.

Treatments	Fruit cracking (%)
	2008
T_1 (Irrigation at 10-15 days interval + FYM (75 kg/tree) + inorganic fertilizer (350 g/tree N) + spray of plain water) control	34.54 (35.97)*
T_2 (Irrigation at 20% ASMD and mulching + FYM (75 kg/tree) + inorganic fertilizer (350 g/tree N) + Azotobacter (18 g/tree) + foliar spray of NAA at 40 ppm)	1.89 (7.82)
T_3 (Irrigation at 20% ASMD and mulching + FYM (113 kg/tree) + inorganic fertilizer (525 g/tree N) + Azotobacter (18 g/tree) + foliar spray of NAA at 40 ppm)	4.42 (12.12)
T_4 (Irrigation at 20% ASMD and mulching + FYM (75 kg/tree) + inorganic fertilizer (350 g/tree N) + Azotobacter (18 g/tree) + foliar spray of K_2SO_4 at 8%)	7.27 (15.62)
T_5 (Irrigation at 20% ASMD and mulching + FYM (113 kg/tree) + inorganic fertilizer (525 g/tree N) + Azotobacter (18 g/tree) + foliar spray of K_2SO_4 at 8%)	9.35 (17.78)
CD at 5%	1.62
CV %	5.89

*Transformed values.

due to boron treatments may be attributed to its physiological role in synthesising of pectin substances in cells. Boron is responsible for increasing the elasticity of the cell membranes and prevents the breakdown of vegetative tissues. Boron also improved the translocation of sugar and synthesis of cell wall material. Thus, this decrease in fruit cracking might be the result of borate bridging with cell wall constituents, thus giving it elasticity, response to it as advocated by Singh et al. (2005) in litchi.

Experiment IV: Effect of treatment combinations on fruit cracking in Baramasi lemon

The minimum fruit cracking to the rate of 1.89% was registered in fruits obtained from T_2 trees kept under optimum orchard management practices. The highest cracking of fruit during experimentation in year 2008 was recorded in control to the rate of 34.54% which is considered a huge loss to fruit growers. The reduction in fruit cracking to maximum extent in treatment T_2 (Table 4) giving negligible loss (1.89%) could have been anticipated due to efficient management of water, nutrients and optimum level of PGRs. The absence of even a single practice in various treatment combinations intensified the fruit cracking percentage. In the present experimentation, the increase of inorganic N and FYM in T_3 negated the proper orchard management leading to increase of fruit cracking over treatment T_2. This may be due to ill-effects of excessive fertilization which caused a burning effect of the citrus rind resulting in a weak rind susceptible to crack. The moisture regulation and conservation created favourable conditions for continued growth of peel for a longer span, making it sufficiently

elastic to keep pace with the internal growth of pulp thereby, resisting pressure to split or crack. The optimized fertilizer application in the form of organic manure, inorganic nitrogen and biofertilizer resulted in higher nutrient uptake partly due to higher nutrient composition in soil and better availability of water as a result of root proliferation. The foliar spray of growth regulator (NAA at 40 ppm) reduced the fruit cracking to a great extent. Thus, the maintenances of adequate soil moisture and nutrients reduced this problem (Measham, 2011). If a single component is avoided, it directly or indirectly influences the cracking percentage.

Conclusion

From the foregoing discussion, it can be concluded that efficient management of water and nutrients is essential to increase the quality of lemon fruits and to prevent the fruit from cracking. Generally, multiple applications consisting of proper water management, good fertilizer programme and good preventive spray programme are necessary. Hence, intelligent management practices such as irrigation at 20% available soil moisture depletion (ASMD), mulching with black polythene, application of FYM (75 kg/tree), inorganic fertilizer (Nitrogen 350 g/tree), azotobacter (18 g/tree) and foliar spray of NAA at 40 ppm in lemon, substantially reduced the cracking losses and resulted in better fruit quality. Thus, to combat this serious problem the consortium of different appropriate treatments is of utmost importance. Therefore, it can be further concluded that fruit cracking can be controlled by a combination of treatments, and inadequate water and nutrient management practices at a critical stage of fruit development adversely affect the fruit quality.

ACKNOWLEDGMENTS

The authors express their thanks to the staff of "Punjab Government Progeny Orchard and Nursery, Attari, Amritsar" for their immense contributions, without which the experiment would not have been successful.

REFERENCES

Amiri NA, Kangarshahi AA, Arzani K (2012). Reducing of citrus losses by spraying of synthetic auxins. Int. J. Agric. Crop Sci. 4:1720-1724

Bar-Akiva A (1975). Effect of potassium nutrition on fruit splitting in Valencia orange. J. Hort. Sci. 50(1):85-89.

Chandra M (1988). Studies on the fruit cracking in lemon (Citrus limon) cv. Pant lemon-1. M.Sc. Thesis, G.B. Pant University of Agriculture and Technology, Pantnagar, India.

Cline JA, Trought M (2007). Effect of gibberellic acid on fruit cracking and quality of Bing and Sam sweet cherries. Can. J. Plant Sci. 87:545–550.

Hoffmann H (2007). Citrus fruit loss in the home garden. Department of Agriculture, Western Australia. Gardennote 38.

Indiamart   (2007). Homepage<http://www.indiamart.com/jayenterprises/fertilizers. html#bio-fertilizer>. Accessed 2007 March, 10.

Lal H, Samra JS, Arora YK (2003). Kinnow mandarin in Doon Valley: I. Effect of irrigation and mulching on growth, yield and quality. Indian J. Soil Conserv. 31:162-67.

Lu PL, Lin CH (2011). Physiology of fruit cracking in wax apple (Syzygium samarangense) J. Plant Sci. 8: 70–76.

Measham PF, Bound A, Gracie J, Wilson SJ (2010). Incidence and type of cracking in sweet cherry (Prunus avium L.) are affected by genotype and season. Crop Pasture Sci. 60:1002-1008.

Measham P (2011). Rain-induced fruit cracking in sweet cherry (Prunus avium L.). Ph.D thesis, School of Agricultural Science, University of Tasmania.

Morgan KT, Rouse RE, Roka FM, Futch SH, Zekri M (2005). Leaf and fruit mineral content and peel thickness of 'hamlin' orange. Proc. Fla. State Hort. Soc. 118:19-21.

Nair SK, Najachandra G (1995). Nitrogen fixing bacteria associated with plantation and orchard crops of Kerala. Final Report - ICAR Ad-hoc Research Project. pp 1-65.

Prahraj CS, Kumar D, Sharma RC (2002). Integrated use of fertilizers and biofertilizers along with plant growth promoting bacteria for higher efficiency in potato (Solanum tuberosum). In: Extended summaries of the 2nd International Agronomy Congress on Balancing Food and Environment Security- A Continuing Challenge, pp 232-235. New Delhi, India.

Sharma RR, Shukla R (2002). Managing fruit cracking in lemon Kagzi Kalan. Indian Hortic. 47(3):14-15.

Sharma SK, Thakur KS (2001). Effect of Azotobacter and nitrogen on plant growth and fruit yield of tomato. Veg. Sci. 28:146-48.

Singh AK, Sharma P, Sharma RM, Tiku AK (2005). Effect of plant bioregulators and micronutrients on tree productivity, fruit cracking and aril proportion of litchi (Litchi Chinesis Sonn.) cv. Dehradun. Haryana J. Hort. Sci. 34(3-4):220-221.

Zhong W, Yuan W, Huang X, Wang H, Li J, Zhang C (2006). A study on the absorption of exogenous calcium and sucrose and their deposit onto the cell walls in litchi pericarp. J. Fruit Sci. 23(3):350-354.

13

Field spread of banana streak virus (BSV)

Kubiriba, J.[1], Tushemereirwe, W. K.[1], Kenyon, L.[2] and Chancellor, T. C. B.[2]

[1]Kawanda Agricultural Research Institute-NARO, P. O. Box, 7065, Kampala, Uganda.
[2]Natural Resources Institute, University of Greenwich, Central Avenue, Chatham, Maritime Kent ME4 4TB, UK.

Musa (banana and plantain) provides a major source of carbohydrates for about 400 million people of whom 20 million are from East Africa. Yet, banana is threatened by number constraints, banana streak virus inclusive. Banana streak virus (BSV) was monitored in Rakai and Ntungamo, Uganda for up to 72 months after planting (MAP) and 29MAP respectively. BSV incidence increase over time was fitted into exponential model and spatial spread analysed by 2DCLASS and 2DCORR. BSV infection was initiated in Rakai 29 months after planting (MAP), but only 6 MAP in Ntungamo. BSV incidence then increased at a rate of 0.10 plants respectively / infected plant / month at a rate 0.23 plants / infected plant / month in Rakai and Ntungamo respectively. In both sites, spatial analysis showed that there were significant aggregated BSV spatial patterns. New infections were more likely to occur within a 10 rows/coloumns from an old infection. Significant edge effects were also detected in Ntungamo, indicating that there was significant spread from the immediate surroundings (infected established field suggesting need for separation of new fields from old infected fields to delay onset of BSV. Roguing should be frequent enough to offset rate of BSV incidence increase. The study shows that BSV is a slow spreading disease; however, there is sufficient time in this perennial cropping system for it to increase to epidemic levels. It is however, possible to check the advance of the BSV epidemic through phytosanitary measures.

Key words: Banana streak virus (BSV), spatial and temporal spread, phytosanitation.

INTRODUCTION

Plantains and bananas (*Musa*) are among the most important fruits, cultivated in over 120 countries (FAO, 2001). In the tropics, the *Musa* provides a major source of carbohydrates for about 400 million people (Swennen et al., 1995) of whom 20 million are from East Africa. Uganda ranks second in the world in banana production with an annual output of about 10.5 million tonnes (FAO, 2001). Over 12 million people including 65% of the urban population depend on the crop as their staple and income.

Although global *Musa* production has increased by 113% from 46 million tonnes in 1968 to 98 million tonnes in 1998, their average yield have risen by only 18% from 8.45 t/ha to 9.96 t/ha during this period (Karamura, 1998).

Yields in Africa, the Carribean and Latin America have not increased over the last 30 years and increase in production is due solely to an expansion in the area under production (IITA, 2004). In Uganda, productivity has been steadily declining due to the effects of pests and diseases, declining soil fertility, poor crop husbandry, and socio-economic and post-harvest problems (Gold et al., 1993).

Banana streak though of recent occurrence, is now one of the major banana diseases in Uganda, where it was first recorded in 1990 (Dabek and Waller, 1990). A severe outbreak of the virus was later reported from Rakai District in the early 1990's (Tushemereirwe et al., 1996). It is now found in many other banana growing areas of Uganda and attacks most of the banana varieties (Kubiriba et al., 1997). Jones and Lockhart (1993) reported up to 90% yield losses on "Poyo" plants with severe BSV symptom. In severely infected areas in Uganda, plantations have had almost 100% loss in saleable yield (Tushemereirwe et al., 1996). Apart from the effect of BSV on the yield (quantity and quality) and plant growth of the banana, it also hinders germplasm exchange (Lockhart, 1996).

Like most other badna viruses, BSV has been shown to be transmitted semi-persistently by mealybugs namely *Planococcus citri*, *Sacharococcus sacchari*, and *Dymicoccus brevipes* under screenhouse conditions (Lockhart and Olszewski, 1993; Su, 1998; Kubiriba et al., 2001b). BSV–infected plants were clustered in banana fields in Uganda and that BSV incidence decreased from the focal infection to the periphery of the clusters of infected plants (Kubiriba et al., 2001a). The study was done in established farmers' fields and the findings were then considered only as circumstantial evidence of field spread of BSV. There appears to be no other information available about the spatiotemporal dynamics of BSV spread within or between banana plantations and yet this should be the basis for effectively controlling BSV.

Temporal and spatial pattern analysis may assist in the development of hypotheses to account for the associations among diseased plants and clues to the spread dynamics of the disease and therefore its control (Thresh, 1974). This paper analyses temporal and spatial spread dynamics of BSV under field conditions in 2 sites in Uganda. The information generated may then provide the basis for designing a more effective management strategy for BSV.

MATERIALS AND METHODS

Trial establishment

The 2 sites (Rakai and Ntungamo) were selected to host the trials because they both have high BSV incidence (> 70%) and well established mealybug populations. The first trial was set up in Rakai using the AAA-EA cv "Kisansa" planted in a 23 plants by 23 plants block in September 1998. The planting material was obtained from a farm where no symptoms of BSV had been observed. The site had been under observation since 1995. The assumed sources of infection were the fields surrounding the trial, which had BSV incidences up to 90%. Another spread trial (12 × 12 plants) was set up in Ntungamo in April, 2002. The planting material used was of the local AAA-EA cv. Kisansa sourced from the same farm earlier described.

Data collection, processing and analysis

The cv. 'Kisansa' spread trial in Rakai was inspected every 3 months but the spread trials in Ntungamo were inspected monthly. Data recorded included date of data collection, position [X, Y] of the plant in the trial [lattice] that is, plant number along the rows and columns. Every plant in each block was inspected for presence of BSV symptoms and recorded (1) for presence and (0) for absence of symptoms.

Temporal spread of BSV incidence data was fitted (P < 0.0001) using the exponential model. The exponential model is suitable for modelling early phases of most polycyclic epidemics (Nutter, 1997). The rate of increase of BSV incidence at the 2 sites was estimated by regressing the natural log of BSV incidence (estimated by exponential model) against time (MAP) thus: $Lny = Lny_0 + rt$, where y is the BSV incidence at time, t and y_0 is the initial BSV incidence. The slope, r is the rate of increase of BSV incidence.

2DCLASS analysis was used to examine the spatial disease patterns of the STCLASS computer programme for personal computers (Nelson et la., 1992). The observed standardised count frequency (SCF) for each [X, Y] distance class was compared to expected SCFs, estimated by 800 computer simulations using the Monte Carlo pseudo-random function and an equal number of symptomatic plants randomly distributed to generate test lattices of the same dimension. 2DCLASS spatial patterns were interpreted as random if the proportion of significant SCFs was less than 5% and aggregated if the proportion of significant SCFs was greater than 5% (Nelson , 1995). The data were considered to have significant edge effects if more than 12.5% of the [X, Y] distance classes within distal row and column [X_{max}, Y_{max}] had significantly greater than expected SCF values (Nelson, 1995). The size of the core cluster was obtained by counting the number of significantly greater than expected SCF values adjacent to the origin [X_0,Y_0] of the distance class matrix that form a discrete group within the area circumscribed by the outer row and column limits of the core cluster. The outer limits of the core cluster are marked by the presence of significantly less than expected SCF values around the core cluster in the distance class matrix. The proximity index is an estimate of the density of the core cluster, which is calculated as the proportion of SCF+ within the area circumscribed by the outer row and column limits of the core cluster.

Spatial patterns were further analysed by 2DCORR in which observed proportions of infected pairs of plants in the field for which plants are separated by distance (r), are generated. 2DCORR also generates the predicted proportion of infected plant pairs for each distance orientation class. The deviation (δ) between observed and expected proportions of infected pairs of plants provides the information over which spatial correlation is significant. The maximum deviation ($δ_{max}$), also estimated as Kolmogorov-Smirnov test statistic were then used to estimate overall significance of deviation of observed BSV spatial patterns from expected random spatial distribution. The spatial correlation is significant if $δ_{max}$ is greater (P < 0.05) than its critical value obtainable from mathematical tables. The corresponding distance (r_{max}) is the maximum radial distance separating infected plant pairs that have significant spatial correlation. It marks the spatial limits of a cluster of infected plants in the field.

RESULTS

BSV symptoms were first observed on cv. Kisansa plants in the spread trial (23 x 23 plants) in Rakai 29 months after planting (MAP). BSV incidence then increased from 1.4%, 37 MAP to 28%, 72 MAP. In the Ntungamo trial, however, the cv. Kisansa plants first showed symptoms 7 MAP and BSV incidence increased from 2.1% to 43.1%, 29 MAP (Figure 1). BSV disease progress curves fitted to an exponential model (*P< 0.0001*). The curves indicated that BSV disease progress was monitored through the linear phases up to progressive phases. BSV incidence increased at an average rate of 0.21 (0.23 plants/infected plant) per month, (P < 0.01) in Ntungamo.

This rate is about two- times greater than the

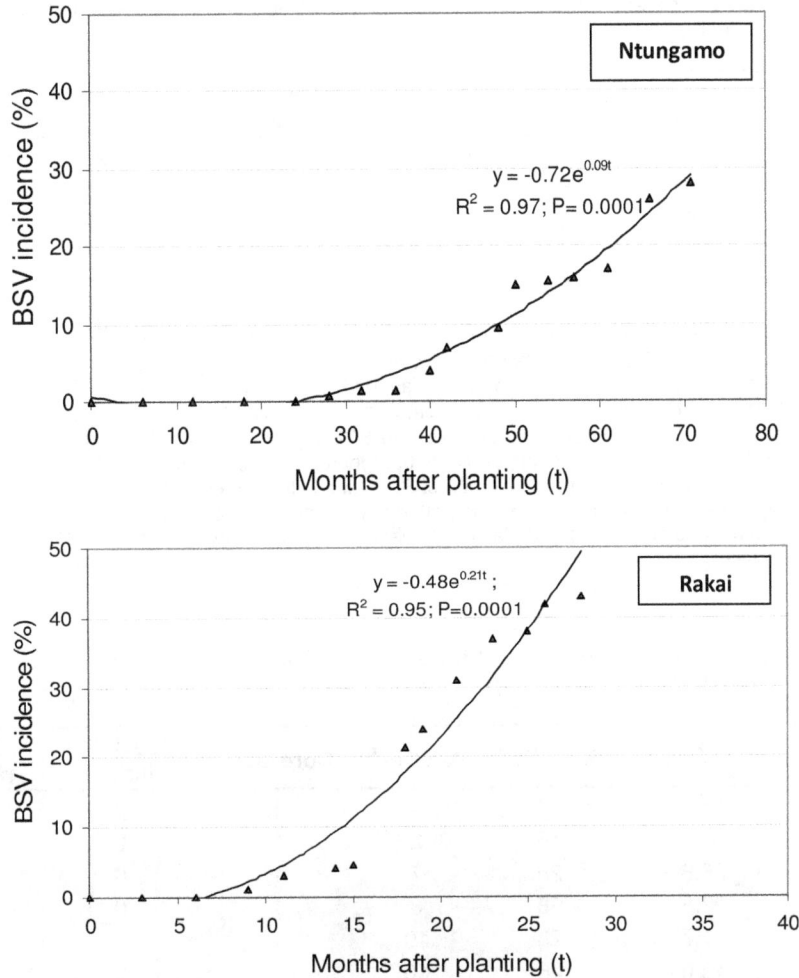

Figure 1. BSV incidence increase over time in Ntungamo and Rakai. Both disease progress curves were fitted with an exponential model (P< 0.0001); BSV incidence rate of increase was estimated by Linearizing the exponential equation using natural log as: Lny = Lny$_0$ + rt, where y= BSV incidence at time t, y$_0$ is the initial BSV incidence and r, is the absolute rate of increase of BSV incidence.

0.09 (0.10 plants/infected plant) per month observed in Rakai.

2DCLASS analysis of spatial data demonstrated that there was aggregation of infected plants on both sites because a high proportion of distance classes with SCF values (SCF+) were significantly greater than expected *(P ≤ 0.05)* in a random spatial structure (Tables 1 and 2) occurred. The SCF + values were relatively smaller in Rakai than in Ntungamo Proportions of distance classes with SCF values (SCF-) significantly less than expected *(P ≥ 0.95)* in a random spatial occurrence were also present. Low SCF- values tend to mark the spatial limits of clusters of infected plants within the field. Edge effects were detected in Ntungamo but not in Rakai. In Ntungamo, distance classes with SCF + values tended to be located in the lower right hand corner [X$_{max}$,Y$_{max}$] region [6-12, 6-12] of the distance class matrix (Figure 2),

however the distance classes with SCF+ values tended to be located in the upper left hand corner [X$_0$,Y$_0$] (Figure 3) for Rakai. This may suggest that there was more spread of BSV from plant to plant into the spread trial from the nearby infection sources in Ntungamo than from within field.

Spatial analysis of data from both sites using 2DCORR revealed that early stages of BSV spread were characterised by random patterns of distribution of infected plants as demonstrated by non significant maximum Kolmogorov-Smirnov test statistic, indicating no spatial correlation between infected plants. Later infected plants in both spread trials had an aggregated spatial structure indicated by a significant maximum Kolmogorov-Smirnov test statistic (P < 0.05) (Tables 3 and 4). Maximum radial distance (r$_{max}$), separating plants of infected plant pairs with significant spatial correlation

Table 1. Incidence, proportion of standardised count frequency (SCF) and spatial pattern from 2DCLASS analysis of BSV spread in 'cv. Kisansa' spread trial in Rakai (R1).

Months after planting (MAP)	% BSV incidence	% SCF+[a]	% SCF-[b]	Core cluster size[c]	Proximity index[d]	Spatial pattern[e]	Edge effect[f]
51	14.6	7	2	23	0.28	A	-[g]
54	14.9	6	2	17	0.21	A	-
57	15.5	6	3	12	0.33	A	-
62	16.5	5	3	17	0.21	A	-
67	25.5	8	5	21	0.33	A	-
72	27.6	6	3	11	0.31	A	-

*October 2001, February 2002, April 2002 and October 2002 are not included in this table because DCLASS could not be performed on data recorded on those dates since BSV incidence was less than 10%, the minimum requirement for 2DCLASS. [a]Proportion of [X,Y] distance classes with standardised count frequency (SCF)greater than expected ($P \leq 0.05$) compared with a random distribution of newly diseased plants. [b]Proportion of [X, Y] distance classes with SCF significantly less than expected ($P \geq 0.95$). [c]The number of significant SCF+ distance classes contagious with the origin [0, 0] of the distance class matrix that form a discrete group. [d]An estimate of the density of the core cluster calculated as the proportion of SCF+ within the area circumscribed by the outer row and column limits of the core cluster. [e]A = Aggregated (non- random) disease pattern; newly infected plants tend to be found near already infected plants. [f]Groups of significant ($P \leq 0.05$) and contagious SCF values for distance classes at the edge of the distance class matrix [X_{max}, Y_{max}]. [g]- = Edge effects were not detected.

Table 2. Incidence, proportion of standardised count frequency (SCF) and spatial pattern from 2DCLASS analysis of BSV spread in 'cv. Kisansa' spread trial in Ntungamo (N5).

Months after planting (MAP)	% BSV incidence	% SCF+[a]	% SCF-[b]	Core size[c]	Proximity index[d]	Spatial pattern[e]	Edge effect[f]
20	23.6	22	4	1	0	A	+[g]
22	31.3	16	2	1	0	A	+
24	36.8	26	7	1	0	A	+
26	37.5	28	8	1	0	A	+
27	41.7	28	10	1	0	A	+
29	43.0	26	10	1	0	A	+

*September 2003 and November 2002are not included in this table because SDCLASS could not be performed of data recorded on those dates since BSV incidence was less than 10%, the minimum requirement for 2DCLASS. [a]Proportion of [X,Y] distance classes with standardised count frequency (SCF)greater than expected ($P \leq 0.05$) compared with a random distribution of newly diseased plants. [b]Proportion of [X,Y] distance classes with SCF significantly less than expected ($P \geq 0.95$). [c]The number of significant SCF+ distance classes contagious with the origin [0,0] of the distance class matrix that form a discrete group. [d]An estimate of the density of the core cluster calculated as the proportion of SCF+ within the area circumscribed by the outer row and column limits of the core cluster. [e]A = Aggregated (non- random) disease pattern; newly infected plants tend to be found near already infected plants. [f]Groups of significant ($P \leq 0.05$) and contagious SCF values for distance classes at the edge of the distance class matrix [X_{max}, Y_{max}]. [g]+= Edge effects were detected.

ranged from 6 to 9 on both sites. This suggests that spatial limits of clusters of infected plants within the field may extend up to 10 plants away from focal infection points in both Ntungamo and Rakai.

Discussion and conclusion

Literature indicate (Campbell and Madden, 1990; Gray et al., 1986) that the random patterns of distribution, as observed in this study, of BSV infected plants at the initial stages of the epidemics in both sites may be due to a number of reasons such as, background contamination or primary spread from a remote source of infection. Background contamination may be caused by use of

pre-infected planting material or mealybugs in the field at the time of planting (Cabaleiro and Segura, 1997). Materials used for establishing the trials should have been indexed for BSV presence but the methods of BSV detection were not reliable at the time. However, the cv. Kisansa plants were obtained from a field that was under observation since 1995 and no symptoms had been observed there. In Rakai, where these materials were planted, first infection was observed 28 MAP. Guidelines for the safe international movement of *Musa* germplasm suggest that indexing and symptom observation should be done for 6 to 9 months (Diekmann and Putter, 1996).

Harper et al. (2002b) reported that there was no evidence that activatable BSV sequences are present in these AAA-EAs. Geering et al. (2000) also reported that

26 MAP PLOT MAP

27 MAP PLOT MAP

29 MAP PLOT MAP

KEY

- = Asymptomatic banana plant

D = Symptomatic banana plant

26 MAP 2DCLASS

27 MAP 2DCLASS

29 MAP 2DCLASS

KEY

+ = Significantly large standardised count frequencies (SCF+)

$ = Significantly small standardised count frequencies (SCF-)

- = SCF Not significant

26 MAP PLOT MAP

27 MAP PLOT MAP

29 MAP PLOT MAP

KEY

- = Asymptomatic banana plant

D = Symptomatic banana plant

26 MAP 2DCLASS

27 MAP 2DCLASS

29 MAP 2DCLASS

KEY

+ = Significantly large standardised count frequencies (SCF+)

$ = Significantly small standardised count frequencies (SCF-)

- = SCF Not Significant

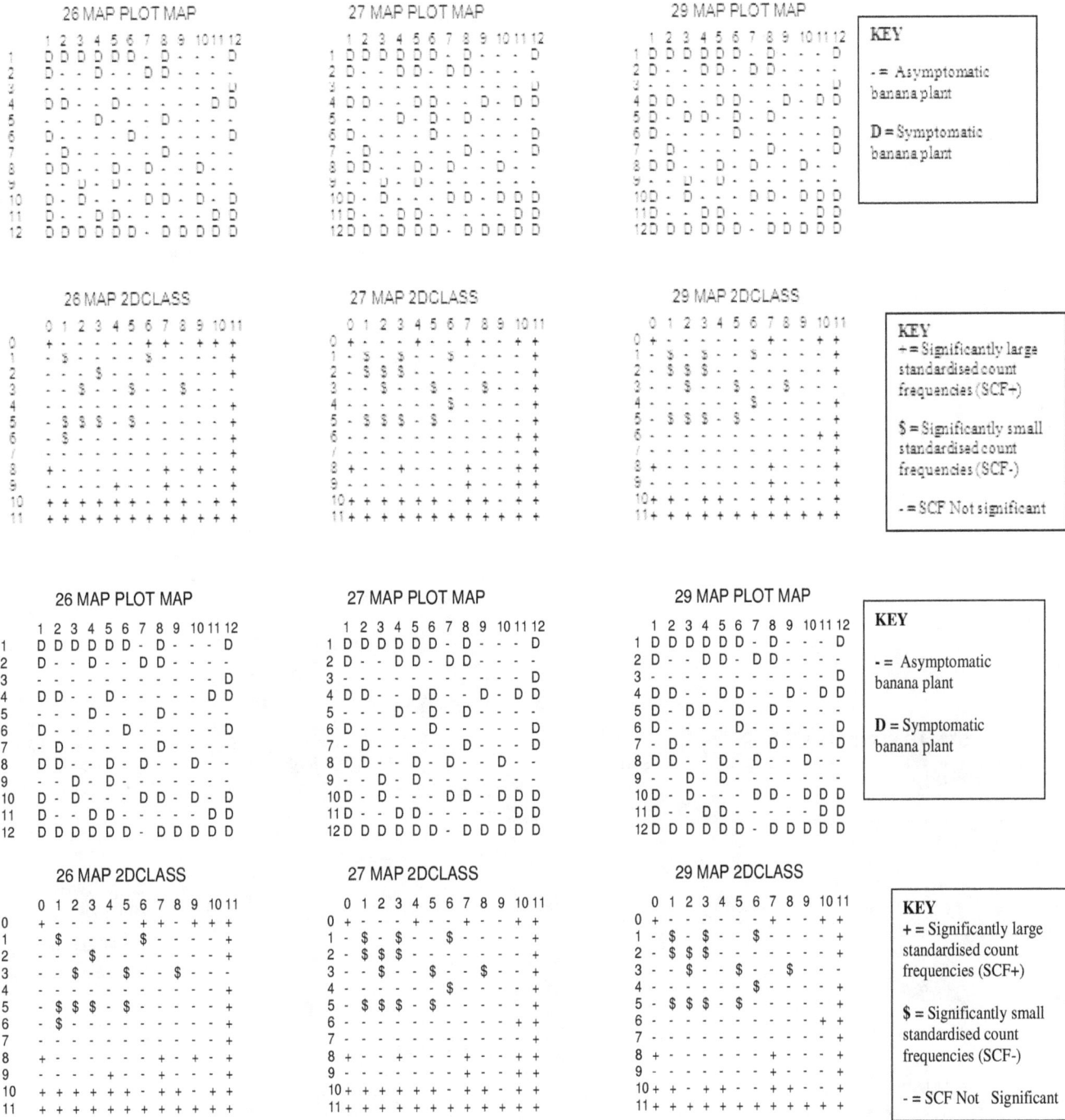

Figure 2. Spatial pattern maps of BSV (first row) with associated 2DCLASS proximity patterns (second row) for the cv. Kisansa (AAA-EA) spread trial in Ntungamo.

episomal expression of the inactive integrated BSV sequences were more associated with the B-genome. BSV in inoculated plants is detectable within 4 to 6 weeks after inoculation (Su, 1998; Kubiriba et al., 2001b). CSSV, a close relative of BSV (Harper and Hull, 1998), showed symptoms within 2 years after inoculation in the field in mature trees (Owusu, 1972). It, therefore does not seem likely, that there was primary infection arising from episomal expression of the inactive integrated BSV sequences in the banana genome, from suckers used for planting were infected at the time of planting, or from mealybugs present in the field at around planting time. BSV infection observed on cv. Kisansa (AAA-EA), therefore, more likely arose from vectored transmission.

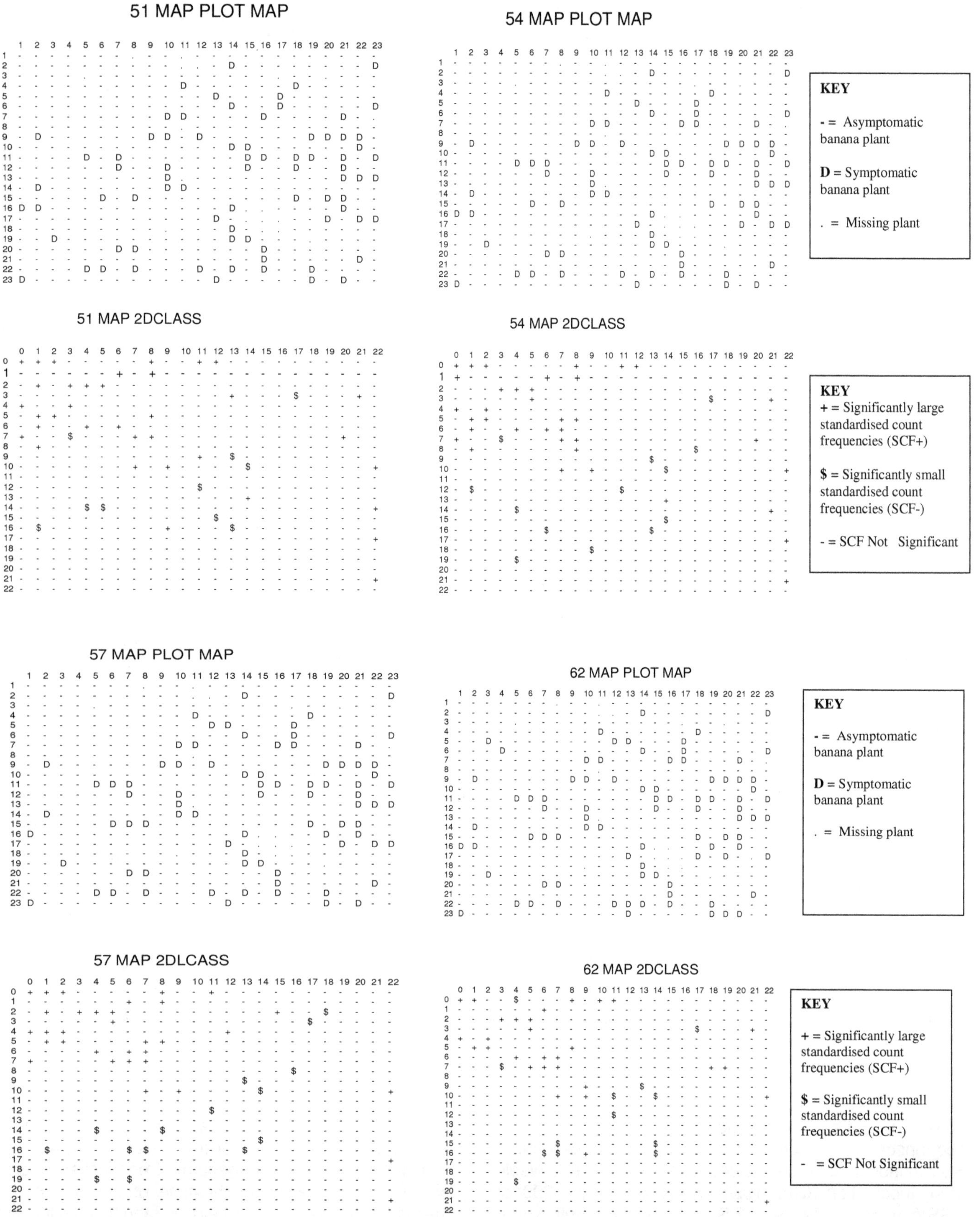

Figure 3. Spatial pattern maps of BSV (first row) with associated 2DCLASS proximity patterns (second row) for the cv. Kisansa (AAA-EA) spread trial in Rakai.

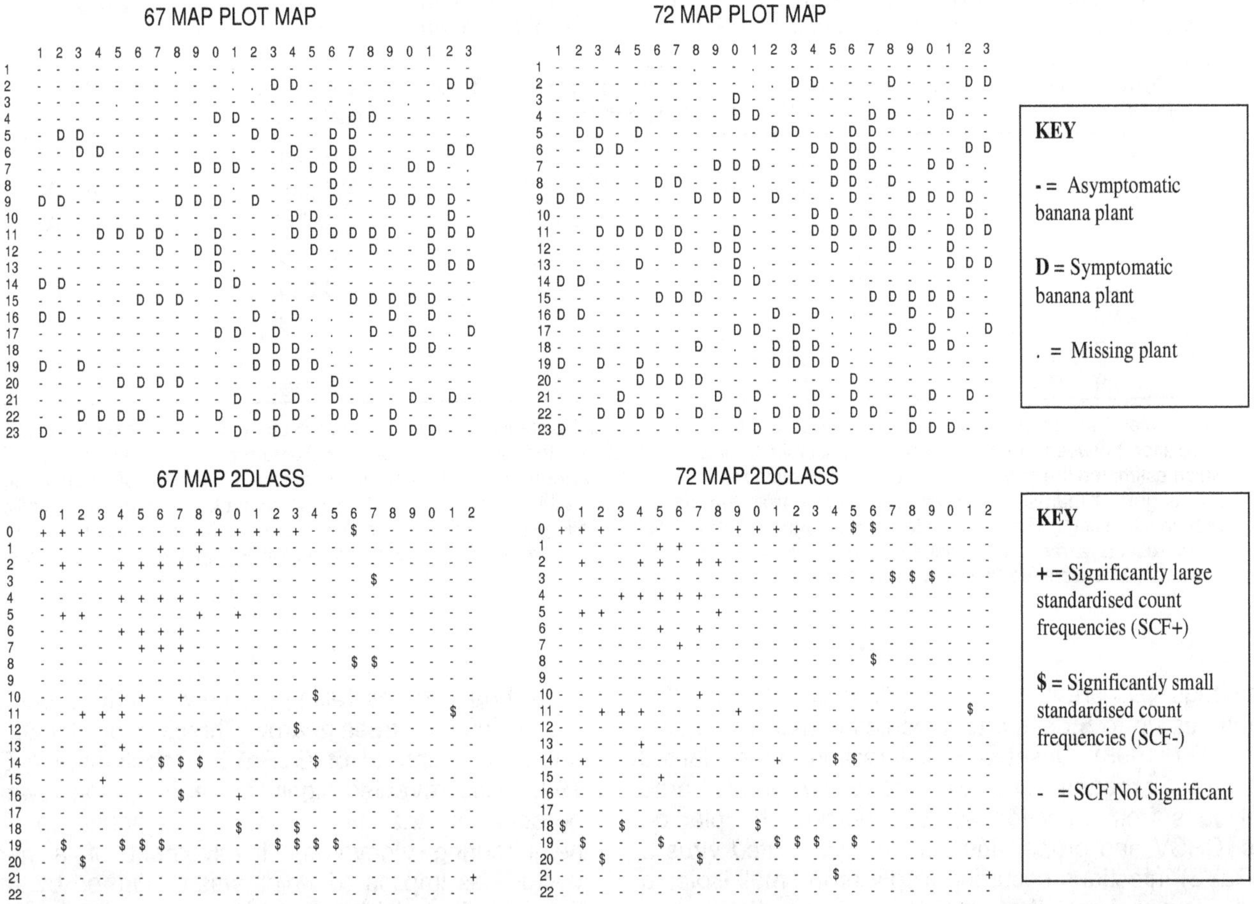

67 MAP PLOT MAP

72 MAP PLOT MAP

KEY

- = Asymptomatic banana plant

D = Symptomatic banana plant

. = Missing plant

67 MAP 2DLASS

72 MAP 2DCLASS

KEY

+ = Significantly large standardised count frequencies (SCF+)

$ = Significantly small standardised count frequencies (SCF-)

- = SCF Not Significant

Figure 3. (Cont'd).

Table 3. Kolmogorov-Smirnov test statistic and corresponding radial distance separating plants of the infected pairs with maximum spatial correlation generated by 2DCORR from BSV spread data in Rakai (R1).

Months after planting MAP	Number of infected plants (N* = 529)	[1]Maximum radial distance (r_{max})	Maximum Kolmogorov-Smirnov test statistic (δ_{max}) [2]	Spatial pattern[3]
37	8	14	0.05ns	-
41	13	9	0.10ns	-
43	30	9	0.04ns	R
49	47	9	0.036ns	R
51	77	9	0.018*	A
54	79	9	0.015*	A
57	81	6	0.014*	A
62	87	9	0.016*	A
67	131	6	0.007*	A
72	144	6	0.006*	A

* N = total number of plants in the trial. [1]r_{max} is the maximum radial distance separating infected plant pairs that have significant spatial correlation between them. It marks the spatial limits of a cluster of infected plants. [2]δ_{max} is the maximum Kolmogorov-Smirnov test statistic, which estimated the overall significance of deviation from random behaviour. The spatial correlation between infected plant pairs is significant if δ_{max} is greater than its critical value (varies with number of infected plants) obtainable from the mathematical tables. Non-significant values are marked with ns and significant ones with an asterisk (*). The critical value of Kolmogorov-Smirnov test statistic generally decreases as number of infected plants increases. [3]R represents a random spatial structure, A an aggregated one and – were not described as random or aggregated because number of infected plants were too few (BSV incidence < 5%).

Table 4. Kolmogorov-Smirnov test statistic and corresponding radial distance separating plants of the infected pairs with maximum spatial correlation generated by 2DCORR from BSV spread data in Ntungamo (N5).

Months after planting (MAP)	Number of infected plants (N*=144)	Maximum radial distance (r_{max})[1]	Maximum Kolmogorov-Smirnov test statistic (δ_{max})[2]	Spatial pattern[3]
17	3	9	0.07ns	-
19	22	9	0.06*	A
20	34	9	0.04*	A
22	45	9	0.035*	A
24	53	9	0.025*	A
26	54	8	0.028*	A
27	60	9	0.019*	A
29	62	9	0.019*	A

* N = total number of plants in the trial. [1]r_{max} is the maximum radial distance separating infected plant pairs that have significant spatial correlation between them. It marks the spatial limits of a cluster of infected plants. [2]δ_{max} is the maximum Kolmogorov-Smirnov test statistic, which estimated the overall significance of deviation from random behaviour. The spatial correlation between infected plant pairs is significant if δ_{max} is greater than its critical value obtainable from the mathematical tables. Non-significant values are marked with ns and significant ones with an asterisk (*). The critical value of Kolmogorov-Smirnov test statistic generally decreases as number of infected plants increases.
[3]R represents a random spatial structure, A an aggregated one and – were not described as random or aggregated because number of infected plants were too few (BSV incidence < 5%).

Mealybugs (*D. brevipes* and *P. citri*) can retain BSV sometimes up to 5 days after inoculation (Kubiriba et al., 2001b; Su, 1998). Onset of BSV infection in the planted trials was probably due to spread by wind borne mealybugs from infection sources outside the planted trials. CSSV and grape vine leaf roll associated virus 3 (GLRaV3) infections occurring singly or in small isolated groups away from the boundary of the field were attributed to the viruliferous windborne mealybugs sources outside the field (Ollennu et al., 1989; Cabaleiro and Segura, 1997). Zadoks and van den Bosch (1994) also contend that the epidemiological implication of spatial independence of infected plants is that, the disease might have spread from infection sources located outside the sampled area.

Clustering nature of BSV infected plants in the trials in both Rakai and Ntungamo indicates secondary spread of disease from a source within the field. The clusters tended to be loosely defined as demonstrated by the low core cluster intensities in Rakai which is characteristic of spread of insect vectored viruses since vector movement is usually not restricted to adjacent plants (Madden et al., 2000). Thresh (1958) reported that new outbreaks tended to develop around existing ones and most CSSV spread was attributed to within field radial spread by mealybug nymphs through the canopy. BSV, like its close relative CSSV, seems to be slow to spread in some locations, and therefore probably amenable to control by phytosanitary measures. BSV management strategy based on phytosanitation would mainly comprise use of clean planting material for establishing new fields and roguing infected plants from the established fields.

BSV spread data demonstrated significant edge effects data in Ntungamo. Similar infections were observed with CSSV in cocoa plantings and were attributed to

mealybug nymphs falling on new plantings below from the old infected trees nearby (Thresh et al., 1988). Spatial analysis of spread of GLRaV3 (Cabaleiro and Segura, 1997) also revealed significant aggregation towards the borders of the field (usually adjacent to infected neighbouring vineyards). Involvement of a non-flying vector was implicated which was confirmed by sampling the fields that yielded *P. citri* that transmitted the virus. It is likely that in Ntungamo, BSV was spread to the peripheral plants of the spread trial by infected mealybug nymphs from the surrounding fields. There was also evidence to show that infected plants were more likely to occur within 10 rows from the boundary in Ntungamo. In Ghana, where CSSV is endemic, spatial limits of CSSV infected plants stretched up to 25 rows from the boundary of the field neighbouring infected plants (Ollenu et al., 1989). This may be important in determining the separation distance between new fields planted with healthy materials and old infected fields where edge effects are important. In cocoa fields in Ghana (Thresh and Owusu, 1986), where a Cordon sanitaire (unplanted band) was maintained between the replanted area and the surrounding diseased area, there was a substantial delay prior to epidemic build up, depending on the width of the Cordon sanitaire. The border effects on BSV spread show that it is important to separate new fields from old infected ones to delay onset of BSV infection in newly planted banana fields established with virus –free suckers.

In Ntungamo, where spread from the borders was more evident, test plants in one of the small spread plots (4 plants × 4 plants) showed first BSV symptoms 6 MAP and the proportion of infected plants increased to 12/16 (75%) at 23 MAP (data not shown). However, in bigger spread trials (12 plants × 12 plants) on the same plantation, BSV incidence increased only to about 28% in the same period. This is in agreement with the findings of Ollennu et al. (1989), there was more risk of spread of CSSV to small or irregular plantations from nearby sources. Since the spread tends to be associated to borders, mainly peripheral plants get infected. These plants form a large proportion of the total stand in small plantings and those of elongate or very convoluted boundaries (Thresh and Owusu, 1986). It may be more beneficial to establish large new fields rather than small ones.

After new fields have been established with non-diseased materials, they together with old plantations need to be inspected regularly so that infected plants can be rogued. Commencement of the roguing activities is site specific since onset of BSV is also site specific and should start earlier in Ntungamo than in Rakai after establishment of new fields. This study also demonstrated that new BSV infections occur nearer to old infections than far away. It is therefore necessary to focus around previous infections when checking for new infected plants, without loosing focus of secondary focal infection, which occurs away from the primary infection focus. It was also shown that the rate of BSV incidence monthly increase was 0.21% in Ntungamo and 0.09% in Rakai. For any suppression programme to succeed in halting the advance of a disease epidemic, the rate of roguing of infected plants should be high enough to offset the rate of increase of the disease (Gowttwald et al., 1996). There are greater chances of success for the control programme, if started at the early stages in the epidemic than later.

An eradication campaign was launched in 1946 though unpopular with the farmers; CSSV was largely under control, by early 1960 and only resurfaced when the government stopped the campaign. Similar campaigns were successfully implemented in controlling semi-persistently transmitted diseases such as Citrus tristeza virus (CTV) in Israel (Fisherman et al., 1983) and BBTV in Australia through visual inspection, eradication and planting with healthy suckers in Australia (Sharma, 1987). The spatiotemporal dynamics of BSV spread on the two fields have been described up to the incidence of 29% and 43% in Rakai and Ntungamo, respectively. Other features of epidemiological importance may be revealed later if the epidemics at the 2 sites are followed to final stages, however, characteristics of the BSV epidemic so far described in Rakai and Ntungamo still provide information that could be the basis for the recommendation of better management practices for BSV. Although BSV is a slow spreading disease, there is sufficient time in this perennial cropping system for it to increase to epidemic levels. However, it is possible to check the advance of the BSV epidemic through phytosanitary measures. This requires the active and well-organised participation of farmers assisted by extension and research staff and local government authority.

REFERENCES

Cabaleiro C, Segura A (1997). Field transmission of grapevine leafroll associated virus (GLRa-V-3) by mealybug, Planococcus citri. Plant Dis. 81:283-287.

Campbell CL, Madden LV (1990). Introduction to plant disease epidemiology. John Wiley and Sons. New York. P. 532.

Dabek AJ, Waller JM (1990). Black leaf streak and viral leaf streak: new banana diseases in East Africa. Trop. Pest Manage. 36(2):157-158.

Diekmann M, Putter CAJ (1996). FAO/IPGR Technical Guidelines for the Safe Movement of Germplasm. Musa. 2nd Edition. Food and Agriculture Organisation of the United Nations, Int. Plant Genet. Resour. Inst. P. 15.

FAO (2001). Production Year book 2001, FAO, Rome Italy.

Fisherman S, Marcus R, Talpaz H, Bar–Joseph M, Oren Y, Solomon R, Zoher M (1983). Epidemiology and economic models for the spread and control of Citrus triciteza virus disease. Phytoparasitica 11:39-49.

Geering ADW, McMichael LA, Dietzgen RG, Thomas JE (2000). Genetic diversity among banana streak isolates from Australia. Phytopathology 90:921-927.

Gold CS, Ogenga-Latigo MW, Tushmereirwe WK, Kashaija IN, Nankinga C (1993). Farmer perceptions of banana pest constraints in Uganda. In: Proceedings of a Research Co-ordination Meeting for Biological and Integrated Control of Highland Banana Pests and Diseases in Africa, Cotonou, 12-14 November 1991. Gold, C.S. and Gemmil, B. (Eds.) International Institute of Tropical Agriculture. Ibadan. pp. 3-24.

Gray SM, Moyer JW, Kennedy GG, Campbell CL (1986). Virus suppression and aphid resistance effects on spatial and temporal spread of watermelon mosaic virus. Phytopathology 76:536-540.

Harper G, Hart D, Moult S, Hull R (2002b). Detection of banana streak virus in field samples of bananas in Uganda. Ann. Appl. Biol. 141(3):247-257.

Harper G, Hull R (1998). Cloning and sequence analysis of banana streak virus DNA. Virus Genes 17:271-278.

IITA (2004). Crops and Farming systems. Bananas and Plantain. P. 72. www.iita.org/crop/plantain.htm.27k

Jones DR, Lockhart BEL (1993). Banana streak virus. Fact sheet No.1. Montpellier. France Int. Netw. Improv. Bananas Plantain.

Karamura D (1998). Numerical taxonomic studies of the East African highland bananas (Musa AAA-East Africa) in Uganda. PhD Thesis. The University of Reading, UK. P. 184.

Kubiriba J, Legg JP, Tushemereirwe W, Adipala E (2001b). Vector transmission of banana streak virus in the screenhouse in Uganda. Ann. Appl. Biol. 139:37-49.

Kubiriba J, Tushemereirwe W, Karamura EB (1997). Distribution of Banana streak virus (BSV) in Uganda. Mus. Afr. 11:17.

Kubiriba, J, Legg JP, Tushemereirwe W, Adipala E (2001a). Disease spread patterns of banana streak virus in farmers' fields in Uganda. Ann. Appl. Biol. 139:31-36.

Lockhart BEL (1996). Virus diseases of Musa in Africa: Epidemiology, detection and control. In: Proceedings of the First International Conference on Banana and Plantain for Africa. Craenen, K., Ortiz, R., Karamura, E.B. and Vuylsteke, D. (Eds.). Pub. Int. Soc. Hortic. Sci. Leuven. Belgium. pp. 355-360.

Lockhart BEL, Olszewski NE (1993). Serological and genomic heterogeneity of banana streak badnavirus: Implications for virus detection in Musa germplasm. In: Breeding Banana and Plantain for Resistance to Diseases and Pests, J. Genry. (Ed.). Montpellier, France. INIBAP. pp. 105-113.

Madden LV, Jeger MJ, van den Bosch (2000). A theoretical assessment of the effects of vector – virus transmission mechanism in plant virus epidemics. Phytopathology 90:576-594.

Nelson SC (1995). STCLASS – Spatiotemporal distance class analysis software for the personal-computer. Plant Dis. 79:643-648.

Nelson SC, Marshal PL, Campbell CL (1992). 2DCLASS, a 2-Dimensional Distance class analysis software for the personal-computer. Plant Dis. 76:427-432.

Nutter FW Jn (1997). Quantifying the temporal dynamics of plant virus epidemics: A review. Crop Prot. 7:608-618.

Ollennu LAA, Owusu GK, Thresh JM (1989). Spread of cocoa swollen shoot disease into recent plantings in Ghana. Crop Prot. 8:251-264.

Owusu GK (1972). Virus research. Acquisition of swollen shoot virus by mealybugs from cocoa plants during the period of latent infection. Report, Cocoa Research Institute. pp. 60-61. Ghana 1969 – 1970.

Sharma SR (1987). Banana bunchy top virus- a review. Int. J. Trop. Plant Dis. 6:19-41.

Su H-J (1998). First occurrence of banana streak badnavirus and studies on vectorship in Taiwan. Pp. 20-25. In: Banana streak virus: a unique virus-Musa interaction? Proceedings of a Workshop of the PROMUSA Virology Working Group. Montpellier, France. January 19-21, 1998.

Swennen R, Vuylsteke D, Ortiz R (1995). Phenotypic diversity and patterns of variation in West and Central African plantains (*Musa* spp., AAB Group Musaceae). Econ. Bot. 49:320-327.

Thresh JM (1958). The spread of virus disease in cacao. Tech. Bull. West Afr. Cocoa Res. Insit. P. 36.

Thresh JM (1974). Temporal patterns of virus spread. Ann. Rev. Phytopathol. 12:111-128.

Thresh JM, Owusu GK (1986). The control of cocoa swollen shoot disease in Ghana; an evaluation of eradication procedures. Crop Prot. 5:41-52.

Thresh JM, Owusu GK, Boamah A, Lockwood G (1988). Cocoa swollen shoot; an archetypical crowd disease. *Zeitschriftfur Pflanzenkrankheiten and Pflanzenschutz.* 95:428-446.

Tushemereirwe WK, Karamura EB, Karyeija R (1996). Banana Streak Virus (BSV) and associated filamentous virus (unidentified) disease complex of highland bananas in Uganda. Infomusa 5:9-12.

Zadoks JC, van den Bosch F (1994). On the spread of plant disease: a theory of foci. Ann. Rev. Phytopathol. 32:503-521.

14

Morphology of Chok Anan mango flower grown in Malaysia

Phebe Ding and Khairul Bariah Darduri

Department of Crop Science, Faculty of Agriculture, Universiti Putra Malaysia, 43400 UPM Serdang, Selangor, Malaysia.

A fundamental understanding of mango flowering in the tropics is essential to efficiently utilize cropping management systems which could extend both the flowering and crop production seasons. However, the information and appreciation of the floral biology of this popular fruit species is still lacking. Therefore, the objective of this work was to observe the morphology of Chok Anan mango flowers using scanning electron microscope. The Chok Anan mango flower is monoecius where both the male and hermaphrodite flowers exist in the same panicle. The fruit develops from the hermaphrodite flower while the male flower contributes the pollen when pollination occurred. The Chok Anan mango flowers contained five sepals and five petals arranged in a whorl. There were not much differences in the structure of sepals and petals between the male and hermaphrodite flowers. The structure of the petals is not flat but billowy. Both sepals and petals consist of ground parenchyma tissue with laticifer and idioblast cells, starch granules and vascular bundle tissues. Trichomes are present in both sepals and petals. The male flower has the same structure as the hermaphrodite flower except that it does not have carpel. Further study is needed to understand the peculiarities of the floral morphology to allow us to predict mango production.

Key words: Pollen, sepal, petal, hermaphrodite, floral biology.

INTRODUCTION

Mango (*Mangifera indica* L.) is one of the most popular fruit of the tropics and belongs to Anacardiaceae family with 60 genera (Abidin, 1991). It is among 15 types of fruits identified in the 3rd National Agricultural Policy to be developed as an export crop. 'Harumanis', 'Chok Anan', 'Masmuda' and 'MAHA' are the popular mango varieties planted in Malaysia. Currently, Malaysia produces mango mainly for the local market and only exports small quantity of fruits. The main export destinations are Singapore, Hong Kong and Brunei. In the year 2001, the export was 4165 metric tons valued at RM9.3 million. Analysis based on the production data from 1997 until 2001 predicted that the export value to be as much as 10,

254 metric tons by the year 2010 (FAMA, 2004).

The ability to reproduce is a unifying and essential characteristic of all organisms. Although there are a number of reproduction strategies found in the plant kingdom, sexual reproduction by way of flower and seed production is one of the most common (Kays, 1991). Species of flowering plants are most reliably identified by their flowers, the sexually reproductive organs. A flower is similar to vegetative short shoot (lacking appreciable internodes) that bears 4 kinds of laterally attached organs in successive whorls: sepals, petals, stamens and carpels (Tucker, 2003). Flowering and the fruit set are the most critical of all events occurring after

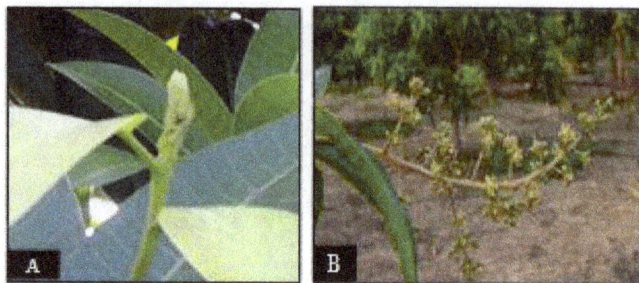

Figure 1. A. Photograph of the initial bud of first week inflorescence where floral shoot arise from a dormant apical bud. B. Photograph of fourth week inflorescence with a few flowers. Male and hermaphrodite flowers are found in the same panicle of inflorescence.

establishment of a tree crop. A fundamental understanding of flowering is essential to efficiently utilize cropping management systems, which would extend both the flowering and crop production seasons (Chacko, 1991). Regulation of flowering allows growers to harvest their crops at the most profitable times.

The information and appreciation of the floral biology of mango is still lacking. Therefore, the objective of this work was to study Chok Anan mango flowers using scanning electron microscope to demonstrate the features which could be useful in the understanding of flower biology. The information obtained from this study can be applied in production of mango especially during pollination.

MATERIALS AND METHODS

The flowers of Chok Anan mango were taken from 4 three-year old plants grown in the Department of Agriculture, Serdang, Selangor. The flowers were tagged from 3 inflorescences of a tree at the bud and flower initiation stage. The development of floral shoot was observed daily for 5 weeks until fruit set and recorded using a digital camera. 12 mature flowers were picked before flower anthesis on the fourth week. The flowers were immediately fixed using formalin-acetic-acid for 4 h under vacuum. The samples were fixed using the following procedure where they were washed in 1% cacodylate buffer; post-fixed in 1% osmium tetroxide in 0.1 M cacodylate buffer for 2 h and finally washed in 0.1 M cacodylate buffer before dehydration through graded series of ethanol to absolute ethanol and critical point dried in a Balzer CD 30 critical point drier. The flowers were then mounted on stubs, sputter coated with gold and viewed in a JOEL 5610 LV scanning electron microscope (Japan) at an acceleration voltage of 15 kV.

RESULTS AND DISCUSSION

Floral shoot development

The floral shoot of Chok Anan mango was initiated terminally from a dormant apical bud of a stem (Figure 1A). The apical bud may turn into leaves or developed

into an inflorescence after 2 weeks. However, factors that determine switching from vegetative to reproductive mode are still poorly understood in mango (Blaikie et al., 2004). The transition to reproductive stage may stimulate by number of different molecules such as carbohydrates, auxins, gibberellins, cytokinin and calcium (Chasan and Walbot, 1993). The initiating bud was green and cone-like shaped. The initiating bud gradually developed into an inflorescence by the third week and flower primordial started to emerge thereafter. The inflorescence formed into a terminal pyramidal green panicle. The structure was rigid and erect and the flower bud developed by the third week. Generally, the panicle colour changed from green to yellow then to pink by the fourth week (Figure 1B). The change in colour was due to the flower colour: green during juvenile, yellow at maturity and pink at anthesis. By the fifth week the fruit set is formed but unpollinated flowers turned brown and aborted.

The panicle of Chok Anan mango inflorescence varies in size. The small panicle has only a few flowers, but the larger panicle could be formed from thousands of flowers; an excess of three thousand flowers were counted in an inflorescence of Chok Anan mango. The flowers were small, monoecius and polygamous. Both male and hermaphrodite flowers with varying stages of anthesis were found in an inflorescence, indicating the flowers were not receptive at the same time. Thus, fruits that are formed have different stages of maturity.

Flower of Chok Anan mango

By the fourth week of flower development, petals started to open exposing the anther and stigma with full anthesis by the end of the fourth week. Both male (Figure 2A) and hermaphrodite (Figure 2B) flowers were found in the inflorescence. The male and hermaphrodite flowers have similar structure except that the male flowers were without carpel. In the hermaphrodite flower of Chok Anan mango, the carpel is round with an ovary, and the style and stigma are supported by a 5-lobed nectary. Nectary is a secretory structure that release nectar with a high sugar content (Evert, 2006). This indicated insects are the agents for pollination in Chok Anan mango flower. Through secretion of nectar, nectary provides a reward to insects and other animals that serve as pollinators. The fruit developed from the hermaphrodite flower while the male flower contributed pollens during pollination.

The perianth consisted of an outer whorl of 5 sepals and an inner whorl of 5 petals. This is followed by 5 stamens where 1 to 2 of the stamens are with fertile anther while the others are staminodes. There are not much of differences in the structure of sepals and petals between the male and hermaphrodite flowers. The sepals are ovate to ovate oblong and highly pubescent while the petals are oblong, billowy and less pubescent. Both sepals and petals consisted of ground parenchyma tissue with laticifers (Figure 3A) and idioblast cells, starch

Figure 2. SEM micrograph showing a male (A) and a hermaphrodite (B) flowers at late fourth week of floral development (x 25). S: sepal, P: petal, Lb: 5-lobed nectary, At: anther, Std: staminode, O: ovary, Stg: stigma, Sty: style.

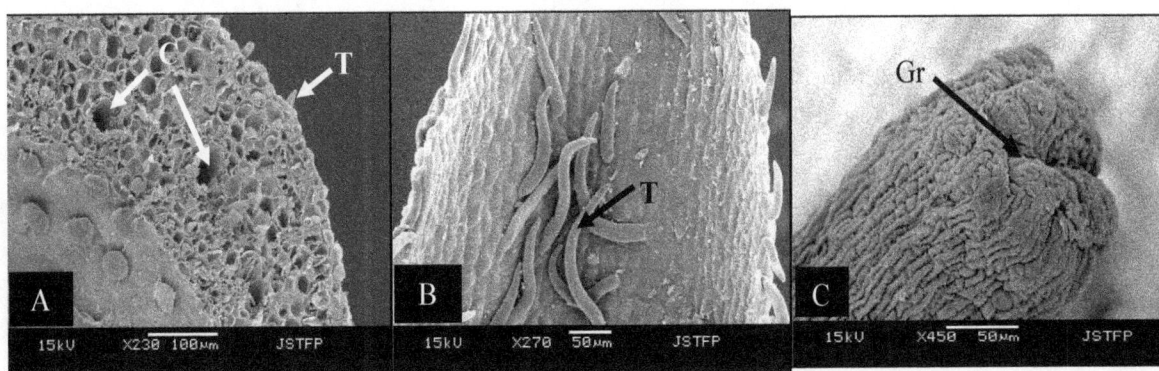

Figure 3. A. SEM micrograph of a sepal. Both sepal and petal contained cavity that believed to be laticiferous cells. C, cavity, T, trichome. B. Trichomes (T) at sepal (× 270). C. The groove (Gr) on stigma of hermaphrodite flower (× 450).

granules and vascular bundle tissues. Laticifers are cells or series of connected cells containing latex and forming systems that permeate various tissues of plant body (Evert, 2006). The laticifers contribute sticky latex when mango flowers being detached. Trichomes were present in both sepals and petals (Figure 3B). Trichomes, in Greek, mean a growth of hair (Evert, 2006). It has a variety of function, however, the functions in sepals and petals of mango flower is unclear.

The style of hermaphrodite flower has a small stigmatic surface grooved as a receptive surface for pollen grains (Figure 3C). In a hermaphrodite flower, the style is longer than the filament of stamen, indicating the existence of some degree of self incompatibility, thus cross pollination is needed. Scholefield (1982) found that 'Haden' mango flower consisted of 1 fertile stamen and 4 short sterile staminodes with the style of carpel shorter than filament of stamen, thus indicating that the floral morphology of mango varies with variety. It is reported that low day/night temperature had a significant effect on floral morphology (style, stigma, ovary and anther size) in

'Nam Dok Mai', 'Kensington', 'Irwin' and 'Sensation' mango cultivars (Sukhvibul et al., 1999).

Conclusions

To our knowledge this is first report on floral morphology Of Chok Anan mango flower. As reported earlier mango floral morphology can be affected by day/night temperature. Further study is needed to understand the peculiarities of the floral morphology during day/night temperature of drought and wet season to allow us to predict mango production.

REFERENCES

Abidin MIZ (1991). Pengeluaran Buah-buahan. Kuala Lumpur: Dewan Bahasa dan Pustaka, pp. 105-119.
Blaikie SJ, Kulkarni VJ, Muller WJ (2004). Effects of morphactin and paclobutrazol flowering treatments on shoot and root phenology in mango cv. Kensington Pride. Sci. Hort. 101:51-68.

Chacko EK (1991). Mango flowering. Still an enigma. Acta Hort. 291:12-21.

Chasan R, Walbot V (1993). Mechanisms of plant reproduction: Questions and approaches. Plant Cell. 5:1139-1146.

Evert FR (2006). Esau's plant anatomy - Meristems, cells, and tissues of the plant body: their structure, function, and development. 3rd ed. New Jersey: John Wiley & Sons, Inc.

FAMA (2004). Analisis Industri Buah Mangga. Kementerian Pertanian. Perpustakaan Negara Malaysia.

Kays JS (1991). Postharvest physiology of perishable plant products. US: Van Nostrand Reinhold.

Scholefield PB (1982). A scanning microscope study of flowers of avocado, litchi, macadamia and mango. Sci. Hort. 16:263-272.

Sukhvibul N, Whiley AW, Smith MK, Hetherington SE, Vithanage V (1999). Effect of temperature on inflorescence and floral development in four mango (*Mangifera indica* L.) cultivars. Sci. Hort. 82:67-84.

Tucker SC (2003). Floral development in Legumes. Plant Physio. 131:911-926.

15

Influence of canopy management practices on fruit composition of wine grape cultivars grown in semi-arid tropical region of India

**Satisha Jogaiah, Dasharath P. Oulkar, Amruta N. Vijapure, Smita R. Maske,
Ajay Kumar Sharma and Ramhari G. Somkuwar**

National Research Centre for Grapes, P. B. No. 3, Manjri Farm, Solapur Road, Pune – 412 307, Maharashtra, India.

Effect of canopy management practices on berry composition of red and white grape cultivars grown in Pune region of India was examined. Cabernet Sauvignon and Sauvignon Blanc vines were selected for the study. Both the cultivars exhibited significant variation in fruit composition parameters in response to various canopy management practices. Combination treatment of leaf removal (LR) either with shoot thinning (ST) or cluster thinning (CT) exhibited high total soluble solids (TSS), lowest acidity (malic acid), lower potassium content and higher anthocyanin content. The vines which received ST+CT+LR treatment and control vines recorded least anthocyanin concentration and phenolic compounds indicating excess light exposure or excess shade to clusters is not congenial for producing better quality fruits. Leaf removal treatment in combination with either shoot thinning or cluster thinning was found to be superior under semi-arid tropical conditions to obtain good quality fruits. Reasons for such variations in fruit composition parameters under different management practices are discussed.

Key words: Anthocyanins, canopy management, organic acids, phenolic compounds, wine grapes.

INTRODUCTION

Wine grape cultivation is gaining strong impetus in tropical climatic conditions of the world. Tropical viticulture has only been practiced commercially, since approximately 50 years. Countries such as Brazil, India, Thailand and Venezuela play a leading role in the tropical grape production. However, it can be noted that there is a trend towards the expansion of tropical viticulture in the world, since there are vineyards being implemented in different countries in America (Bolivia, Colombia, Peru, Guatemala), in Africa (Madagascar, Namibia, Tanzania) and Asia (Vietnam, China). The production technology in the tropical regions differs significantly from the one employed in the traditional temperate regions. It is necessary to break the bud dormancy in order to foster bud burst, and special management techniques have to be employed to overcome problems of low fertility and to control vigor.

It is generally opined that wine grapes require a temperate climate that includes predominantly winter rainfall, frost-free late spring, and warm to hot summers to ensure ripening, i thus the global wine industry has been analysed predominantly in terms of Old World and New World wines from regions characterised by those criteria (McLennan, 1996). However, this largely ignores the nascent frontier of new climate wines, including the new altitude wines of tropical zones. Between 1996 and 2006 the area under commercial grape cultivation in tropical zones in Africa, Asia and Central and

South-America, north and south of the equator, between the Tropic of Cancer (23.27°N) and the Tropic of Capricorn (23.27°S) increased by 155% from 55,000 ha to over 140,000 ha. The increase was most rapid in Asian countries like India, Thailand, Myanmar and Vietnam where new vineyards for table grape and wine production are established every year (http://estructuraehistoria.unizar.es/gihea/documents/GwynCampbell.pdf)

Canopy management practices in wine grape cultivation have been developed with an aim of optimizing sunlight interception, photosynthetic capacity and fruit microclimate to improve fruit yield and wine quality, especially in vigorous and robust growing varieties with dense canopies. For wine making, significant benefits have been obtained from comprehensive approaches, to control shoot vigour through the use of different methods of trellis system, training systems, pruning methods, deficit irrigation, rootstocks and canopy management practices (Smart, 1985; Smart et al., 1990). Canopy management practices like leaf removal improved the bunch and berry characteristics with respect to reduced incidence of botrytis incidence and increased anthocyanin and reduced malic acid content in Graciano and Carignan grapes (Tardaguila et al., 2010). Many workers have shown positive effects of canopy management practices on composition of wine grapes in recent years. Fruit zone shading reduced total soluble solids and anthocyanin accumulation in Nebbiolo grapes. Excessive sunlight exposure caused sun burn damage and did not increased TSS or anthocyanin concentration (Chorti et al., 2010). Similarly cluster thinning increased TSS by 25 brix and showed positive impact on wine anthocyanin, berry skin tannins and seed tannins in Corot Noir grapes (Sun et al., 2012). Gatti et al. (2012) in their studied on effect of cluster thinning and pre-flowering leaf removal on fruit composition of Sangiovese grapes observed high brix level corresponded to the highest TA in defoliated vines and conversely the lowest TA and high pH in early cluster thinning and lag phase cluster thinned vines.

Initially, wineries in tropical climate used to follow production technologies similar to traditional old world wine producing countries. But, these production technologies did not work well and hence new and specialized techniques and equipments are being used in tropical wine grape production. The quality of grapes has been improved tremendously, after the establishment of two pruning and single cropping cultivation practices. Though sunlight is not a limitation in semi-arid tropics of India, excess sunlight can harm the production and thereby reduce the wine quality. The other major drawback in tropical climate is the more vigorous nature of vines which needs to be curtailed to improve fruit composition, especially in wine grapes. No systematic research on canopy management practices to improve fruit composition of wine grapes has been attempted in major wine grape growing regions of India. Hence, this preliminary investigation was undertaken, to study the influence of important canopy management practices on fruit composition of Sauvignon blanc and Cabernet Sauvignon grapes.

MATERIALS AND METHODS

This experiment was conducted at the experimental vineyard of National Research Centre for Grapes, Pune during two growing seasons of 2010-2011 and 2011-2012. Pune is located in Midwest Maharashtra state (India) at an altitude of 559 m above the mean sea level. It lies in 18.32° N latitude and 7.51° E longitude. The vines were grown on calcareous black cotton soil (clay content was 44.5%) exhibiting swelling and shrinkage properties. The average bulk density of the root zone up to a depth of 30 cm was 1.25 g/cm^3. The average electrical conductivity (EC) of the irrigation water during the experimentation was 1.98 dS/m with an average pH value of 7.78. The rainfall during 2010-2011 and 2011-2012 was 484 and 540 mm respectively.

Four year old Cabernet Sauvignon and Sauvignon Blanc grapes grafted on 110R rootstock were selected for this study. The vines were planted at a spacing of 2.5 m between rows and 1.2 m between vines within a row. The row orientation was in the direction of North – South. The vines were trained to double cordon small T system. The pruning biomass of the vines was in the range of 1 to 1.25 kg. The concept of balanced pruning is not in practice in tropical viticulture of India, where double pruning and single cropping is being practiced. Hence, approximately 40 to 45 shoots are encouraged per vine in a spacing mentioned above.

Canopy management practices such as shoot thinning (ST), leaf removal (LR) and cluster thinning (CT) were imposed either singly or in different combinations. Shoot thinning was done at 45 days after pruning, wherein approximately 32 shoots (eight shoots on either side of double cordon) were retained per vine by removing weak and non-bearing shoots. Cluster thinning was done after fruit set stage (3 to 4 mm stage) to maintain 40 basal one or two clusters per vine and leaf removal was performed during version stage by removing two leaves above and two below the cluster to expose bunches. The 7 treatments were ST, CT, LR, ST+LR, CT+LR, ST+CT, ST+CT+LR along with control as eighth treatment. Each treatment was replicated thrice with three vines per replication. Except shoot thinned vines, the vines with other canopy management treatments had approximately 40-45 shoots per vine oriented towards east and west side of the cordon.

Fruit composition parameters

After harvesting, about 250 berry samples were collected from each treatment replication wise. Half of the samples were utilized immediately for analysis of basic fruit composition parameters such as total soluble solids (TSS), titratable acidity (TA) and juice pH. The fruit samples were also analyzed for total proteins, total phenols, and potassium content. The remaining half of the berry samples was stored in -20°C for analysis of organic acids and phenolic compounds using high performance liquid chromatography (HPLC) and LC-MS/MS respectively. The frozen samples were analyzed for organic acids and phenolic compounds within 20 days after harvest.

Estimation of total phenols, proteins and potassium content

The total phenol content of the fruit extract was determined using the Folin- Ciocalteu method (Singleton and Rossi, 1965), using gallic acid as the standard. The total protein content was estimated

Influence of canopy management practices on fruit composition of wine grape cultivars grown...

93

as per the procedures of Lowry's method (Lowry et al., 1951). Both these estimations were done using UV spectrometer. Juice potassium content was estimated using flame photometer method. In Cabernet Sauvignon grapes, anthocyanin and phenolic concentration was determined as explained below.

Spectrometric analysis anthocyanins

Frozen berries were removed from cold storage and thawed overnight under refrigerated conditions (4°C) and approximately 100 berries were weighed and homogenized in a grinder. One gram of homogenate was taken in 10-ml plastic centrifuge tubes and 10 ml of 50% (v/v) aqueous ethanol was added and the mixture was agitated for 1 h at 400 rpm. Then the mixture was thereafter centrifuged at 1800 rpm for 10 min. The supernatant (extract) was used for estimation of anthocyanins and total phenols. For analysis of anthocyanins and total phenols, about 200 µl of extract was transferred to acrylic cuvettes and 3.8 ml of 1.0 M HCl was added and covered with paraffin film and mixed by inverting. The mixture was incubated for 3 h at room temperature and the color was measured at 520 nm for anthocyanins and 280 nm for total phenols.

Estimation of organic acids

A new method was developed for estimation of organic acids in grapes and wines using ultra HPLC 1260 Series (Communicated for publication). The method was developed based on the common organic acids present in grape must such as tartaric acid, malic acid, citric acid and lactic acid.

Standard solutions of organic acids

All acids and reagents used were of analytical grade. The standard organic acids were purchased from Thomas Baker. All organic acids used for standards were dissolved in double distilled water. For method development both D and L tartaric acid were used. The concentrations of organic acids varied from 1 to 100 mg/L. The prepared standard solutions of organic acids were stored at 4°C.

Solvents

The mobile phase consisted of acidified water of pH 2 adjusted with Othophosphoric acid and 100% methanol (Volume ratio of 95.0:5.0). Prior to use, the solvent was filtered through vacuum filter and then sonicated for 5 to 10 min in an ultrasonic bath to remove air bubbles.

Equipment

The HPLC used was 1260 Agilent Series with EZ chrome software for data acquisition and analysis.

Chromatographic conditions for determination of organic acids

A Zorbax Eclipse plus C18 column (4.6 × 100 mm × 5 µ), with an injection volume of 10 µl, pressure 45-46 Bar, temperature 25°C, wavelength 210 nm, flow rate 0.80 ml /min. For precision and accuracy validation, grape extract were spiked with organic acids to such an amount that the final concentration of the added acid varied from 20, 40, 60 mg/L. From stock solution of 10,000 mg/L of different organic acids, aliquot of 2, 4 and 6 µl was added to 10 µl of

samples to get final concentration of 20, 40 and 60 mg/L respectively. Three replicates of the spiked samples were prepared and injected and assay was calculated to measure the repeatability of retention times, peak area of standard and samples.

Similarly, an aliquot of grape extract was diluted with mobile phase and 10 µl of the obtained solution was injected to the system. Before injection, all standards and sample solutions were filtered through 0.45 µm nylon membrane filter units. The chromatogram of standard organic acids with their intensity and retention time is shown in Figure 1. The concentration of L- tartaric acid was very negligible in fruit samples and hence, only D- tartaric acid was estimated in all samples.

Estimation of phenolic compounds

The estimation of phenolic compounds was performed as per the procedures of Patil et al. (2011). The stock solutions of individual standards were prepared by weighing 10 ± 0.01 mg of each analyte and dissolving them in 10 ml methanol and were stored in glass vials under refrigerated conditions. The concentration of each compound was calculated using the weight of the standard and weight of solvent and also the purity of the compound. A 25 ppm intermediate working standards was prepared by diluting the above stock solutions in 10 ml. The chromatogram of standard phenolic compounds showing their intensity and retention time is shown in Figure 2.

Sample extraction for LC-MS/MS analysis

One hundred representative berries were homogenized in a mixer grinder. One gram of homogenized sample was extracted in 5 ml of 0.1% formic acid in methanol. The sample was vortexed for 2 min and centrifuged at 5000 rpm for 5 min. Supernatant (1 ml) was injected to LC-MS/MS by passing through 0.2 µm nylon membrane filter paper.

LC-MS/MS analysis

The LC-MS/MS analysis was done with an Agilent Technologies 1200 series hyphenated to API 4000 Qtrap (ABS Sciex) mass spectrometer equipped with electrospray ionization (ESI +) probe. The separation of the phenolic compounds was done on a Precenton SPHER-60 C_{18} 60 A° (150 × 2 mm × 5 µm) with mobile phase A-1% formic acid in water, B-1% formic acid in water: acetonitrile (1:1) and C- acetonitrile. Oven temperature was 35°C and injection volume 10 µl. Mass parameters were curtain gas 20 psi, ion spray voltage 5500 V and source temperature 550°C. Flow rate: 0.400 ml/min.

Statistical analysis

The experiment was conducted as randomized block design with three replications and the data was analysed using SAS Version 9.3. Tukey's test was used for comparing treatment means.

RESULTS

Sauvignon blanc

Significant difference was recorded for berry weight and basic fruit composition parameters among different canopy management practices. Maximum 100 berry

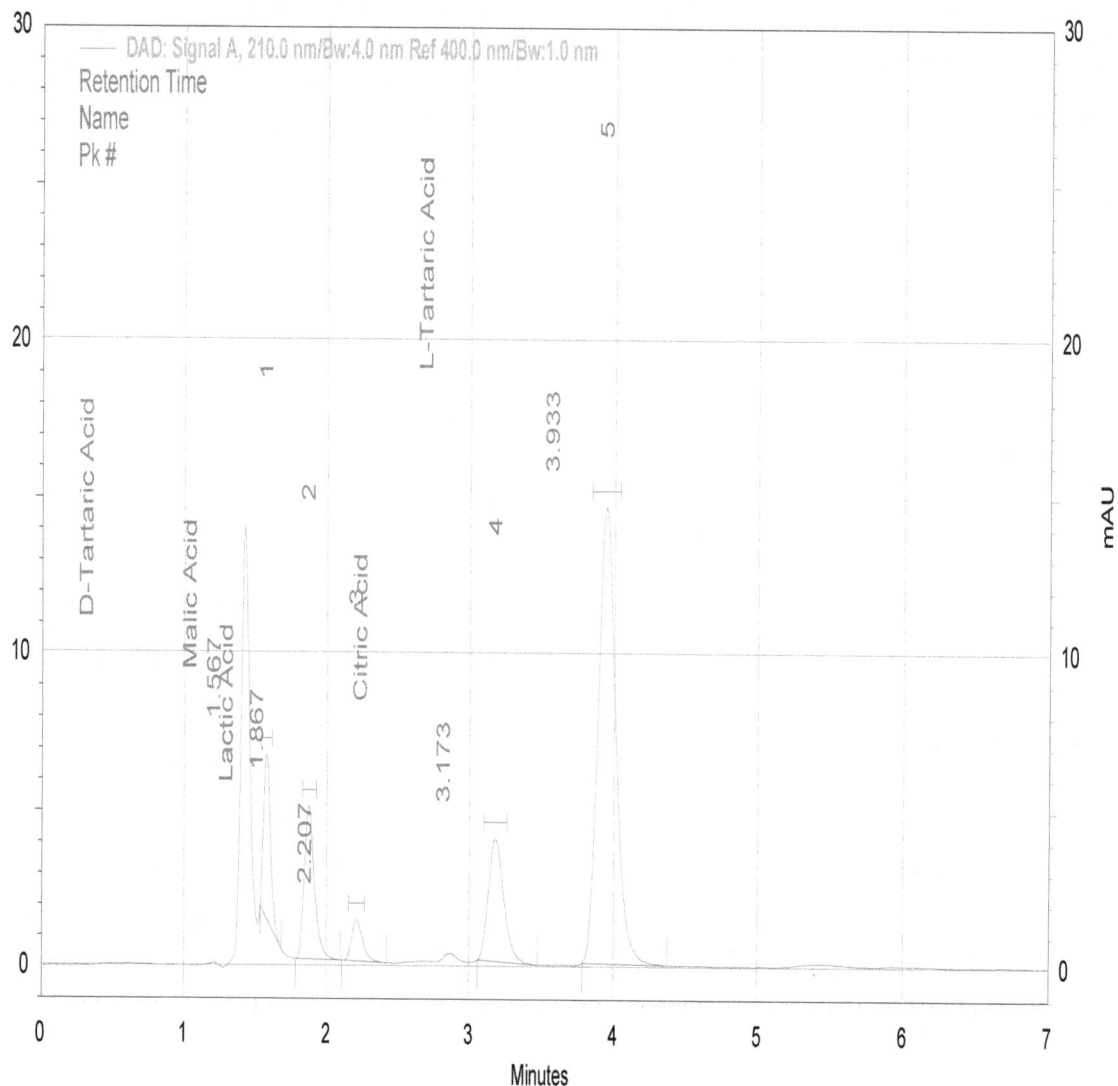

Figure 1. Chromatogram showing intensity of organic acid standards using HPLC.

weight (115.12 g) was recorded on cluster thinned vines. Highest TSS was recorded on CT, LR and ST combination vines (22°B). Least TSS was on control vines (19.83°B) and shoot thinned vines (19.10°B). Highest titratable acidity (TA) was recorded on ST or control vines (1.18%), while least was on LR vines or LR+ST vines (0.90%). Maximum total phenol concentration was recorded on CT+LR vines (6.01 mg/g) while it was least on ST or control vines (3.29 mg/g). Maximum potassium concentration in juice was recorded on control vines (0.17%) while it was least on vines which received LR treatment in combination with ST and / or CT treatments (0.10%) (Table 1). Among organic acids, highest Tartaric acid (1.32 g/L) was recorded on vines which received combination of all the three treatments while it was least on cluster thinned vines. Malic acid was highest (1.51 g/L) on control vines, while it was least on vines which received leaf removal (1.03 g/L) or the

combination of cluster thinning and leaf removal and / or shoot thinning (Table 1).

Among phenolic compounds analyzed, maximum concentration of phenols was in flavonoid category. Catechin concentration was highest which varied significantly among treatments with highest being recorded on either CT+ST or CT + LR vines. Least was on control vines. Significant difference was recorded in major flavonol compound quercetin with highest being recorded on CT+ST and CT+LR vines with least on control vines and ST+LR treated vines. Among hydroxybenzoic acids, gallic acid was in higher concentration with highest being recorded on CT+LR and CT+ST vines. Maximum ellagic acid was recoded on ST+LR and ST+CT+LR vines with least on either ST or LR vines. Among non-flavonoid compounds, concentration of chlorogenic acid was highest in Sauvignon Blanc followed that of cafteric acid and gallic

XIC of +MRM (57 pairs): 169.0/125.0 Da from Sample 54 (Std Phenolic mix 2.5 ppm) of Data27.03.2011....   Max. 4.7e4 cps.

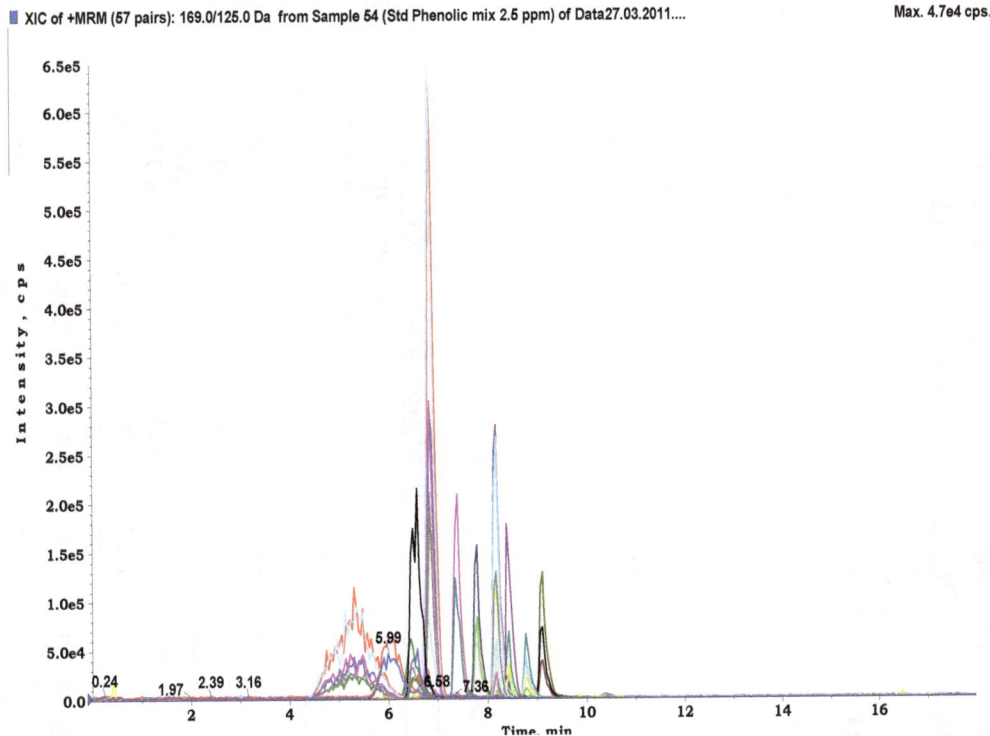

Figure 2. Chromatogram showing intensity of phenolic compound standards using Lc-MS/MS.

acid. But no definite trend could be seen among canopy management practices in concentration of non-flavonoid phenolics. Highest concentration of resveratrol was recorded on vines which received CT+ST treatment (Table 2).

Cabernet Sauvignon

The berry weight and basic fruit composition parameters of Cabernet Sauvignon grapes in response to canopy management practices is shown in Table 3. Maximum berry weight was recorded on vines which received cluster thinning treatment (86.68 g) while it was least on CT+ST vines (59.29 g). Highest TSS was recorded on vines which received combination of all the three practices viz., CT+ST+LR (23.03°B) while it was least on control vines (20.13°B). Highest pH was recorded on vines which received LR treatment (3.85) while it was least on CT +LR, CT+LR+ST and control vines. The TA was highest on ST+LR vines (1.13%) followed by control (1.07%) vines, while it was least on CT+LR vines (0.87%). Vines which received either LR treatment or in combination with either CT or ST recorded maximum anthocyanin and phenol content. Significant difference in juice potassium content was recorded with highest potassium content on either shoot thinned vines (0.199%) or on control vines (0.221). Least potassium was recorded on LR vines (0.145%) followed by those vines which received combined treatment of ST+CT+LR

(0.160%). Least anthocyanin concentration was recorded on vines which received ST (1.66 mg/g) treatment followed by those on control vines (1.76 mg/g). Significant differences in organic acid concentration were recorded among canopy management treatments (Table 3). Highest tartaric acid was recorded on vines which received LR treatments (1.65 g/L) followed by the vines which received all the three treatments (1.49 g/L). Highest malic acid content was recorded on control vines (1.84 g/L) while it was least on ST+CT+LR (1.31 g/L) vines.

As expected, a highest concentration for most of the phenolic compounds was recorded in Cabernet Sauvignon grapes as compared to Sauvignon Blanc grapes (Tables 2 and 4). Significant differences were recorded for most of the flavonoid phenolic compounds of which, catechin concentration was highest followed by that of quercetin and epicatechin contents. In both the cultivars, least catechin concentration was recorded on control vines and vines which received CT+ST+LR treatment compared to those vines which received either single treatment or combination of any two treatments. Highest quercetin was recorded in vines which received either ST or CT treatments followed by the vines which received combination of either ST+LR and / or CT+LR treatment. No definite trend was observed for other flavonoid compounds such as rutine hydrate and kaempferol. The major non-flavonoid compound measured was ellagic acid with highest on vines which received CT+ST treatment (3.84 mg/L) followed by those

Table 1. Influence of canopy management practices on basic fruit composition parameters and organic acids in Sauvignon Blanc grapes (values are average of two years).

Treatment	100 Berry wt. (g)	TSS (°B)	pH	Acidity (%)	Phenol (AU/g)	Protein (mg/g)	Potassium (%)	Tartaric acid (g/L)	Malic acid (g/L)	Citric acid (g/L)
ST	100.33	19.10	3.19	1.18	3.82	181.95	0.126	1.10	1.19	0.042
CT	115.12	20.33	3.17	1.00	4.46	228.7	0.119	0.97	1.23	0.043
LR	99.21	20.16	3.06	0.90	4.86	211.44	0.109	1.00	1.03	0.025
CT+ST	108.03	20.60	3.17	0.98	4.98	253.68	0.121	1.32	1.43	0.045
CT+LR	99.64	21.33	3.19	1.05	6.01	257.26	0.115	0.99	1.15	0.039
ST+LR	90.94	21.13	3.21	0.93	4.56	229.98	0.111	1.37	1.21	0.043
ST+CT+LR	102.32	21.26	3.16	0.93	4.51	241.65	0.117	1.32	1.27	0.045
Control	98.26	19.83	3.27	1.14	3.29	202.73	0.173	1.20	1.51	0.039
SEM ±	3.747	0.379	0.0136	0.0467	0.332	24.621	0.0107	0.042	0.047	0.0015
Significance*	0.0148	0.012	<0.001	0.0051	0.0016	0.406	0.0149	<0.0001	<0.0001	<0.0001

*:Values below 0.05 are significant at p≤0.05 and above are not significant.

Table 2. Influence of canopy management practices on phenolic compounds (mg/L) in Sauvignon Blanc grapes (Values are average of two years).

	Flavonoid phenolics					Non flavonoid phenolics								
	Flavan -3-ols		Flavonols and Flavonl algycons			Hydroxy benzoic acids				Hydroxy cinnamates				Stilbene
Treatments	Catechin	epicatechin	Quercetin	Rutine hydrate	Kaempferol	Gallic acid	Vanillic acid	Ellagic acid	Cafteric acid	Coumaric acid	Chlorogenic acid		Resveratrol
ST	2.22	0.875	0.79	0.615	0.020	0.562	0.293	0.351	0.566	0.021	0.760		0.28
CT	1.98	0.865	2.25	0.600	0.018	0.545	0.231	0.348	0.549	0.024	0.763		0.25
LR	1.81	0.839	2.12	0.644	0.011	0.544	0.112	0.177	0.563	0.068	0.745		0.18
CT+ST	2.29	0.893	2.43	0.644	0.014	0.564	0.319	0.583	0.559	0.087	0.774		0.36
ST+LR	1.86	0.803	1.67	0.676	0.016	0.545	0.230	2.08	0.563	0.086	0.724		0.18
CT+LR	2.21	0.860	2.34	0.644	0.008	0.573	0.194	0.35	0.545	0.106	0.760		0.10
CT+ST+LR	1.70	0.771	2.26	0.654	0.038	0.555	0.214	1.69	0.581	0.032	0.724		0.15
Control	1.52	0.776	0.98	0.624	0.005	0.538	0.250	0.524	0.549	0.060	0.680		0.23
SEM ±	0.165	0.0295	0.199	0.0164	0.0095	0.0055	0.052	0.129	0.0059	0.370	0.0241		0.0304
Significance*	0.0447	0.0641	<0.0001	0.099	0.4215	0.0032	0.2481	<0.001	0.0118	0.645	0.2018		0.0005

*: Values below 0.05 are significant at p≤0.05 and above are not significant.

on control (2.27 mg/L) vines and on CT+ST+LR vines (2.83 mg/L). Concentration of gallic acid, vanillic acid, cafteric acid and coumaric acid were in same proportion with no definite trend among different canopy management practices (Table 4).

Table 3. Influence of canopy management practices on basic fruit composition parameters and organic acids in Cabernet Sauvignon grapes (Values are average of two years).

Treatment	100 Berry wt. (g)	TSS (°B)	pH	Acidity (%)	Anthocyanin (mg/g)	Phenol (AU/g)	Protein (mg/g)	Potassium (%)	Tartaric acid (g/L)	Malic acid (g/L)	Citric acid (g/L)
ST	20.20	3.65	0.97	1.66	5.98	219.76	0.199	83.96	1.23	1.80	0.035
CT	21.33	3.52	0.94	2.39	6.17	233.28	0.191	86.68	1.40	1.55	0.038
LR	21.60	3.45	1.06	2.70	6.92	593.55	0.145	85.29	1.65	1.54	0.034
CT+ST	21.63	3.51	0.89	2.31	6.28	251.92	0.197	59.29	1.40	1.83	0.034
CT+LR	20.96	3.43	0.87	3.28	6.81	260.15	0.163	76.53	1.34	1.55	0.024
ST+LR	20.45	3.45	1.13	3.08	5.98	219.76	0.162	76.08	1.15	1.66	0.033
ST+CT+LR	23.03	3.43	0.95	2.71	6.13	226.34	0.160	78.3	1.49	1.31	0.32
Control	20.13	3.85	1.07	1.76	3.67	299.51	0.221	78.12	1.17	1.84	0.034
SEM ±	0.364	0.0863	0.0433	0.143	0.608	0.0133	0.012	2.890	0.070	0.0847	0.0011
Significance*	0.0007	0.0382	0.006	<0.001	0.0463	0.0415	0.0142	0.065	0.0019	0.0003	<0.0001

*: Values below 0.05 are significant at p≤0.05 and above are not significant.

Table 4. Influence of canopy management practices on phenolic compounds (mg/L) in Cabernet Sauvignon grapes (values are average of two years).

Treatments	Flavonoid phenolics					Non flavonoid phenolics						Stilbene
	Flavan -3-ols		Flavonols and Flavonol algycons			Hydroxy benzoic acids			Hydroxy cinnamates			
	Catechin	epicatechin	Quercetin	Rutine hydrate	Kaempferol	Gallic acid	Vanillic acid	Ellagic acid	Cafteric acid	Coumaric acid	Chlorogenic acid	Resveratrol
ST	6.32	1.109	1.18	1.046	0.0003	0.685	0.510	5.45	0.548	0.616	0.715	ND
CT	6.16	1.100	1.23	0.354	0.0023	0.696	0.509	3.23	0.549	0.619	0.739	ND
LR	3.99	0.687	1.73	0.270	0.0250	0.565	0.490	3.35	0.537	0.597	ND	ND
CT+ST	3.80	0.800	2.57	ND	0.0613	0.611	0.485	3.84	0.573	0.598	ND	ND
ST+LR	4.10	0.811	2.57	ND	0.0006	0.654	0.506	3.43	0.548	0.619	0.716	0.0006
CT+LR	4.47	0.825	2.90	ND	0.0140	0.670	0.503	3.22	0.560	0.594	0.716	0.001
CT+ST+LR	3.07	0.604	3.39	ND	0.0176	0.614	0.506	2.83	0.559	0.602	0.00	0.0006
Control	2.06	0.679	1.85	ND	0.0001	0.610	0.504	2.27	0.572	0.643	0.00	0.004
SEM ±	0.149	0.0325	0.0930	0.114	0.0095	0.0049	0.0074	0.497	0.0026	0.0056	0.0057	0.0008
Significance*	<0.001	<0.000	<0.000	<0.0001	0.0045	<0.001	0.224	0.0206	<0.000	0.0002	<0.0001	0.0004

*: Values below 0.05 are significant at p≤0.05 and above are not significant. ND: Not detected.

DISCUSSION

Canopy management practices along with balanced pruning, training and trellising are primarily focused on altering canopy components and cluster microclimate during fruit development mostly in favour of improved light distribution in the canopy (Kliewer and Smart, 1989). Altering the physical appearance of canopy, by judicious canopy management practices also has physiological implications that virtually always comprise a change in source: sink relationships in the grapevine through improvement in photosynthetic activity and translocation of photosynthates from leaves to sinks such as berries (Johnson et al., 1982; Hunter and Visser, 1988; Candolfi – Vasconcelons and Koblet, 1990; Hunter et al., 1995; Koblet et al., 1996). These canopy management practices include a range of techniques which can be applied in a vineyard to alter the position or amount of leaves, shoots and fruits in space to harness maximum benefits of microclimate. The main objective of this pilot study was to examine the impact of such canopy management practices on changes in fruit composition of Cabernet Sauvignon and Sauvignon Blanc grapes especially in sub-tropical climate of India.

Berry weight was highest in vines which received single treatment of cluster thinning in both the varieties, while it was least on control vines or vines which received shoot thinning treatments. The increased berry weight in cluster thinned vines may be due to diversion of photosynthates in to remaining clusters on the vine. Bunches developed on control vines showed least berry weight. Shoot thinning must have reduced the total photosynthetic capacity of the vines which resulted in reduced accumulation of photsynthates in the developing clusters. The present observation on reduced berry weight on control vines is in accordance to findings of Ristic et al. (2007), where shading (control vines) reduced the size of the berry by 20%. The increase of cluster weight in cluster thinned vines is related to increase in the availability of nutrients to retained clusters on the vines as compared with the un-thinned vines which have more number of clusters. The control vines produced fruits of least weight; the reason for this may be due to competition between the high number of leaves in the shoots and more number of clusters.

Both TA and juice pH were highest in control vines suggesting increased malic and potassium levels. No canopy management practices on those vines might have resulted in more shade inside the canopy especially in bunch zone. This may also suggest delayed fruit ripening on such vines. Similar findings of increased TA in shaded conditions and reduced TA in leaf removed vines were observed by Wolf et al. (1986) and Kliewer and Lider (1970). In contrast, reduced TSS and increased acidity and juice potassium on the same vines explain the importance of exposed canopy to harness sufficient sunlight into the canopy. The decrease in pH in control vines may be due to increased shading of bunches which resulted in bunches to remain cooler, leading to lower pH (Bergqvist et al., 2001) and due to decreased malic acid metabolism (Lakso and Kliewer, 1978). Shaded berries accumulated more potassium, malic acid, and sometimes anthocyanin content (Kliewer and Smart, 1989).

Importance of canopy/cluster exposure to optimum sunlight is evident with respect to most of the biochemical composition which differed significantly between treatments. Although it is worth to notice that leaf removal decreased fruit weight in Gewurztraminer, Seyval, Cabernet Sauvignon, Sangiovese and Trebbiano grapes, many investigators found that sunlight exposed fruits are generally rich in total soluble solids and reduced titratable acidity, compared to non-exposed or canopy shaded (Kliewer and Lakso, 1968; Ferree et al., 2004; Kliewer and Dokoozlian, 2005; Santesteban and Royo, 2006; Main and Morris, 2004). But, in contrast some workers found that defoliation had no effect on soluble solids and titratable acidity (Vasconcelos and Castagnoli, 2000; Howell et al., 1994; Poni et al., 2006). In our study, though we could not observe significant differences in soluble solid concentration among treatments, the TSS was considerably highest on vines which received LR treatment either alone or in combination. The increased TSS on such vines might be either due to remobilization of stored carbohydrates, an increase in photosynthetic activity of remaining leaves and improvement in the light microclimate of remaining leaves and an increase in sink strength as explained by Kliewer and Antcliff (1970), and increased fruit temperature and changes in pattern of assimilate movement (Bledsoe et al., 1988) which needs to be confirmed by measuring light intensity, leaf area index and berry temperature under tropical climate.

Influence of leaf removal performed at different stages of berry development was studied by different workers. Leaf removal is usually performed in the fruit zone during vegetative season between fruit set and ripening (Poni et al., 2006). If it is done at veraison stage, it affects synthesis of primary and secondary metabolites and this effect is directly related to leaf layer number, photosynthetic rate and canopy surface area. Several experiments have shown increased sugars, flavor, flavonoids and decreased acidity when leaf removal was done at veraison stage (Percival et al., 1994; Poni et al., 2006; Zoecklein et al., 1992). In contrast, leaf removal at veraison on plants with low canopy density does not affect grape sugar, acidity and color (Reynolds et al., 1986). The more vigorous nature of vines induced in tropical climate may be benefitted by partial defoliation to improve grape composition and wine quality as suggested by Hunter et al. (1991) that partial defoliation in vigorous varieties is an endeavour to alter grape composition. Some of the investigations have revealed that partial defoliation increases total soluble solids and reduce titratable acidity, malic acid, pH and K level in the fruits (Kliewer and Bledsoe, 1987; Kliewer et al., 1988). In contrast, some of the workers have failed to demonstrate these alterations in fruit compositions (Koblet, 1984; Williams et al., 1987) or in some cases they could see reduced sugar accumulation (Sidahamed and Kliewer, 1980).

The goal of shoot thinning is to reduce canopy density, although the ideal shoot number per meter of cordon is dependent on the cultivar, spacing and site. When shoot thinning is optimized, the vine is more efficient in radiation interception (Smart, 1985). Appropriate shoot thinning can improve fruit composition in *Vitis vinifera* cultivars (Smart, 1988). In the present study, vines which received treatment shoot thinning alone or control vines had lesser tartaric acid content while leaf removal along with cluster thinning resulted in higher tartaric acid and lower malic acid concentration. The shoot thinning alone may increase the vegetative growth of remaining shoots, leading to diminished leaf and cluster exposure as explained by Reynolds et al. (1996) resulting in lesser tartaric acid content and increased malic acid and potassium content resulting in increased juice pH.

Juice potassium content displayed significant difference among treatments in both the varieties wherein, vines which received leaf removal in combination with ST and CT recorded least potassium content than on control vines and ST vines. Between the two varieties studied, Cabernet Sauvignon grapes recorded more potassium concentration compared to Sauvignon Blanc grapes. The increased potassium concentration in control vines of both the varieties is in accordance with the findings of Smart et al. (1985); Bledsoe et al. (1988) and Rojas-Lara and Morrison (1989). In addition, Boulton (1980) identified potassium as a major factor in determining the pH of wines and grapes. He could establish positive correlation between potassium concentration and juice pH. In the present study, this relationship could be observed in Sauvignon Blanc grapes where control vines recorded highest pH and higher potassium content, while similar relationship could not be established in Cabernet Sauvignon grapes. This might be due to the concentration of malic acid which also determines juice pH. Highest concentration of malic acid was recorded in control vines and shoot thinned vines in Cabernet Sauvignon grapes which also recorded highest juice pH. The increased juice pH may be due to degradation of malic acid by respiratory enzymes as it is weaker than tartaric acid. According to Philip and Kuykendall (1973), combination of higher tartaric and lower malic acid is considered as superior grape quality. Thus in present study the acidity balance was therefore apparently changed favourably by leaf removal. As the ratio of malic acid and tartaric acid determines total titratable acidity, there was significant variation in titratable acidity with different canopy management treatments. The current finding of reduced malic acid in leaf removal treatment in combination with either cluster thinning and/or shoot thinning is in accordance with the findings of Kliewer (1967); Wolf et al. (1986); Kliewer and Bledsoe (1987) and Kliewer et al. (1988).

Reduced anthocyanin concentration was recorded on both control vines and on vines which received ST and ST+CT+LR treatments. Vines in control treatment might have developed higher number of leaves with maximum shade inside the canopy. It is likely that both light and berry temperature (either in excess or reduced quantity)

may be the factor in accumulation of anthocyanins. This is in accordance to reduced anthocyanins in control (shaded) vines and fully exposed clusters in Shiraz (Haselgrove et al., 2000), where fully exposed clusters recorded relatively higher berry temperature due to more exposure to sunlight which might have reduced anthocyanin accumulation or increased degradation. The increased anthocyanin accumulation in clusters on vines which received LR, ST+LR and CT+LR treatments suggest that, if light conditions within the canopy are such that bunches / cluster zone receives sufficient sunlight of moderate intensity, then light is not necessarily limiting factor for anthocyanin accumulation (Keller and Hrazdina, 1998). However, these effects may also be temperature dependent as explained by Mabrouk and Sinoquet (1998).

Influence of canopy management practices on berry composition is more pronounced with respect to anthocyanin concentration in Cabernet Sauvignon grapes. In control vines, it is likely that light is a limiting factor for anthocyanin accumulation. Similarly on vines which received combination of all the practices, the sunlight falling on clusters may be high thus increasing berry temperature to inhibit anthocyanin synthesis and/or to increase anthocyanin degradation. This hypothesis is supported by earlier literatures that synthesis of anthocyanin is directly regulated by both light exposure and temperature condition to which grape is subjected (Pirie and Mullins, 1980; Crippen and Morrison, 1986; Smart et al., 1988).

As far as phenolic compounds are concerned, significant difference could be seen in Cabernet Sauvignon than on Sauvignon Blanc. Concentration of major phenolic groups flavan – 3 – ols (catechin and epicatechin) and flavonols (quercetin and myricetin) were least in both control vines and those which received CT+ST+LR treatment, while it was highest in vines which received combination treatment of LR+ST or LR+CT or CT+ST. In Cabernet Sauvignon grapes, highest quercetin was recorded in vines which received either ST or CT treatments followed by the vines which received combination of either ST+LR and / or CT+LR treatment. Among non-flavonoid compounds, ellagic acid was highest with maximum concentration recorded on either ST or CT vines followed by ST+CT vines. Control vines and vines which received single treatments (ST, LR or CT) recorded lesser concentration of quercetin in both the varieties compared to those vines which received combination treatments. This is in accordance to the findings of Ristic et al. (2007), wherein shaded clusters of Syrah could accumulate only trace quality of flavonols such as quercetin compared to exposed clusters. This clearly explains the benefit of canopy management practices to expose clusters to light as quercetin accumulation may be a light dependent process. The increased anthocyanin concentration in those treatments may be also attributed to quercetin concentration as quercetin is important for co-pigmentation.

Light exposure through canopy modification has been shown to significantly influence flavonol accumulation in grapes and wine (Goldberg et al., 1980; McDonald et al., 1998; Haselgrove et al., 2000; Spayd et al., 2002). These observations report that fruit exposed to light mainly via changes in canopy structure have greater levels of flavonols, particularly quercetin glucosides, than shaded fruit. An increase in flavonols from sun exposed fruit may have implications with the stability of the wines particularly if flavonols act as co-pigments for anthocyanins.

Vineyards in semi-arid tropical climate with heavy black soils exhibit excessive vegetative growth. This result in disturbances in source: sink relationship leading to denser canopies and an inferior canopy microclimate for the continuous maximum photosynthetic activity of leaves. These factors may detrimentally affect grape and wine composition in particular resulting in reduced soluble sugars, tartaric acid and anthocyanin and higher concentration of malic acid, potassium and must pH. Though this preliminary study showed improved fruit composition with respect to TSS, acidity, juice pH, anthocyanins (in Cabernet Sauvignon) and few phenolic compounds in both the varieties under study, still detailed study with respect to intensity and time of leaf removal, bunch load, bunch exposure in different canopy sides (viz, east or west) etc. needs to be taken up to derive final conclusions. Measuring other parameters such as light intensity in different canopies, berry temperature, leaf area index, photosynthetic rate etc, will help to understand relationship between canopy management practices and canopy microclimate thus helps to follow, appropriate management practices to improve wine grape composition in semiarid tropical climate.

ACKNOWLEDGEMENTS

Authors are greatly thankful to the Director, NRC Grapes, Pune for providing all the field and laboratory facilities for conducting this experiment and necessary support from the Mr. Praveen Taware, Farm Manager for maintaining experimental vineyard.

REFERENCES

Bergqvist J, Dokoozlian N, Ebisuda N (2001). Sunlight exposure and temperature effects on berry growth and composition of Cabernet Sauvignon and Grenache in the central San Joaquin valley of California. Am. J. Enol. Vitic. 52:1–7.

Bledsoe AM, Kliewer WM, Marois JJ (1988). Effects of timing and severity of leaf removal on yield and fruit composition of Sauvignon Blanc grapevines. Am. J. Enol. Vitic. 39:49-54.

Boulton R (1980). The relationship between total acidity, titratable acidity and pH in wines. Am. J. Enol. Vitic. 31:76-80.

Candolfi-Vasconcelos MC, Koblet W (1990). Yield, fruit quality, bud fertility and starch reserves of the wood as a function of leaf removal in Vitis vinifera – Evidence of compensation and stress recovering. Vitis 29:199-221.

Chorti E, Guidoni S, Ferrandino F, Novello V (2010). Effect of different cluster sunlight exposure levels on ripening and anthocyanin accumulation in Nebbiolo grapes. Am. J. Enol. Vitic. 61:23-30.

Crippen DD, Morrison JC (1986). The effects of sun exposure on the compositional development of Cabernet Sauvignon berries. Amer. J. Enol. Vitic. 38:235-242.

Ferree DC, Scurlock DM, Steiner T, Gallander J (2004). 'Chambourcin' grapevine response to crop level and canopy shade at bloom. J. Am. Pomology Soc. 58:135-141.

Gatti M, Bernizzoni F, Civardi S, Poni S (2012). Effect of cluster thinning and prefloweing leaf removal on growth and fruit composition in cv. Sangiovese. Am. J. Enol. Vitic. 63:325-332.

Goldberg DM, Karumanchiri A, Tsang E, Soleas E (1980). Catechin and epicatechin concentration of red wines: Regional and cultivar related differences. Am. J. Enol. Vitic. 49:23-34.

Haselgrove L, Botting D, Van Heeswijck R, Høi PB, Dry PR, Ford C, Iland PG (2000). Canopy microclimate and berry composition: the effect of bunch exposure on the phenolic composition of Vitis vinifera L. cv. Shiraz grape berries. Aust. J. Grape Wine Res. 6:141-149.

Howell GS, Candolfi-Vasconcelos MC, Koblet W (1994). Response of Pinot noir grapevine growth, yield, and fruit composition to defoliation the previous growing season. Am. J. Enol. Vitic. 45:188-191.

Hunter JJ, Visser JH (1988). The effect of partial defoliation, leaf position and developmental stage of the vine on the photosynthetic activity of Vitis vinifera L. cv. Cabernet Sauvignon. S. Afr. J. Enol. Vitic. 9:9–15.

Hunter JJ, Visser JH, De Villiers OT (1991). Preparation of grapes and extraction of sugars and organic acids for determination by high-performance liquid chromatography. Am. J. Enol. Vitic. 42:237-244.

Hunter JJ, Ruffner HP, Volschenk CG, Le Roux DJ (1995). Partial defoliation of Vitis vinifera cv. Cabernet Sauvignon/99 Richter: Effect on root growth, canopy efficiency, grape composition, and wine quality. Am. J. Enol. Vitic. 46:306-314.

Johnson JO, Weaver RJ, Paige DF (1982). Differences in the mobilization of assimilates of Vitis vinifera L. grapevines as influenced by an increased source strength. Am. J. Enol. Vitic. 33:207 – 213.

Keller M, Hrazdina G (1998). Interaction of nitrogen availability during bloom and light intensity during veraison. II. Effects on anthocyanin and phenolic development during grape ripening. Am. J. Enol. Vitic. 49:341–349.

Kliewer MW, Lakso LA (1968). Influence of cluster exposure to the sun on the composition of Thompson Seedless fruits. Am. J. Enol. Vitic. 119: 175-184.

Kliewer WM, Dokoozlian NK (2005). Leaf area/crop weight ratio of grapevines influence on fruit composition and wine quality. Proceedings of the ASEV 50th Anniversary Annual Meeting. Am. J. Enol. Vitic. 56:170- 181.

Kliewer WM, Antcliff AJ (1970). Influence of defoliation, leaf darkening and cluster shading on the growth and composition of sultana grapes. Am. J. Enol. Vitic. 21:26-36.

Kliewer WM, Bledsoe A (1987). Influence of hedging and leaf removal on canopy microclimate, grape composition, and wine quality under California conditions. Acta Hort. 206:157-168.

Kliewer WM, Marois JJ, Bledsoe AM, Smith SP, Benz MJ, Silvestroni O (1988). Relative effectiveness of leaf removal, shoot positioning, and trellising for improving winegrape composition. Proc. 2nd Int. Cool Climate Vitic. And Oenol. Symp., Jan. 1988, Auckland, New Zealand. pp. 123- 126.

Kliewer WM, Lider, LA (1970). Effects of day temperature and light intensity on growth and composition of Vitis vinifera L. fruits. J. Am. Soc. Hort. Sci. 95:766-769.

Kliewer WM, Smart RE (1989). Canopy manipulation for optimizing vine microclimate, crop yield and composition of grapes. In Manipulation of Fruiting. C.J. Wright (Ed.), Butterworth, London. pp. 275-291.

Kliewer WM (1967). Concentration of tartrates, malates, glucose and fructose in the fruits of the genus Vitis. Am. J. Enol. Vitic. 18:87-96.

Koblet W (1984). Influence of light and temperature on vine performance in cool climates and application to vineyard. Proc. International Symposium on Cool Climate Viticulture and Enology, Eugene, USA. Oregon State University, pp. 139-157.

Koblet W, Keller M, Candolfi-Vasconcelos MC (1996). Effects of training system, canopy management practices, crop load and rootstock on grapevine photosynthesis. In: Poni, S., Peterlunger, E., Iacono, F. & Intrieri, C. (eds). Proc. Workshop Strategies to Optimize Wine Grape Quality, Conegliano, Italy. pp. 133–140.

Lakso AN, Kliewer WM (1978). The influence of temperature on malic acid metabolism in grapes. II. Temperature responses of net dark CO_2 fixation and malic acid pools. Am. J. Enol. Vitic. 29:145–149.

Lowry OH, Rosenbrough NJ, Farr, AL, Randall RJ (1951). Protein measurement with the Folin Phenol Reagent. J. Biol Chem. 193:265-275.

Mabrouk H, Sinoquet H (1998). Indices of light microclimate and canopy structure of grapevine determined by 3D digitising and image analysis and their relationship to grape quality. Aust. J. Grape Wine Res. 4:2-13.

McDonald MS, Hughes M, Burns J, Lean ME, Matthews D, Crozier A (1998). Survey of the free and conjugated myricetin and quercetin content of red wines of different geographical origins. J. Agric. Food Chem. 46:368–375.

McLennan W (1996). Year Book Australia, (Canberra: Australian Bureau of Statistics, 1996), P. 414.

Main GL, Morris JR (2004). Leaf-removal effects on Cynthiana yield, juice composition, and wine composition. Am. J. Enol. Vitic. 55:147-152.

Patil SH, Banerjee K, Oulkar, DP, Jogaiah S, Sharma AK, Dasgupta S, Adsule PG, Deshmukh MB (2011). Phenolic composition and antioxidant activity of Indian wines. Bull. de l'OIV. 84:517-546.

Percival DC, Fisher KH, Sullivan JA (1994). Use fruit zone leaf removal with Vitis vinifera L cv. Riesling grapevines. II. Effect on fruit composition, yield, and occurrence of bunch rot (Botrytis cinerea Pers). Am. J. Enol. Vitic. 45:133-140.

Philip T, Kuykendall JR (1973). Changes in titratable acidity, °B, pH, potassium content, malate and tartarate during berry development of Thompson Seedless grapes. J. Food Sci. 38:874-876.

Pirie A, Mullins MG (1980). Changes in anthocyanins and phenolic content of grapevine leaf and fruit tissues treated with sucrose, nitrate and abscissic acid. Plant Physiol. 58:468-472.

Poni S, Casalini L, Bernizzoni F, Civardi S Intrieri C (2006). Effects of early defoliation on shoot photosynthesis, yield components, and grape composition. Am. J. Enol. Vitic. 5:397-407.

Reynolds AG, Pool RM, Mattick LR (1986). Influence of cluster exposure on fruit composition and wine quality of Seyval Blanc grapes. Vitis 25:85-95.

Reynolds AG, Wardle DA, Naylor AP (1996). Impact of training system, vine spacing, and basal leaf removal on Riesling. Vine performance, berry composition, canopy microclimate, and vineyard labor requirements. Am. J. Enol. Vitic 47:63-76.

Ristic R, Downey M, Iland P, Bindon K, Francis L, Hederich M, Robinson S (2007). Exclusion of sunlight from Shiraz grapes alters wine colour, tannins and sensory properties. Aust. J. Grape Wine Res. 13:53–65.

Rojas-Lara BA, Morrison JC (1989). Differential effects of shading fruit or foliage on the development and composition of grape berries. Vitis 28:199–208.

Santesteban LG, Royo JB (2006).Water status, leaf area and fruit load influence on berry weight and sugar accumulation of cv. 'Tempranillo' under semiarid conditions. Scientia Horticult. 109:60-65.

Sidahamed OD, Kliewer MW (1980). Effect of defoliation, gibberellic acid and 4-chlorophenoxy acetic acid on growth and composition of Thompson Seedless grape berries. Am. J. Enol. Vitic. 31:149-153.

Singleton VL, Rossi JA (1965). Colorimetry of total phenolics with phosphomolybdic phosphotungstic acid reagent. Am. J. Enol. Vitic. 16:144-158.

Smart RE (1985). Principles of grapevine canopy microclimate manipulation with implications for yield and quality- A review. Am. J. Enol. Vitic. 35:230-239.

Smart RE, Robinson JB, Due GR., Briew CJ (1985). Canopy microclimate modification for the cultivar Shiraz, I. Definition of canopy microclimate.Vitis 24:17-31.

Smart RE (1988). Shoot spacing and canopy light microclimate. Am. J. Enol. Vitic. 39:325-333.

Smart R, Smith SM, Winchester RV (1988). Light quality and quantity effects on fruit ripening for Cabernet Sauvignon. Am. J. Enol. Vitic. 39:250-258.

Smart RE, Dick JK, Gravett IM, Fisher BM (1990). Canopy management to improve grape yield and wine quality – principles and practices. South Afr. J. Enol. Vitic. 11:3- 17.

Spayd SE, Tarara JM, Mee DL, Ferguson JC (2002). Separation of sunlight and temperature effects on the composition of Vitis vinifera cv. 'Merlot' berries. Am. J. Enol. Vitic. 53:171-182.

Sun Q, Sacks GL, Lerch SD, Vander Heuvel JE (2012). Impact of shoot and cluster thinning on yield fruit composition and wine quality of Corot Noir. Am. J. Enol. Vitic. 63:49-56.

Tardaguila J, Martinez de Toder F, Poni S, Diago MP (2010). Impact of early leaf removal on yield and fruit and wine composition of Vitis vinifera L Graciano and Carignan. Am. J. Enol. Vitic. 61:372-281.

Vasconcelos MC, Castagnoli S (2000). Leaf canopy structure and vine performance. Am. J. Enol. Vitic. 51:390-396.

Williams LE, Biscay PJ, Smith RJ (1987). Effect of interior canopy defoliation and potassium distribution in Thompson Seedless grapevines. Am. J. Enol. Vitic. 38:287-292.

Wolf TK, Pool RM, Mattick LR (1986). Responses of young Chardonnay grapevines to shoot tipping ethephon, and basal leaf removal. Am. J. Enol. Vitic. 37:263–268.

Zoecklein BW, Wolf TK., Duncan ND, Judge JM, Cook MK (1992). Effect of fruit zone leaf removal on yield, fruit composition, and rot incidence of Chardonnay and White Riesling (Vitis vinifera L.) grapes. Am. J. Enol. Vitic. 43:139-148.

http://estructuraehistoria.unizar.es/gihea/documents/GwynCampbell.pdf

16

Influence of diverse pollen source on compatibility reaction and subsequent effect on quality attributes of sweet cherry

K. K. Srivastava*, N. Ahmad, Dinesh Kumar, Biswajit Das, S. R. Singh, Shiv Lal, O. C. Sharma, J. A. Rather and S. K. Bhat

Section of Crop Improvement, Central Institute of Temperate Horticulture , Old Air Field, Srinagar– 190007, Jammu and Kashmir, India.

Sweet cherry is self incompatible due to having a gametophytic self – incompatibility system. S alleles in the style and pollen determine possible crossing relationship and ultimate fruit set. Complete knowledge of the s allele constitution of cultivars is import out for sweet cherry growers and breeders. Natural pollination in cultivar Van resulted in high fruit set as compared to controlled crossing but fruit size recorded high in crossed than natural pollinated. However, cultivar Stella recorded high fruit set in controlled crossing. In Lapinus also, very few crosses set fruits. Furthermore, the fruit set percent are also low and diverse pollen source has no effect on the fruit weight but total soluble solids (T.S.S) was recorded higher in crossed fruits. Similarly, Lambert in controlled pollination resulted in poor fruit set as compared to natural pollination. Guigne Pourpera Precoca exhibited high fruit set with pollen of Lambert and Bing; but other combinations resulted in poor compatibility. Bigarreau Noir Grossa with all the pollen resulted in poor compatibility, whereas fruit set in natural (controlled) pollination recorded high fruit set. Bigarreau Napoleon with Guigne Pourpera Precoca and Lapinus showed good cross compatibility; when used as male with Bigarreau Napoleon, Lapinus exhibited increased fruit weight and T.S.S. as compared to natural pollinated fruits.

Key words: Cherry, pollen, compatibility, quality attributes, fruit set.

INTRODUCTION

The sweet cherry botanically known as *Prunus avium* L., belongs to Family Rosaceous, a deciduous tree of large stature, occasionally reaching almost 20 m in height with attractive peeling bark. Primary centre of origin of cherry is Caspian and Black seas from where it spread by birds Westwood (1993). Cherries are cultivated almost continuous of the world, offering suitable congenial conditions. Hungary consume almost twice of cherry than Germany and sour cherries are consumed most in the Yugoslavs, Germans etc. Turkey ranks first accounts, 59751.00 ha area and 338361.0 tones production with 5.66 t/ha, productivity (Anonymous, 2009). Since most major cherry cultivars are self incompatible, require cross pollinations to obtain fruits. Maximum pollinating efficiency in sweet cherry can be achieved by planting equal numbers of each cultivars and alternating their location down the rows. Each tree should be surrounded by compatible pollinizer. For good orchard pollination, three to five strong beehives per hectare is ideal, comprising of 20,000 to 30,000 adult' bees per beehives. In sweet cherry flower remain open for 7 to 8 days and stigma is receptive at the opening of flowers. Anther starts dehiscing shortly after flower open and continues the second day (Srivastava and Singh, 1970). Ovule longevity is of 4 of 5 days (Roversi, 1994). Maximum stigma receptivity exhibits for five days after anthesis and few ovules remain functioning even after 13 days of

---

*Corresponding author. E-mail: kanchanpom@gmail.com.

anthesis. Most varieties of sweet cherries are self sterile and need cross pollination. All commercial cultivar have viable pollen, but all varietal combinations are not useful. In cherry upto 25 different S-alleles (Boskovic and Tobutt, 1996, 2001; Boskovic et al., 1997; De Cuper et al., 2005; Sonneveld et al., 2001, 2003; Taq et al., 1996; Vaughan et al., 2008) and 40 incompatibility groups have been reported so far in sweet cherry using different methods (Marchese et al., 2007; Schuster et al., 2007; Tobutt et al., 2004). There are many incompatible cross groups of sweet cherry; hence, the cultivars within a group should not be planted together without a pollinizer (Childers, 1995). Incompatibility in sweet cherry is gametophytic type, results in the inhibition of pollen tube growth in the style (Hurter et al., 1979; Vasilakakis and Porlingis, 1985). Mechanism of incompatibility is genetically controlled by multiple alleles at a single locus. The sweet cherry cultivars with sterility alleles were reported by Crane and Brown (1937), Tehrani and Brown (1992). Black Heart, Van, Venus, Winds have S1, S3 alleles, Bing, Lambert, Napoleon, S3, S4 alleles, whereas Vic, Stella, Vista cultivars are universal doner (Tao et al., 1999b). In sweet cherry cultivars, RFLP profiles have been used to assign self incompatibility alleles in sweet cherry genotypes (Hauck et al., 2002). The introduction of molecular methods for determining self incompatibility alleles in sweet cherry led to the rapid confirmation of the S-allele, on this basis incompatibility groups of many cultivars were reported previously. Self and cross (in) compatibility between cultivars have traditionally been determined by monitoring the fruit set percentage under field condition. The only disadvantage of this method is that fruit set varies from year to year, depending on weather condition. In order to establish high yielding sweet cherry orchards it is essentially required to have the knowledge of incompatibility relationship of the varietal profile available in India. In cherry orchardists are not aware about the incompatibility reactions resulting poor orchard yield. In order to find out best pollen source for most of the commercially grown varieties, this present experiment was initiated. Hence, this present experiment has been designed to determine compatibility relationship among the sweet cherry cultivars.

MATERIALS AND METHODS

This present studies were carried over a period of three years in 2009, 2010 and 2011 on eight cherry varieties buded on *Prunus cerasus* (sour cherry) root stock and planted at 3 × 3 m spacing. The cherry varieties used for this study was Van, Stella, Lapinus, Lambert, Guigne Purpera Precoca, Bigarreau Napoleon, CITH-Cherry-01, Bing and Bigarreau Noir Grossa. The experimental site is located at Karewa Belt of Kashmir situated at latitude 34°, 45°N and longitude of 74°50 E, and elevation is 1649 m masl, area experienced average minimum and maximum temperate 6.52 19.63°C. Area receives the amount of rainfall 650 to 1000 mm with relative humidity 58.35%. The specified branches were tagged and covered with muslin cloth bags to prevent any contamination from the foreign pollen well in advance. To collect the pollen from the designated male parents for the pollination, the flowers were collected at balloon stage just before the petals expand and before anther dehisce. Before crossing the flowers of the seed parents were emasculated at balloon stage by flicking off the sepals, petals and stamens with scissors and pollination was done immediately. In cherry orchardists are not aware about the incompatibility reactions resulting poor orchard yield. In order to find out best pollen source for most of the commercially grown varieties, this present experiment was initiated. In case of bad weather re-pollination was done.

All possible crosses were made during this study including di-allele crosses. The specified branches were tagged and covered with net bag to prevent infection from foreign pollen. To collect pollen, flowers from the netted branch were collected at the balloon stage just before the petals expand and anther dehiscence though receptivity remains for five days. After 30 days of hand pollination fruit set per cent was recorded and calculated by total fruit set divided by total flowers pollinated multiplied by 100. The same crossing was done on three different branches to replicate the treatment. The data were recorded on fruit set percent, fruit diameter, fruit length, fruit height, pulp weight, stone weight and TSS. The Digital Vernier Caliper (*Mitutoyo, Japan*), Hand refractometer (*Atago*) and electronic balance was used to record the observations. The impact of diverse pollen source on compatibility reaction as well as on fruit quality attributes was recorded. In order to compare the quality attributes of crossed fruits, one tree of each variety was left to set fruits through natural pollination for check.

RESULTS

The fruit set percent and quality attributes of cherry from cross combinations are presented in Table 1. Fruit set and quality attributes were recorded in cultivar Van. Highest fruit set 80.74, 68.32 and 61.30% were recorded during 2009, 2010 and 2011, respectively under naturally pollinated condition, however, fruit set percent from the specific pollen source were poor. Van and Guigne Pourpera Precoca resulted in poor fruit set in all the years; Van X Lambert resulted in high fruit set in three years (Figure 1). Similar trend in fruit diameter was also observed. Fruit diameter obtained from crossing was higher than the fruits obtained from the control. Fruit diameter registered highest 23.4, 22.26, 23.59 mm in Van × Guigne Pourera Precoca and lowest in Van × Lambert. Fruit length was recorded highest (22.10 mm) in Van × Guigne Pourpera Precoca (Table 1). Significant variations in TSS were obtained in different cross combinations, highest TSS. (17.46%) was noted in Van × Stella, followed by Van × Guigne Pourpera Precoca, whereas lowest (11.2% in Van × Lapinus during 2011, while fruits obtained from control (natural) pollination had low TSS (11.26%) during 2009 and 2010 (Figure 3).

The cultivar Stella exhibited high fruit set as compared to natural pollination. Fruit set 54.10%, 61.99 and 66.40% during 2009, 2010 and 2011, respectively were recorded. Stella × Bigarreau Napoleon recorded highest fruit set 75.73, 82.23 and 74.83%, respectively for the 3 studied years, Stella with all the pollen source resulted high fruit set except Stella and Lambert (Figure 1). Significant difference of pollen source on the fruit weight was noted.

Table 1. Fruit set and quality attributes as influenced by diverse pollen source in cherry.

Cross combinations	2009				2010				2011			
	Fruit dia. (mm)	Fruit length (mm)	Fruit pulp weight (g)	Stone weight (g)	Fruit dia. (mm)	Fruit length (mm)	Fruit pulp weight (g)	Stone weight (g)	Fruit dia. (mm)	Fruit length (mm)	Fruit pulp weight (g)	Stone weight (g)
Van x Stella	21.3	20.9	4.90	0.4	20.83	20.6	4.26	0.36	21.13	21.45	4.36	0.33
Van x Lapinus	20.53	20.0	4.86	0.5	19.93	21.84	4.5	0.36	20.77	23.90	5.16	0.26
Van x Lambert	16.92	17.33	2.70	0.3	19.58	20.18	4.1	0.26	20.64	21.83	4.53	0.26
Van x Guigne Pourpera Precoca	23.4	22.1	5.63	0.53	22.26	21.66	5.0	0.4	23.59	22.38	5.76	0.46
Van	20.00	21.91	4.8	0.5	20.28	19.57	4.6	0.4	19.41	21.57	3.9	0.26
CD (P = 0.05)	1.21	1.10	0.53	0.11	1.13	1.75	0.97	0.33	1.85	1.91	6.53	0.45
Stella x van	21.46	21.56	5.26	0.26	21.87	22.13	4.64	0.6	22.68	22.47	6.3	0.36
Stella X Lapinus	21.5	19.93	3.93	0.4	20.21	19.33	4.70	0.26	21.87	20.43	4.03	0.36
Steela x Guigne Pourpera Precoca	21.36	21.6	5.63	0.73	22.00	20.16	5.66	0.5	21.81	22.74	6.1	0.36
Stella x Bigarreau Napoleon	20.6	21.5	5.6	0.46	20.63	20.65	4.96	0.26	20.88	21.95	5.83	0.36
Stella x Biggareau Napoleon	20.53	21.26	5.03	0.5	19.98	20.3	5.13	0.5	21.29	22.41	5.36	0.36
Stella x Bigarreau Noir Grossa	20.83	20.46	4.4	0.56	19.5	20.26	4.33	0.46	21.08	21.32	4.93	0.26
Stella x Lembert	18.56	17.6	2.23	0.16	18.5	18.56	3.20	0.4	19.71	18.9	3.5	0.26
Stella	19.90	20.85	4.0	0.5	19.99	17.61	4.13	0.36	19.99	20.94	4.2	0.36
CD (P = 0.05)	1.72	1.15	0.57	0.23	1.72	1.81	1.02	0.45	1.73	1.83	0.51	0.71
Lapinus x Van	21.16	20.33	4.66	0.53	20.53	19.33	5.1	0.4	21.52	20.29	4.43	0.5
Lapinus x Guigne Pourpera Precoca	21.26	21.63	3.96	0.6	19.68	19.57	4.13	0.36	21.20	21.31	4.2	0.5
Lapinus x Bigarreau Napoleon	21.66	20.76	5.16	0.5	21.16	20.63	4.83	0.4	21.75	22.13	5.26	0.43
Lapinus	19.89	20.80	4.3	0.5	19.35	20.45	5.36	0.46	20.73	22.04	4.1	0.53
CD (P = 0.05)	1.25	1.17	0.49	0.13	1.11	1.20	0.43	0.13	1.30	1.18	0.33	0.21
Lambert x Van	18.4	17.03	3.03	0.36	23.75	21.21	5.76	0.5	18.93	17.63	3.06	0.33
Lambert x Stella	20.43	20.26	4.76	0.46	21.16	21.76	4.03	0.43	21.25	20.56	4.8	0.43
Lambert x Lapinus	18.53	18	3.9	0.5	19.06	17.50	3.9	0.36	20.77	19.29	3.96	0.26
Lambert x Bigarreau Noir Grossa	18.96	18.42	3.6	0.53	21.28	20.02	4.35	0.47	20.37	20.50	4.06	0.53
Lambert	19.85	21.00	4.6	0.4	23.16	21.21	5.63	0.5	21.7	21.61	4.7	0.4
CD (P = 0.05)	0.92	1.11	0.38	0.12	1.20	1.80	0.82	0.21	2.30	2.17	0.51	0.32
Guigne Pourpera Precoca. x Lambert	21.15	18.6	3.96	0.4	20.27	16.71	4.00	0.43	21.39	19.10	3.83	0.63
Guigne Pourpera Precoca. x Bing	22.36	18.76	4.4	0.5	19.16	15.87	2.99	0.4	22.58	19.25	4.33	0.76
Guigne Pourpera Precoca. x Biggareau Napoleon	21.2	16.66	3.9	0.5	22.24	17.86	3.97	0.47	21.06	17.73	3.8	0.73
Guigne Pourpera Precoca. x Bigarreau Noir Grossa	21.8	17.83	4.03	0.6	22.1	18.2	4.03	0.46	21.43	17.35	3.66	0.73
Guigne Pourpera Precoca. x Van	21.86	18.33	4.56	0.53	19.3	16.52	2.99	0.4	22.11	19.11	4.86	0.43
Guigne Pourpera Precoca x Steela	21.43	18.4	4.16	0.6	21.15	16.76	3.18	0.41	21.86	19.15	4.3	0.5
Guigire Pourpera Precoca	18.03	19.05	5.1	0.5	20.79	17.9	5.5	0.53	19.87	17.87	3.2	0.63

Table 1. Contd.

CD (P = 0.05)	2.11	1.02	0.52	0.21	2.10	0.70	0.71	0.21	2.11	2.02	0.87	0.17
CITH-Cherry-01 x Stella	21.3	20.43	5.53	0.4	20.93	21.66	5.36	0.5	21.25	22.35	5.86	0.43
CITH-Cherry-01 x Lambert	19.6	18.1	3.7	0.4	18.83	19.66	3.66	0.43	17.47	20.53	4.2	0.23
CITH-Cherry-01 X Bing	20.23	22.43	4.96	0.56	20.13	22.33	4.86	0.43	20.8	22.50	5.23	0.26
CITH-Cherry-01 x Bigarreau Noir Grossa	20.33	21.3	5.23	0.6	19.43	20.66	4.76	0.43	20.74	21.88	5.53	0.26
CITH-Cherry-01 x Lapinus	19.4	21.5	4.5	0.5	20.5	20.83	4.23	0.53	20.78	21.57	4.86	0.23
CITH – Cherry- 01	19.20	20.01	4.2	0.5	14.55	20.20	4.1	0.5	20.04	21.64	4.33	0.6
CD (P = 0.05)	1.12	1.51	0.51	0.11	2.20	0.79	0.79	0.24	2.30	2.05	0.91	0.21
Bing x Stella	21.5	24.66	6.73	0.43	21.7	22.92	5.04	0.31	22.13	26.7	6.96	0.33
Bing x Lambert	21.86	21.5	6.13	0.53	22.65	25.00	6.31	0.47	21.41	22.68	6.26	0.33
Bing x Guigne Pourpera Precoca	20.76	23.83	5.63	0.5	22.88	24.07	5.95	0.48	20.89	24.13	6.0	0.43
Bing x Bigarreau Napoleon	20.43	19.66	4.7	0.53	23.15	24.41	5.35	0.37	24.37	26.15	7.6	0.43
Bing	20.00	21.10	4.8	0.5	20.84	21.14	5.03	0.43	22.01	23.4	5.10	0.5
CD (P = 0.05)	1.82	2.01	0.58	0.31	2.00	0.68	0.19	0.90	1.91	2.02	0.91	0.27
Bigarreau Noir Grossa x Van	23.52	21.54	6.26	0.6	23.47	22.73	4.63	0.46	21.32	21.61	5.53	0.26
Bigarreau Noir Grossa x Stella	20.86	23.16	4.1	0.56	20.14	20.76	3.27	0.37	21.39	23.6	4.5	0.26
Bigarreau Noir Grossa x Lambert	22.43	23.06	6.43	0.53	19.23	19.50	4.36	0.5	23.15	23.63	6.7	0.33
Bigarreau Noir Grossa x Guigne Pourpera Precoca	19.33	18.83	3.86	0.36	18.5	19.83	4.3	0.53	19.59	21.06	4.33	0.36
Bigarreau Noir Grossa x Bigarreau Napoleon	19.55	19.5	3.8	0.4	26.47	24.24	6.83	0.5	19.61	20.14	3.76	0.33
Bigarreau Noir Grossa x Bing	18.16	18.66	3.56	0.43	18.5	19.0	3.3	0.4	19.4	19.66	3.63	0.3
Bigarreau Noir Grossa x Bigarreau Noir Grossa	17.5	18.7	3.06	0.43	20.33	20.06	3.3	0.3	18.17	18.19	3.43	0.43
Bigarreau Noir Grossa	21.80	22.51	4.4	0.4	22.53	22.99	5.83	0.43	20.26	22.50	3.6	0.36
CD (P=0.05)	1.00	1.91	0.51	0.31	3.01	0.71	0.18	0.89	1.81	2.03	1.12	0.31
Bigarreau Napoleon x Lapinus	20.86	22.43	5.73	0.6	19.99	18.71	3.63	0.56	22.31	22.84	6.06	0.23
Bigarreau Napoleon x Guigne Pourpera Precoca	19.87	17.45	4.3	0.5	20.81	18.42	4.0	0.46	19.02	19.80	3.36	0.33
Bigarreau Napoleon x Van	18.73	20.93	3.63	0.46	18.26	20.43	4.06	0.4	20.46	23.06	4.33	0.26
Bigarreau Napoleon	16.62	17.30	4.1	0.5	20.44	19.44	4.4	0.53	19.61	21.86	4.9	0.4
CD (P = 0.05)	1.61	1.18	0.45	0.11	1.36	0.54	0.71	0.11	1.64	1.17	0.43	0.11

High fruit weight were recorded in combination of Stella × Guigne Pourpera Precoca, Stella × Bigarreau Napoleon and Steel × Van with low in Stella × Lambert was recorded during all the three years (Figure 2). TSS was recorded at par with natural pollination among the entire cross combinations except Stella × Guigne Pourpera Precoca (14.1, 13.76 and 14.2% during 2009, 2010 and 2011, respectively. However, during 2010, TSS in Control was significantly higher than fruits obtained from cross combinations (Figure 3). In poor fruit set. Bigarreau Noir Grossa with all the

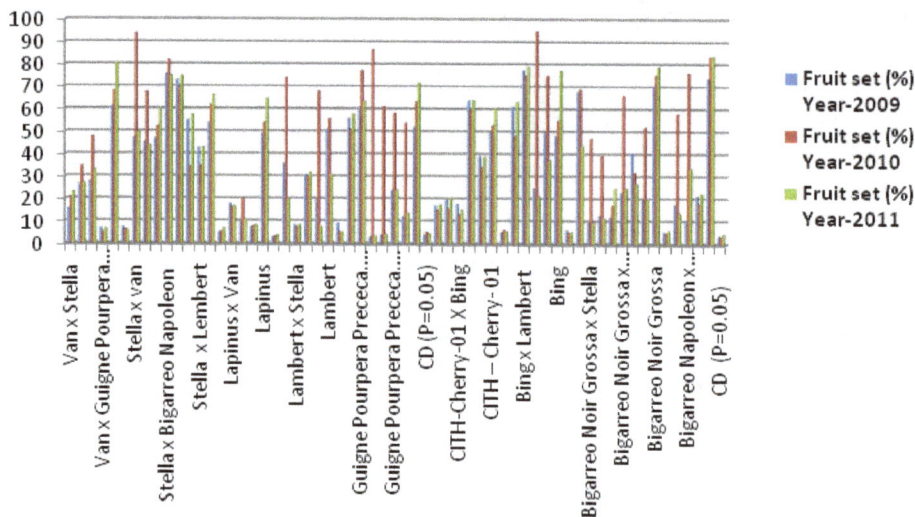

Figure 1. Impact of diverse pollen source on fruit set of sweet cherry.

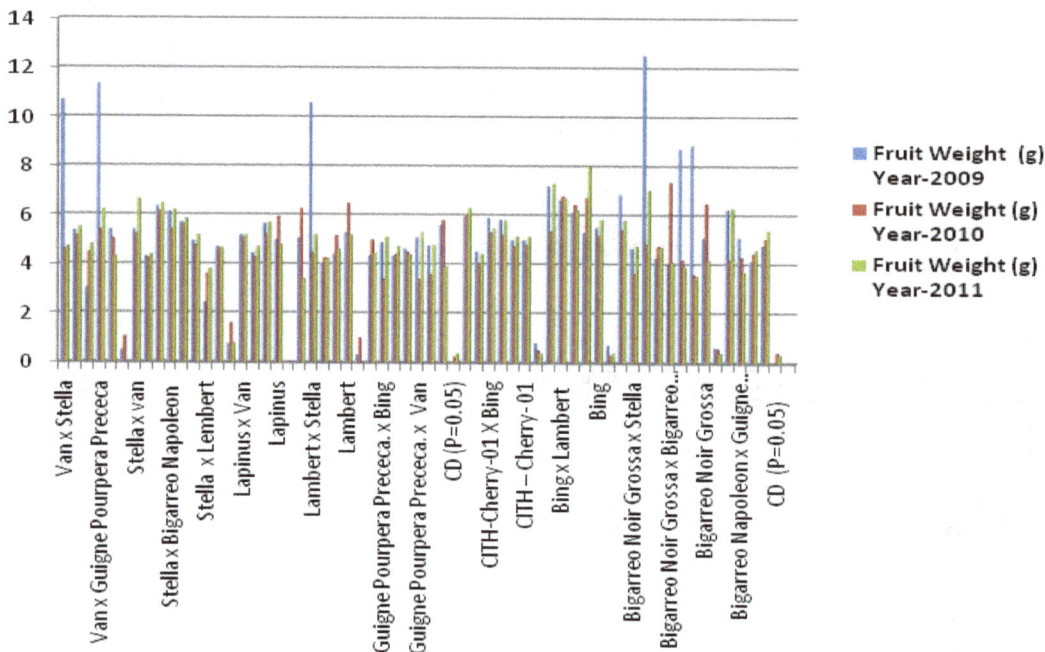

Figure 2. Impact of diverse pollen source on fruit weight of sweet cherry.

combinations resulted poor fruit set as compare to control, higher fruit set 70.50, 75.43 and 79.07% recorded during 2009, 2010 and 2011, respectively in control. However, Bigarreau Noir Grossa with Van, Bing in 2009 and with Van, Stella and Bigarreau Napoleon in 2010 resulted in high fruit set, where as all the combinations resulted in poor fruit set during 2011 except with Van (Figure 1). When Lambert was used as pollen source with Bigarreau Noir Grossa resulted high TSS (Figure 3). Bigarreau Napoleon with all the cross combinations resulted in low fruit set as compared to control except Bigarreau Napoleon with Guigne Pourpera Precoca and Lapinus, 76.34 and 58.19%, respectively, similarly pollen of Lapinus resulted increased in fruit weight and TSS case of Lapinus, very few cross combinations set fruits, further, the fruit set percent were also found lowest in comparison to control (49.25%) (Figure 1). No significant variations in fruit weight was recorded however, Lapinus × Bigarreau Napoleon and Lapinus × Van recorded comparatively higher fruit weight, but during 2010, fruit weight among crossed fruits were found less than control (Figure 2). TSS recorded

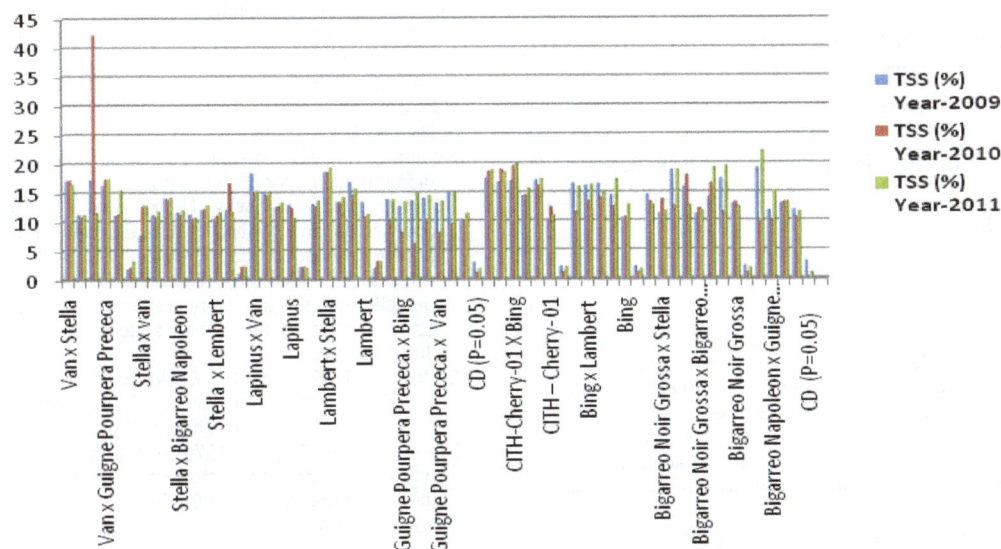

Figure 3. Impact of diverse pollen source on quality of resultant fruits.

significantly higher in the crossed fruits than naturally set fruits. High TSS 15.13 and 18.3, 14.53 and 15.3 and 15.1 and 15.33 were noted in Lapinus × Van, Lapinus × Guigne Pourpera Precoca during 2009, 2010 and 2011 respectively (Figure 3).

Lambert cultivar were crossed with different male parents, the fruit set were recorded less than control with 51.20, 55.76 and 30.98% during 2009, 2010 and 2011 respectively except Lambert x Van (74.0%) and Lambert X Bigarreau Noir Grossa. When Lambert crossed with Stella resulted high fruit weight and TSS for the three consecutive years.

Fruit set pattern among the cross combinations of Guigne Pourpera Precoca with cross combination were found poor except with Lambert and Bing which were recorded higher as compared to other combinations.

Guigne Pourpera Precoca recorded high fruit set during 2010 with Bigarreau Napoleon and Van (86.9, 61.17 and 58.27%), however, other cross combinations resulted in poor fruit set for the two years, that is, 2009 and 2011(Figure 1). Diverse pollen source had significant impact on quality attributes of fruits, fruit length, fruit diameter and TSS (Table 1).

Fruit set pattern was recorded low as compared to control, except CITH-Cherry - 01 × Bigarreau Noir Grossa, 64.16, 60.03 and 64.13% during 2009, 2010 and 2011 respectively (Figure1). Similarly, fruit weight and TSS also recorded high in CITH-Cherry-01 × Bing during three consecutive years.

Fruit set was recorded significantly higher as compare to control during three consecutive years, Bing with Stella and Lambert resulted in high fruit set (Figure 1), TSS and fruit weight constantly. However, Bing × Guigne Pourpera Precoca resulted 95.0% during 2010, rest two years, this combination resulted constantly.

DISCUSSION

Most of the cherry varieties are self sterile and need cross pollination. All commercial cultivars have viable pollen, but not all varietal combinations are useful. Childers (1995) suggested 18 cross compatible groups of sweet cherries. Incompatibility in sweet cherries is gametophytic in nature. The incompatibility in cherry and pear appears to function via glycoprotein's which correlate with the incompatibility alleles (Raff et al., 1981) for Bing, Lambert and Napoleon the suitable polinizers are Van, Black Republican, Corum, Stella and Black Tartarian . The poor fruit set among the crossed combinations might be due to incompatibility group cultivars have same geographical origin and are genetically closely related which showed incompatibility reaction. In cherry, stigma of all species and varieties are receptive as soon as flower open, but maximum receptivity of sweet cherry has been recorded for 4 to 5 days after anthesis (Stosser and Anvari, 1982; Guerrero et al., 1985). Ganopoulos et al. (2010) recorded same trends in cross compatibility in sweet cherry at Greece, Similarly at Himachal Pradesh, Ananda and Verma (1992), recorded that 75 crosses of almond, 53 were cross compatible, 12 crosses partially cross compatible and 10 crosses fully cross incompatible. Reciprocal combinations gave a good amount of fruit set and were partially to completely cross compatible. Stella is a self compatible cultivar has $S_3 S_4$ alleles but the S_4 is mutated to the 4^1 type. Ganopoulos et al. (2010) reported most frequent genotypes were $S_3 S_4$ (51%) and $S_4 S_9$ (9%) alleles. Hence, in order to increase the potential for cross fertilization in Kashmir, it is essential that additional cross compatible cultivars should be planted for obtaining high yield in cherry.

Based on this findings, we concluded that in sweat cherry, self and cross incompatibility are both prevalent. Hence, cultivars of same group should not be planted together. Cultivars Bigarreau Noir Grossa, Lambert and Lapinus exhibited poor cross compatibility, however, Stella, Guigne Pourpera Precoca and CITH Cherry 01 showed good cross compatible relation with crossed parents in this study.

REFERENCES

Ananda SA, Verma RL (1992). Studies on pollination requirements in almond. *Proceeding of Natural Seminar on Emerging* trends in temperate fruit production in India. In NHB Tech. Communication. 1, pp. 68-73.

Anonymous (2009). District wise Ares and protection and productivity in Jammu and Kashmir, Department of Horticulture, Govt. of Jammu and Kashmir.

Boskovic R, Tobutt KR, Nicoll FJ (1997). Interspecific of isoenzymes and their linkage relationships in two interspecific cherry progenies. Euphytic 93:129-143.

Boskovic R, Tobutt KR (1996). Correlation of stylar ribonuclease zymograms with incompatibility alleles in sweet cherry. *Euphytica* 90:245-250

Boskovic R, Tobutt KR, (2001). Genotyping cherry cultivars assigned to incompatibility groups, by analyzing stylar ribonucleases. Theoret. Appl. Genet. 103:475-485.

Childers NF (1995). Modern Fruit Science. Horticultural Publications, 3909NW31 place Gainasville , Florida 32606. p. 85.

Crane MW, Brown AG (1937). Incompatibility and sterility in sweet cherry. J. Pomol. Hort. Sci. 15:86-116.

De Cuper B, Sonneveld T, Tobutt KR (2005). Deterring self-incompatibility genotypes in Belgian wild cherries. Mole. Ecol. 14:945-955.

Ganopoulos IV, Argiriou A, Saftaris AS (2010). Determination of self incompatible genotypes in 21 cultivated sweet cherry cultivars in Greece and implications for orchard cultivation. J. Hort. Sci. Biot. 85(5):444-448.

Guerrero Prieto VM, Vasilakakis MD, Lombard PB (1985). Factors controlling fruit set of 'Napoleon' sweet cherry in Western Oregon. Hort. Sci. 20(5):913-14.

Hauck NR, Yamane H, Tao R, Iezoni AF (2002). Self- compatibility and incompatibility in tetraploid sour cherry (*Prunes cerasus* L.). Sexual Plant Reprod. 15:39-46.

Hurter N, Van Tonder MJ, Bester CW (1979). Cross pollination of the plum cultivar Red Gold. Decid. Fruit Grower 29:152-155.

Marchese AKR, Raimondo A, Motisi A, Boskovic RI, Caruso, T (2007). Morphological characteristic, microsatellite fingerprinting and determination of incompatibility genotypes of Siclian sweet cherry cultivars. J. Hort. Sci. Biotechnol. 82:41-48

Raff JW, Knox RB, Clarke AE (1981). Style antigens of *P. avium* L. Planta 153:125-129.

Roversi A. (1994). Effective pollination period for sweet cherry. Rivista di fruit e di Ortoflor 56(6):53-55.

Schuster M, Flachowsky H, Kohlar D (2007). *Prunus avium* L. accessions and cultivar of the German Fruit Gene bank and from pravite collections. Plant Bread. 126:533-540.

Sonneveld T, Tobutt KR, Robbins TP (2003). Allele - specific PCR detection of sweet cherry self-incompatibility (S) alleles S_1 to S_{16} using consensus and allele- specific primers. Theoret. Appl. Genet. 107:1059-1070.

Sonneveld T, Robbins TP, Boskovic R, Tobutt KR, (2001). Cloning of six cherry self-incompatibility alleles and development of allele – specific PCR detection. Theoret. Appl. Genet. 102:1046-1055.

Srivastava RP, Singh I (1970). Floral biology, fruit get, Fruit drop and physic-chemical, characters of sweet cherry (Prunus avium L.). Indian J. Agric. Sci. 40:400-420.

Stosser R, Anvari SF (1982). Pollen tube growth and fruit set as inflanced by senescence of stigma, style and ovules. XXIst International Horticultural Congress 1:1138.

Tao R, Yamane H, Sugiura A, Murayama H Sassa H, Mori H (1999b) . Molecular typing of S-alleles through identification, characterization and cDNA cloning for S -RNases in sweet cherry. J. Am. Soc. Hort. Sci. 124:224 -233.

Tehrani G, Brown SK (1992) pollen compatibility and self fertility in sweet cherry. Plant breed. Rev. 9:367-388.

Tobutt KR, Sonneveld TZ, Befeki T, Boskvick R (2004) .Cherry (in) compatibility genotypes – an updated cultivar table. Acta Hort. 663:667-672.

Vasilakakis M, Porlingis IC (1985). Effect of temperate on Pollen germination , pollen tube growth, effective pollination period and fruit set. Hort. Sci. 20:733-735.

Vaughan SP, Boskovick RI, Gisbert-Climent A, Russell K Tobutt KR (2008) .Characterisation of novel S-allele from cherry (Prunus avium L.) .Tree Genet. Genomes 4:531-541.

Westwood MN (1993). Temperate Zone Pomology, physiology and culture, 3rd ed. Timber Press, Portland, Oregon.

17

Development of an efficient protocol for micropropagation of pineapple (*Ananas comosus* L. var. smooth cayenne)

Inuwa Shehu Usman[1], Maimuna Mohammed Abdulmalik[1], Lawan Abdu Sani[2] and Ahmed Nasir Muhammad[3]

[1]Department of Plant Science, Ahmadu Bello Univeristy, Samaru, Zaria, Nigeria.
[2]Department of Plant Science, Bayero Univeristy Kano, Kano, Nigeria.
[3]School of Engeneering Science Technology, Federal Polytechnic, Kazuare, Jigawa state, Nigeria.

The aim of this study was to establish an efficient micropropagation protocol for pineapple. Axillary buds were excised from the crown and inoculated on a liquid basal culture Murashige and Skoog (MS) medium supplemented with sucrose (3%), benzylaminopurine (BA) (2.5 μM) and naphthaleneacetic acid (NAA) (0.62 μM) for shoot induction. Shoot multiplication and elongation was on MS basal medium supplemented with 3% sucrose, 0.8% agar and different concentrations of BA (5, 7.5, 10 and 12.5 μM) and NAA (2, 2.5 and 3 μM). Result showed that MS supplemented with BA (5 μM) and NAA (3 μM) gave the highest number of plantlets of 11.5 and 14.4 and the highest mean plant height at shoot elongation of 5.8 and 7.6 cm, respectively. Regenerated plantlets were hardened on different media. Non-acid washed riverside sand gave the highest recovery rate of 87.4%. Use of riverside sand as substrate for hardening will serve as cost-effective substitute for perlite or vermiculite.

Key words: pineapple, micropropagation, benzylaminopurine (BA), naphthaleneacetic acid (NAA), acclimatization.

INTRODUCTION

Pineapple is the third most important tropical fruit in the world, after the banana and citrus. The fruits are important source of vitamin A and B_1 and contain a protein digesting enzyme bromelain. Pineapple fruits are consumed fresh or processed into canned fruit, juice, or jam. Potentials exist for commercial production and processing of crop in Nigeria and other deveolping tropical countries. Conventionally, pineapple is propagated by the use of slips arising from the stalk below the fruit, suckers originated from leaf axils or leaves, crowns of the fruits or ratoons that arise from underground part of the stems. This conventional method of propagation is slow and allows for transfer of pineapple

requires large volume of planting materials, which are hardly obtained by conventional method of propagation. Micropropagation technique (Plant tissue culture) offers an opportunity for large scale production of uniform pineapple planting material in a relatively short period of time (Escalona et al., 1999; Ika and Mariska, 2003). Rapid multiplication of pineapple through axillary buds culture was reported (Dal Vesco et al., 2001; Be and Debergh, 2006; Danso et al., 2008; Zuraida et al., 2011; Yapo et al., 2011). However, low multiplication rate and poor survival rates during acclimatization have been identified as some of the problems affecting the micropropagation technique (Escalona et al., 1999)

Table 1. Optimizing *in vitro* shoot proliferation in pineapple using different combinations of Benzyl-6-aminopurine (BA) and Napthaleneacetic acid (NAA).

Parameter	Crop vigour score (1-5) at shoot elongation	Mean plantlet height at shoot elongation (cm)	Mean plantlet height at shoot proliferation (cm)	Number of plantlets at multiplication
BA (µM)				
5.00	4.0[a]	5.8[a]	2.2[b]	11.5[a]
7.50	3.7[b]	5.8[a]	2. 9[a]	11.1[a]
10.00	3.7 [b]	7.0[a]	2.5[b]	11.0[a]
12.50	2.5[c]	5.0[a]	1.7[c]	7.8[b]
SE(±)	0.08	0.72	0.09	0.95
CV (%)	6.70	36.7	12.1	27.50
NAA (µM)				
2.00	4.1[a]	5.1[b]	3.1[a]	7.6[b]
2.50	2.3[b]	4.9[b]	1.9[b]	9.1[b]
3.00	4.0[a]	7.6[a]	1.9[b]	14.4[a]
SE (±)	0.07	0.62	0.08	0.82
CV (%)	6.70	36.70	12.10	27.5
BA x NAA	**	*	**	**

Means with the same letter in a column are not significantly different using Lysergic acid diethylamide (LSD) (*significant at $P<0.005$, **significant at $P<0.001$).

In addition, the choice of medium used for tissue culture depends upon the species, cultivar and culture conditions and adjustments in growth medium are determined by experimentation (Usman et al., 2011). Therefore, this study was conducted with the primary objective of establishing an efficient rapid micropropagation protocol for large scale propagation of pineapple.

MATERIALS AND METHODS

The study was carried out in the Biotechnology Laboratory of Department of Plant Science, Ahmadu Bello University Zaria, Nigeria. The plant material of pineapple (*Ananas comosos* L. var. smooth ceyanne) was obtained from a fruit market in Zaria. Crowns of young fruits were taken as a source of explant. The leaves were then removed and the crown was thoroughly washed with detergent under running tap water. The crown was surface sterilized by sequential treatment for 5 min in 70% alcohol, 20 min in 10% NaOCl (commercial bleach) plus 2 to 3 drops of tween 20, rinsed thrice with sterile distilled water, 10 min in 5% NaOCl plus 2 to 3 drops of tween 20 washed three times with sterile distilled water and with occasional stirring. All sterilization work was done under laminar air-flow cabinets.

Axillary buds were excised from the crown and inoculated in test tubes over paper bridges on a liquid basal culture Murashige and Skoog (1962) medium supplemented with sucrose (3%), BA (2.5 µM) and NAA (0.6 µM) for shoot induction. Media was adjusted to pH 5.8 before autoclaving for 15 min at 12°C and 15 psi. At every stage the cultures were kept in the growth room at a temperature of 27 ± 2°C, with light provided by cool white florescent tube lights for a 16 h photoperiod. After four weeks of inoculation, shoots were subcultured in kidney jars on MS basal medium plus 3% sucrose, solidified with 0.8% agar and supplemented with different concentrations of growth regulators; BA (5, 7.5, 10 and 12.5 µM) and NAA (2, 2.5 and 3 µM) for shoot multiplication.

The shoots multiplied from each treatment were then separated and rooted on hormone free MS basal medium. Subsequently, the rooted plantlets were acclimatized in the screen house during which they were grown on different media that is, acid washed river side sand and non acid river side sand to evaluate hardening response. Data were recorded on percentage of plantlets showing transplanting shock sign at 1 week after transplanting (WAT) and fully recovered plantlets at 4 WAT. The experimental design used was completely randomized design replicated three times. For the multiplication study a 4 x 3 factorial experiment was used. Data collected were subjected to analysis of variance and means were separated using Lysergic acid diethylamide (LSD) (SAS Institute, 1988).

RESULTS AND DISCUSSION

Experiments using phytohormones has for long established the importance of relative ratio of cytokinin and auxin in plant development. Regeneration of shoots and roots from tissues and cells culture can be induced by increasing or decreasing the relative cytokinin-to-auxin ratio in the culture mediun. Since crop species have different response to treatment with exogenous phytohormones, determintion of optimum cytokinin-to-auxin ratio that will induce maximum level of morphogenesis is essentail in the establishment of efficient micropropagation system. The effects of different concentrations of benzylaminopurine (BA) and naphthaleneacetic acid (NAA) on *in vitro* morphogenesis and plantlets development in pineapple are presented on Table 1. The presence of exogenous BA and NAA in culture medium was found to promote morphogenesis and plantlets development. Supplementing Murashige and Skoog (1962)

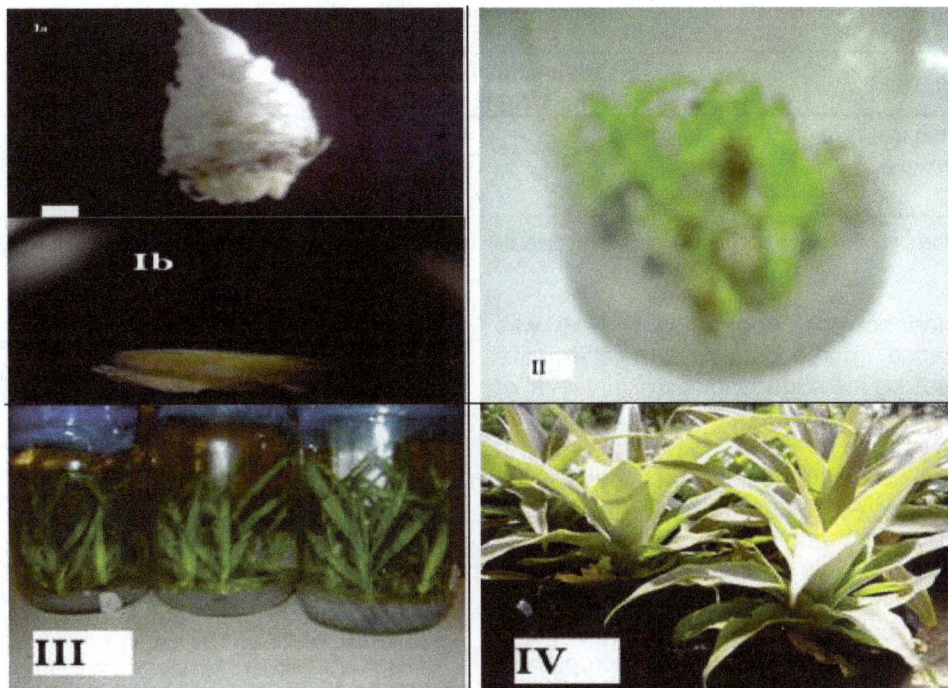

Figure 1. I to IV; I: IA = pineapple crown and IB = excised bud, II: multiplication on ms medium portified with 5.0 µm BA, III: plantlets elongation on ms supplented with 5.0 µm naa, IV: acclimatized pineapple seedlings ready for transplanting.

with 5.0 µM BA proved to provide the best condition for morhpogenesis. MS portified with 5.0 µM BA provided the highest number of plantlets (11.5) compared to other concentrations used in this study.

Our observation is consistent with the optimum range (4.0 to 5.0 µM BA) earlier reported for pineapple (Dal Vesco et al., 2001; Danso et al., 2008; Zuraida et al., 2011; Yapo et al., 2011). Further increase in BA concentration from 5.0 µM BA resulted in a progressive decline in the number of plantlets. However, the decline become statistically significant only when the concentration of BA was increased to 12.5 µM. The effect of cytokinin in apical dominance is antagonistic to that of auxin. While auxin are known to promote apical dominanace by suppressing the activity of auxiliary buds, elevated levels of cytokinin in the auxiliary buds were reported to release them from apical dominance (Shimizu-Sato et al., 2009). Increase in the ratio of BA-to-auxin in the medium portified with BA correlated with prolific induction of shoots. This could be as the result of direct absorption and subsquent accumulation of BA in the auxiliary buds of the cultured plantlets. Over production of cytokinin in nicotiana and cucumber tissues transformed with 35S-ipt resulted in development of stunted plantlets and proliferation of undifferenciated cells (Smigocki and Owens, 1989). It could therefore be possible that 5.0 µM of BA is plateau for *in vitro* proliferation in pineapple and any increase in BA concentration will result in the decline in proliferation of

shoots until it reachs a point at which the effect become detrimental to the cell's ability to differenciate.

The ability of different concentrations levels (2.0, 2.5 and 3.0 µM) of NAA to induce plantlets proliferation *in vitro* was also evaluated in this study. Application of exogenous NAA was observed to significantly influenced *in vitro* morphogenesis in pineapple (Figure 1). The highest plantlets number (14.4) was obtained when MS was fortified with 3.00 µM NAA. Similar observation was also reported by Dal Vesco et al. (2001). Reduction in the concentration of NAA to 2.5 µM resulted in significant decrease in the number of plantlets. Plantlets height at elongation was also influenced by NAA concentration. Supplementing MS with 3.0 µM NAA gave the highest plantlets height (7.6 cm) at elongation phase. There was a significant decrease in the plantlets height with decrease in the concentration of NAA. Auxins are generally known for thier ability to induce rhizogenesis at low concentration and callogenesis at rised concentration.

However, NAA has been widely used to induce direct shoots formation and embryogenesis under *in vitro* condition. The ability of NAA to induce *in vitro* morhpogenesis has been reported in pineapple (Dal Vesco et al., 2001; Danso et al., 2008; Zuraida et al., 2011; Al-Saif et al., 2011; Yapo et al., 2011) and other species (Garcia et al., 2007; Ebrahimie et al., 2007; Usman et al., 2011; Sani et al., 2012).

The effect of exogenous NAA on *in vitro* morphogenesis

Table 2. Hardening response of pineapple plantlets grown on different media.

Treatment	Plantlets showing transplanting shock signs at 1 WAT (%)	Fully recovered plantlets (%) at 4 WAT
Acid washed river side sand	14.7	15.1
Non-acid washed river side sand	11.3	87.4
LSD (5%)	7.31	22.47

Means with the same letter in a column are not significantly different using Lysergic acid diethylamide (LSD) at 5% probability level.

seems to be an indirect one. Application of NAA was reported to influence morphogenesis via alternative pathways, by increasing the endogenous ctokinin level or increasing the levels of endogenous NAA (Ebrahimie at el., 2006). Both cytokinin and NAA were reported to actively participate in determining the pattern of differenciation in plants. Auxin are known for their ability to induce cell elongation and could be the possible reason for efficient plantlets elongation in the NAA portified media used in this study. The BA and NAA interraction was significant for both Number of plantlet at multiplication, plant hieght at elongation and crop vigor (factor of nuber of leaves and general greeness of the plantlets) during elongation phase. Interaction however, was not significant for plantlets height during proliferation (Plates I to IV).

The response to hardening of pineapple plantlets was also evaluated on two different harening media (Table 2). No significant difference was observed among the media in terms of plantlets showing transplanting shock signs at 1 wk after transplanting (1 WAT). However, non-acid washed river side sand significantly produced more fully recovered plantlets (87.4%) at 4 wks after transplanting (4 WAT). The poor recovery observed by the acid-washed river sand could be attributed to the pH level of the soil.

Conclusion

The result obtained in this study suggests that an efficient micropropagation protocol has been achieved. BA (5 μM) and NAA (3 μM) could be use for plantlets multiplication, shoot elongation and good crop vigor. While hardening of plantlets during acclimatization can be, done using non-acid washed riverside sand. The relatively very high recovery rate of 87.4% obtained with riverside sand is conceived as a cost saving measure that could substitute for the more expensive and difficult to source substrates like vermiculite, perlite and ghyphy under our laboratory conditions.

REFERENCES

Al-Saif Adel MA, Sharif Hossain BM, Rosna MT (2011). Effects of benzylaminopurine and naphthalene acetic acid on proliferation and shoot growth of pineapple (Ananas comosus L. Merr) in vitro. Afr. J. Biotech. 10(27):5291-5295.
Be LV, Debergh PC (2006). Potential low-cost micropropagation of pineapple (Ananas comosus). South Afr. J. Bot. 72:191-194.
Dal Vesco LL, Pinto AA, Zaffari GR, Nodari RO, Sedrez dos Reis M, Guerra MP (2001). Improving Pineapple Micropropagation Protocol through Explant Size and Medium Composition Manipulation. Fruits 56:143-154.
Danso KE, Ayeh KO, Oduro V, Amiteye S, Amoatey HM (2008). Effect of 6-Benzylaminopurine and Naphthalene Acetic Acid on in vitro Production of MD2 Pineapple Planting Materials. World Appl. Sci. J. 3(4):614-619.
Ebrahimie E, Mohammad RN, Abdolhadi H, Mohammad RB, Mohammadie-Dehcheshmeh M, Ahmad S, German S (2007). Induction and Comparison of Different in vitro Morphogenesis Pathways using Embryo of Cumin (Cuminum cyminum L.) as a Model Material. Plant Cell. Tiss. Organ. Cult. 90:293-311.
Ebrahimie E, Habashi AA, Mohammadie-Dehcheshmeh M, Ghannadha M, Ghareyazie B, Yazdi-Amadi (2006). Direct shoot Regeneration from Mature Embryo as a Rapid and Genotype-independent Pathway in Tissue Culture of Heterogeneous Diverse sets of Cumin (Cuminum cyminum L.) Genotypes. In vitro Cell Dev. Biol-Plant. 42:455-460.
Escalona ML, Gonzalez JC, Daquinta BM, González JL, Desjardins Y, Borroto CG (1999). Pineapple (Ananas comosus L. Merr) Microprogapation in Temporary Immersion Systems. Plant Cell.Reports 18:743-748.
Garcia RDC, Castellar A, Andrea L, Claudia M, Catia CM, Mansur E (2007). In vitro Morphogenesis Patterns from Shoot Apices of Sugarcane are Determined by Light and Type of Growth Regulator. Plant Cell. Tiss Organ Cult. 90:181-190.
Ika RT, Mariska I (2003). In vitro Culture of Pineapple by Organogenesis and Somatic Embryogenesis: Its Utilization and Prospect. Buletin AgroBio. 6(1):34-40.
Murashige T, Skoog F (1962). A Revised Medium for Rapid Growth and Bioassays with Tobacco Tissue Cultures. Physiol. Plant. 9:8-11.
Sani LA, Mustapha Y, Usman IS (2012). In Vitro Regeneration of Commercial Sugarcane (Saccharum spp.) Cultivars in Nigeria. J. Life Sci. 6:454-459.
Shimizu-Sato S, Tanaka M, Hitoshi M (2009). Auxin-Cytikinin Interaction in the Control of Shoot Branching. Plant. Mole. Biol. 69:429-435.
Smigocki AC, Owens LD (1989). Cytikinin-Auxin Ratio and Morphology of Shoots and Tissue Transformed by Chemeric Isopentenyl Transferase Gene. Plant. Physio. 91:808-811.
SAS Institute (1988). SSA/STAT Users' Guide, Release 6.03 Eds SAS Inst. Carry, NC.
Usman IS, Ado SG, Ng SY (2011). Media appraisal for somatic embryo genesis of elite inbred lines of maize. J. Life Sci. 5(37):360-363.
Yapo ES, Tanoh HK, Mongomaké K, Justin YK, Patrice K, Jean-Michel M (2011). Regeneration of Pineapple (Ananas comosus L.) Plant through Somatic Embryogenesis. J. Plant Biochem. Biotech. 20(2):196-204.
Zuraida AR, Nurul Shahnadz AH, Harteeni A, Roowi S, Che Radziah CMZ, Sreeramanan S (2011). A novel Approach for Rapid Micropropagation of Maspine Pineapple (Ananas comosus L.) Shoots Using Liquid Shake Culture System. Afri. J. Biotech. 10(19):3859-3866.

18

Monitoring population density and fluctuations of *Xyleborus dispar* and *Xyleborinus saxesenii* (Coleoptera: Scolytıdae) with red winged sticky traps in hazelnut orchards

Islam Saruhan[1] and Hüseyin Akyol [2]

[1]Department of Plant Protection, Faculty of Agriculture, Ondokuz Mayis University, Samsun, Turkey.
[2]Bleak Sea Agricultural Research Institute Samsun, Turkey.

Bark and ambrosia beetles (Coleoptera: Curculionidae:Scolytinae) include many important pest species of forest and fruit trees. They usually prefer the physiologically stressed trees for colonization but also it is known that they attack healthy trees. Bark and ambrosia beetles are consisted of two main ecological groups, bark beetles grow in bark and ambrosia beetles in sapwood. Especially ambrosia beetles are very detrimental in Turkish hazelnut orchards. This study was carried out between 2005 and 2007 to monitor populations of *Xyleborus dispar* and *Xyleborinus saxesenii* (Col.: Scolytidae), causing considerable damages in hazelnut (*Corylus avellana* L.) orchards in Ordu and Samsun Provinces. The populations of the bark beetles were monitored using sticky traps with red wings that are registered and used to capture these pests. Ethyl alcohol (96%) was used as the attractant in the traps. The results of the three-year study indicated that both pest species emerged in different times in Ordu and Samsun. *X. dispar* emerged in large numbers in springs (March - May) as overwintered adults; however, *X. saxesenii* emerged in large numbers in summers (June - August). Population density is usually the *X. dispar* was found to be more. *X. dispar* on 05.15.2007 (398 adult/trap), *X. saxesenii* on.09.01.2005 (383 adult/trap) the highest catch was recorded in Samsun province.

Key words: Hazelnut, monitoring population, *Xyleborus dispar*, *Xyleborinus saxesenii*, red winged sticky traps.

INTRODUCTION

Hazelnut is one of the most important agricultural products of Turkey; it is cultivated over approximately 550 thousand hectares (84% of global production area), producing 500 thousand tonnes (69% of global production), of which approximately 300 thousand tones are exported (Yavuz, 2007).

Approximately 150 insect species have been detected in hazelnut orchards. However, only 10 to 15 of these species result in economic losses, varying between years

and the region of hazelnut (Işık et al., 1987). According to various studies conduced in Turkey, Hazelnut weevil (*Balaninus nucum* L.) is the most significant hazelnut pest (Işık et al., 1987; Ecevit et al., 1995; Tuncer and Ecevit 1996 a, b; Saruhan and Tuncer, 2001; Tuncer et al., 2002). Bark beetles (*Scolytidae*) comprise another pest group of hazelnut (Ak et al., 2005a, b, c). These pests are a risk for stone or pome fruits, kiwi and forests, and have recently been shown to harm hazelnut orchards. Serious

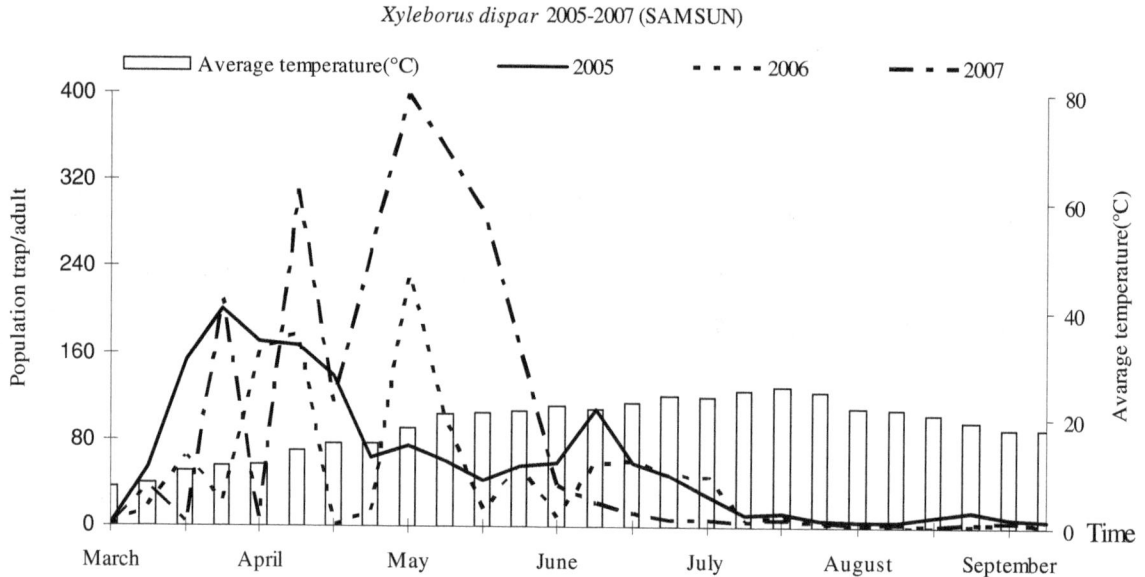

Figure 1. Population density and fluctuation of *X. dispar* in Samsun province (2005-2007).

damage of these species were observed on hazelnut plants in low and middle altitude of region in last years (Mani et al., 1990; Raulder, 2003; Kaya, 2004; Ak et al., 2006b, 2010). While other pests directly or indirectly affect the quality and yield in hazelnut orchards, bark beetles (*Scolytidae*) cause product losses by draining young or old hazelnut branches. Additionally, as these pests spend most of their lives in the woody tissue of their host, they are extremely difficult to eradicate. Therefore, chemical control must be supported by cultural and biological control.

Controlling bark beetles depends on the emergence time of the adult females. Therefore, Red Winged Sticky Traps (ethyl alcohol baited) are used to detect their emergence and achieve mass capture.

Population monitoring of pest species is of great importance in determining the emergence time of adult individuals in order to control them. Population monitoring enabled the determination of emergence times and population fluctuations of pest species in hazelnut orchards.

Due to the high population of pests in hazelnut orchards and the importance of determining the emergence time of adults in controlling them, this study examined population fluctuation of *X. dispar* and *X. saxesenii*. Both species are important in terms of both presence and density in hazelnut orchards in Turkey. They were monitored at different locations in two provinces (Samsun and Ordu) for three years and the emergence time of adults was determined.

MATERIALS AND METHODS

The main materials of the study consist of hazelnut orchards,

Scolytidae (bark beetles) species (*X. dispar* and *X. saxesenii*), Red Winged Sticky Traps and 96% ethyl alcohol as the attractant.

The trap used in the study consists of four red-colored sticky plates as a wing and a 1-liter plastic bottle hung just below them. Each wing of the sticky trap has an area of 148.9 cm^2 (14.6 cm height and 10.2 cm width). The total area of the sticky part of the trap is 0.12 m^2. The plastic bottle has four holes to enable the alcohol to evaporate.

Population monitoring of *X. dispar* and *X. saxesenii* used red-winged sticky traps licensed for use against bark beetles, which were located in hazelnut orchards of Samsun (Terme) and Ordu (Central) provinces between 2005 and 2007. Three traps were hang at each location used in the study. Traps were placed 1.5 m above the ground and spaced 20 m from each other. Population fluctuations of *X. dispar* and *X. saxesenii* were monitored in Samsun province (41° 12' 37" N - 36° 59' 32" W) and Ordu province (40° 58' 48" N - 37° 55' 44" W) in 2005 to 2007 (Figures 1 to 4). The numbers of trapped *X. dispar* and *X. saxesenii* were monitored weekly between March and October; traps were cleaned after each count and this process continued throughout the year. In the first year, traps were hang in Samsun location on 03.15.2005, and removed on 10.25.2005. In Ordu location, they were hang on 03.31.2005 and removed on 10.20.2005. In the second year, they were hang in Samsun location on 03.15.2006 and gathered on 10.26.2006, and in Ordu location they were hang on 03.16.2006 and gathered on 10.27.2006. In the last year of the study, the traps were hang in Samsun location on 03.21.2007 and gathered on 11.01.2007. In Ordu location, they were hung on 03.23.2007 and gathered on 10.15.2007.

RESULTS

Population density and fluctuation of *X. dispar* in Samsun Province (2005-2007)

The first *X. dispar* individuals were trapped during the last week of March (03.25.2005) in Samsun in 2005. The highest catch was recorded in the third week of April

Xyleborus dispar 2005-2007 (ORDU)

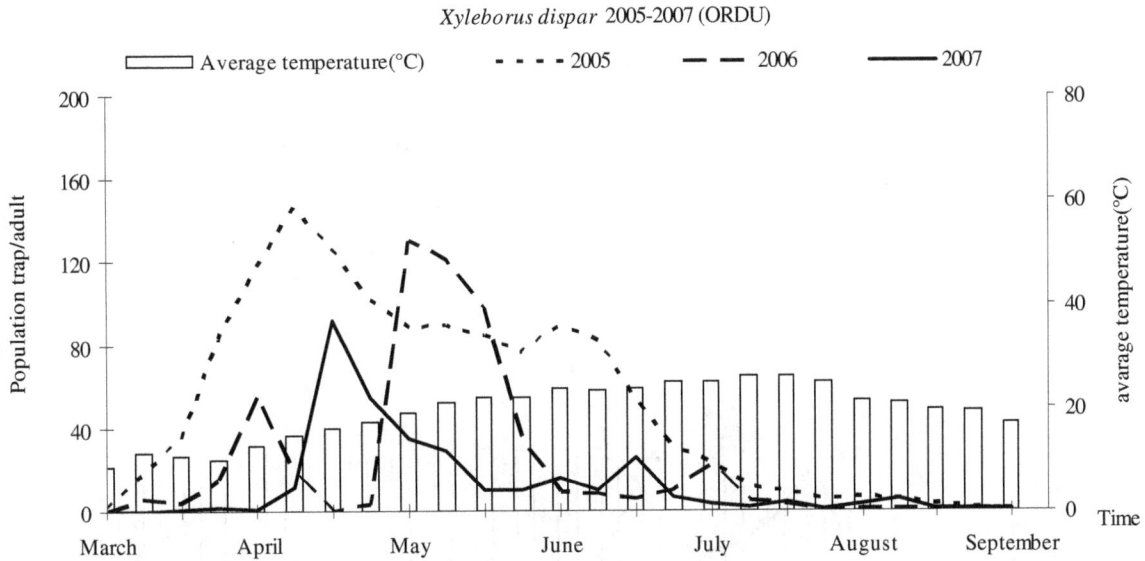

Figure 2. Population density and fluctuation of *X. dispar* in ordu province (2005-2007).

Xyleborinus saxesenii 2005-2007 (SAMSUN)

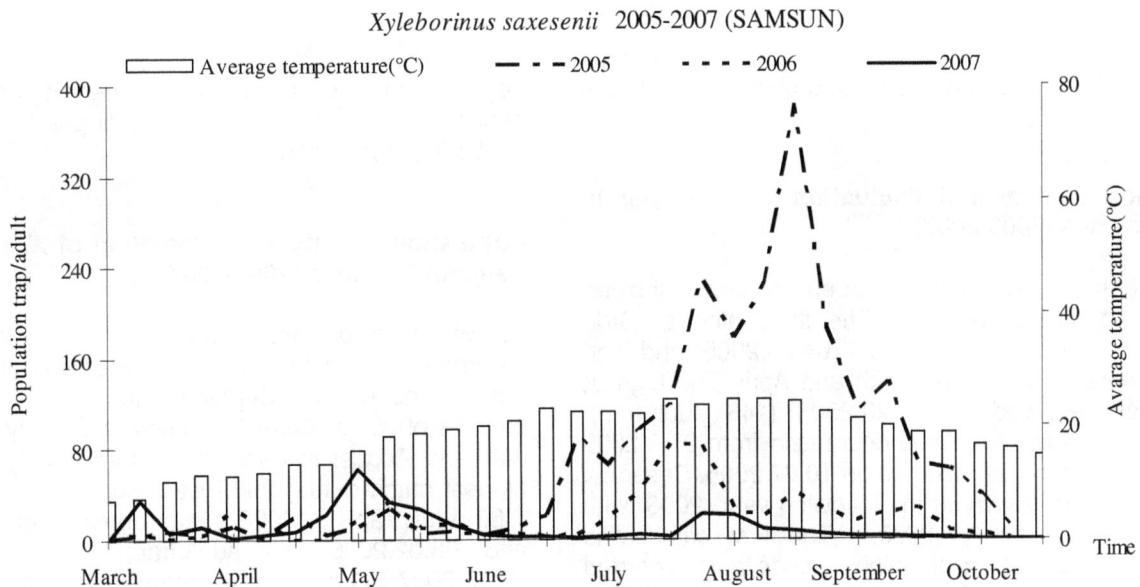

Figure 3. Population density and fluctuation of *X. saxesenii* in Samsun province (2005-2007).

(04.20.2005) (201 adult/trap). The number of individuals caught in traps decreased from this date, but an increase was observed again at the end of June and the beginning of July. The last adult individuals were trapped on 10.25.2005. In 2006, similarly to the previous year, the first catch was observed in the third week of March (03.21.2006). High numbers of individuals were trapped during May and June. The highest catch was recorded on 05.18.2006 (232 adult/trap). Catches declined from 07.13.2006 and the last catch was observed on 10.26.2006. In 2007, the first catch occurred in the last

week of March (03.27.2007). High catches were observed during April. The highest catch was recorded on 05.15.2007 (398 adult/trap). Catches declined from 06.12.2007 and the last catch was recorded on 10.02.2007. High catches were observed during April. The highest catch was recorded on 15.05.2007 (398 adult/trap) (Figure 1). Data from Samsun Province for the three study years reveals that pests generally begin to emerge by the end of March, depending on the season, and reach their highest density emergence in April. Therefore, biological control for the pest should be

Xyleborinus saxesenii 2005-2007 (ORDU)

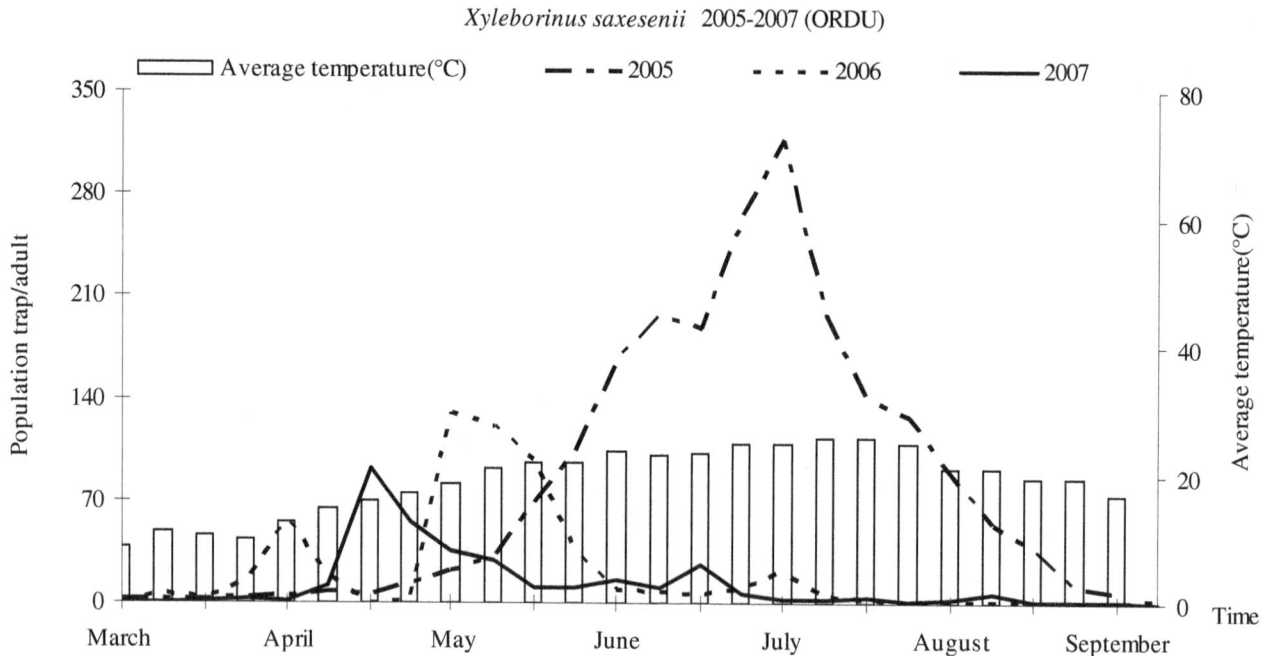

Figure 4. Population density and fluctuation of *X. saxesenii* in Ordu Province (2005-2007).

commenced at the end of March and chemical control in the first week of April.

Population density and fluctuation of *X. dispar* in Ordu Province (2005-2007)

The pattern for *X. dispar* trapped in Ordu showed similarity to that of Samsun. The first catch at Ordu occurred in the first week of April (04.08.2005) and high catches were observed in March and April. The highest catch was recorded on 05.20.2005 (146 adult/trap). Catches of *X. dispar* at Ordu decreased from 07.9.2005 and the last catch was recorded on 10.07.2005. Seasonal differences were observed the following year, 2006, with earlier catches than in 2005. The first catch was observed on 03.23.2006 and high catches were recorded in April and May 2006, as in the previous year. The highest catch of 2006 was recorded on 05.29.2006 (130 adult/trap). From 07.04.2006, catches decreased and the last catch was observed on 10.27.2006. In Ordu during 2007, *X. dispar* was first trapped on 04.04.2007 and high catches were observed between 05.08.2007 and 07.12.2007. The highest catch was recorded on 05.16.2007 (92 adult/trap). The number of individuals trapped decreased from 07.20.2007 and ended on 10.05.2007 (Figure 2). The data indicate that the emergence of *X. dispar* depends on seasonal factors and that traps should begin to be used one week later than those at the Samsun site, and that chemical control should be started by the first week of April, as in Samsun province site. Both provinces showed a population decline at the beginning of May and an increase at the

beginning of June. Therefore, commencing chemical pest control by the first week of June is of great importance in controlling populations.

Population density and fluctuation of *X. saxesenii* in Samsun Province (2005-2007)

As with *X. dispar*, individuals of *X. saxesenii* were first recorded in traps in the last week of March and lasted until the last week of September in Samsun. The highest catches of *X. saxesenii* in Samsun generally occurred in July and August in each of the three study years. The highest catch during 2005 was recorded on 09.01.2005 (383 adult/trap); in 2006 it was detected on 07.26.2006 and 08.02.2006 (83 adult/trap), and in 2007 on 05.15.2007 (63 adult /trap). Even though the emergence of *X. saxesenii* in the summer season was detected in the first week of May 2007 and the first week of June during 2005 and 2006, the dense emergence of *X. saxesenii* in summer season was determined as being between the end of June and beginning of August. In 2007, adult emergence in May- June might result from the low population and high temperature since May (Figure 3). According to the data acquired, management of *X. saxesenii* should commence at the beginning of July in Samsun province, unlike *X. dispar*.

Population density and fluctuation of *X. saxesenii* in Ordu Province (2005-2007)

In 2005, catches of *X. saxesenii* in Ordu were recorded

between 03.31.2005 and 10.20.2005. The highest catch was recorded on 08.05.2005 (317 adult/trap). And also high catches were observed in the dates between 07.29.2005 and 08.19.2005. In 2006, catches were recorded between 03.16.2006 and 10.27.2006. High catches were recorded between 07.20.2006 and 09.22.2006. The highest catch was recorded on 08.04.2006 (98 adult/trap). In 2007, the first and last catches were recorded on 03.23.2007 and 10.15.2007, respectively. The highest caught was seen on 07.20.2007 (58 adult/trap), and the time period with high catches was between 07.06.2007 and 07.27.2007 (Figure 4).

DISCUSSION

In both locations in all three study years, the emergence of adult *X. dispar* was determined to be between March and June, depending on temperature. The emergences during these periods were found to be non-continuous and occurred as a result of nestling of adult in different periods.

Adult *X. saxesenii* were determined to emerge densely in the summer period between the end of June and August at both locations. Additionally, adult individuals nestling in the spring period were found in traps.

According to population monitoring over the three years in Samsun and Ordu provinces, adult individuals of both pests species emerged at different periods. The population progress of *X. dispar and X. saxesenii* were similar to each other at both locations.

As a result of population monitoring in Samsun and Ordu provinces, the dense emergence of adult *X. dispar* was found to occur in spring and adult *X. saxesenii* emerged in summer. Apart from these dense emergences, each species was found to make instantaneous (non-continuous) emergences. These findings support those reported by Ak (2004) and Ak et al. (2005a, b, c, 2006a).

It was found that *X. dispar* had higher population densities than *X. saxesenii* in both locations during 2006 and 2007 but not 2005.

Population monitoring of *X. dispar* showed that adult emergence increased during the spring season when the temperature was approximately 18 to 20°C. Similarly, a study by Kaya (2004) of mixed fruit trees (apple, pear, plum, peach etc.) found that, from 1997 to 1999, the first emergence of adult *X. dispar* was on May 8th, April 26th, and May 6th, respectively. Schultz et al. (2002), using ethyl alcohol baited Lindgren and Japanese traps, reported that the first emergence of adult *Xyleborus crassiusculus* was at the end of March and the beginning of April. Mani et al. (1990), using ethanol traps, stated that *X. dispar* was caught in spring when the temperature was 20°C and that catches lasted for 3 to 4 weeks. In a study using ethyl alcohol funnel traps in fruit orchards in Canada (Creston), White (1992) reported that *X. dispar* emerged at the end of March and beginning of April and

had two peaks in April and June. Ciglar and Boric (1998) stated that 98% ethyl alcohol diluted 1:1 with water could be used as baiter in winged traps; *X. dispar* emerged in spring when the temperature was 20°C and the emergence lasted from the last week of April to mid June. The results of population monitoring of *X. saxesenii* showed the emergence of adults in the spring season when the temperate reached 18 to 20°C, as in *X. dispar,* and the dense emergence was detected in June- August. Similarly, a study by Markalas and Kalapanida (1997) examined flight models of some *Scolytidae* using an ethyl alcohol baited slot tarp in an oak forest in Greece between 1992 to 1993; it was reported that *X. dispar* emerged in high numbers between March and June, and *X. saxesenii* from the end of April to the end of August. In a study of flight dynamics using alcohol traps, Raulder (2003) found that *X. saxesenii* and *X. dispar* began to fly in spring (end of March or the first week of April) when the daily temperature was 18°C and above, lasting until autumn; and that the period of dense flight occurred at the end of April to mid June. A survey of *Scolytidae* in Oregon forest (Cramer, 2005) observed the first emergence of *X. saxesenii* in mid February when the temperature reached 18°C and lasted until the end of autumn, with the period of highest emergence being from the beginning to the end of June.

X. dispar was found to have higher population density than *X. saxesenii* and the emergence of its nestled adults was observed in spring (March - April - May). This result highlights the importance of determining the appropriate period to commence the control of especially *X. dispar.* In this period, the control (mechanical, biotechnical and chemical) of nestled adults can give effective results. During the summer period, *X. dispar* showed low emergence from the end of June to mid August. *X. saxesenii* was found to emerge in low density in spring, depending on temperature, with dense emergence observed in the summer season (June - August). Therefore, control of this species should be conducted during this period.

REFERENCES

Ak K (2004). Giresun, Ordu ve Samsun İllerinde Fındık Bahçelerinde Zarar Yapan Yazıcıböcek (Coleoptera: Scolytidae)Türlerinin Tespiti ve Kitlesel Yakalama Yöntemi Üzerinde Araştırmalar. Selçuk Üniversitesi Fen Bilimleri Enstitüsü (Basılmamış) Doktora tezi, Konya. P. 92.

AK K, Uysal M, Tuncer C (2005a). Giresun, Ordu ve Samsun illerinde fındık bahçelerinde zarar yapan yazıcıböcek (Coleoptera: Scolytidae) türleri, kısa biyolojileri ve bulunuş oranları. Ondokuz Mayıs Üniversitesi Ziraat Fakültesi Dergisi. 20(2):37-44.

AK K, Uysal M, Tuncer C (2005b). Giresun, Ordu ve Samsun illerinde fındık bahçelerinde zarar yapan yazıcıböceklerin (Coleoptera: Scolytidae) zarar seviyeleri. Gaziosmanpaşa Üniversitesi Ziraat Fakültesi Dergisi. 22(1):9-14.

AK K, Uysal M, Tuncer C, Akyol H (2005c). Orta ve Doğu Karadeniz Bölgesinde fındıklarda zararlı önemli yazıcıböcek (Colepotera: Scolytidae) türleri ve çözüm önerileri. Selçuk Üniversitesi Ziraat Fakültesi Dergisi. 19(37):37-39.

AK K, Uysal M, Tuncer C (2006a). Yazıcı Böceklerin Samsun ili fındık bahçelerindeki populasyon değişimi ve kitle yakalama yöntemi üzerinde araştırmalar. Selçuk Üniversitesi Ziraat Fakültesi Dergisi. 20(39):15-22.

AK K, Uysal M, Tuncer C (2006b). "Karadeniz Bölgesinde kivilerde zararlı yazıcıböcek (Coleoptera: Scolytidae) türleri ve mücadelesi, 365-370". Gaziosmanpaşa Üniversitesi, Ziraat Fakültesi, II. Ulusal Kivi ve Üzümsü Meyveler Sempozyumu (14-16 Eylül 2006, Tokat) Bildirileri. P. 380.

AK K, Güçlü Ş, Tuncer C (2010). Kivide yeni bir meyve zararlısı: *Lymantor coryli* (Perris, 1853) (Coleoptera: Scolytidae). Türkiye Ento. Dergisi. 34(3):391-397.

Ciglar I, Boric B (1998). Bark beetle (Scolytidae) in Croation orchards. Acta Hortic. 525:299-305.

Cramer EE (2005). A survey of three key species of Ambrosıa Beetles (Coleoptera: Curculionidae: Scolytinae: *Monarthrum scutellare*, *Xyleborus dispar*, *Xyleborinus saxesenii*) in Oregon's Wıllamette Walley Nursery Industry. A Project, Oregon State University, Bioresource research and University Honors College. P. 27.

Ecevit O, Tuncer C, Hatat G (1995). Karadeniz Bölgesi bitki sağlığı problemleri ve çözüm yolları. O. M. Ü., Ziraat Fakültesi Dergisi. 10(3):191-206.

Işık M, Ecevit O, Kurt MA, Yücetin T (1987). Doğu Karadeniz Bölgesi fındık bahçelerinde Entegre Savaş olanakları üzerinde araştırmalar. Ondokuz Mayıs Üniversitesi Yayınları. No: 20, Samsun. P. 95.

Kaya M (2004). Bursa ilinde değişik meyve ağaçlarında *Xyleborus dispar* (F.) (Coleoptera: Scolytidae)'ın ergin populasyon değişimi üzerinde araştırmalar. Yüzüncü Yıl Üniversitesi, Zir. Fak., Tarım Bilimleri Dergisi. 14(2):113-117s.

Mani E, Remund U, Schwaller F (1990). Der Ungleiche Holzbohrer, *Xyleboryus dispar* F. (Coleoptera: Scolytidae) im Obst und Weinbau. Landwirtschaft Schweiz Band. 3(3):105-112.

Markalas S, Kalapanida M (1997). Flight pattern of some Scolytidae attracted to flight barier traps baited with ethanol in an oak in Greece. Anz. Schadlingskde., Pflanzenschutz, Umweltschutz. 70:55-57.

Raulder H (2003). Observation on the flight dynamics of Bark Beetle (*Xyleboruus saxeseni* and *Xyleboruus dispar*). Gesunde Pflanzen. 55(3):53-61.

Saruhan İ, Tuncer C (2001). "Population densities and seasonal fluctiations of Hazelnut pests in Samsun, Turkey. 419-429". Proc. V. Int. Congress on Hazelnut. Acta. Hortic. P. 556.

Schultz PB, Dills MS, Whitaker CS (2002). Managing Asian Ambrosia Beetle in Virginia. Sna Res. Conf. 47:167-169.

Tuncer C, Ecevit O (1996a). "Fındık Zararlıları ile mücadelede entegre model tasarımı, 40-53". Fındık ve Diğer Sert Kabuklu Meyveler Sempozyumu (10-11 Ocak 1996, Samsun) Bildirileri. P. 419s.

Tuncer C, Ecevit O (1996b). "Samsun ili fındık üretim alanlarındaki zararlılarla savaşım faaliyetlerinin mevcut durumu üzerinde bir araştırma, 286-292". Fındık ve Diğer Sert Kabuklu Meyveler Sempozyumu (10-11 Ocak 1996, Samsun) Bildirileri P. 419s.

Tuncer C, Saruhan İ, Akça İ (2002). Karadeniz Bölgesi fındık üretim alanlarındaki önemli zararlılar. Eko-kalite. Samsun Ticaret Borsası Yayın organı yıl:2, Sayı 2:43-54.

White KJ (1992). Scolytid pests in fruit tree orchards. Simon Fraser University. Master Thesis. P. 52.

Yavuz GG (2007). Fındık. Tarımsal Ekonomi Araştırma Enstitüsü-BAKIŞ. Sayı: 9, Nüsha: 8

19

The role of avocado production in coffee based farming systems of South Western Ethiopia: The case of Jimma zone

Berhanu Megerssa

Department of Agricultural Economics, Rural Development and Value Chain Management, Jimma University College of Agriculture and Veterinary Medicine, P. O. Box 307, Jimma, Ethiopia.

Empirical data collected from smallholder farmers (N = 99) residing in three Woredas of Jimma zone, was analyzed to investigate production issues pertaining to avocados, and the pros and cons of producing the crop was investigated. To address these tasks, field and desk research was accompanied by interviews and discussions with focus groups. Despite the crop's potential, data suggest that avocados have not yet been fully exploited in Jimma zone where the crop remains an elemental component of farming systems. Results highlighted a shortage of large commercial processors, weak financial institutions and monetary support, and a lack of enabling infrastructure, especially such as research, and extension to support farmers, and production and marketing chains. Despite large production volumes, the potential of avocado production is not being realized because of these identified shortcomings. Consequently, intervention points and socio-economic constraints are outlined in an attempt to improve avocado production capabilities in Jimma zone.

Key words: Avocado, Jimma, production, smallholder farmers.

INTRODUCTION

Avocado (*Persea americana* Miller) is native to Mexico. Because of its high calorific value, the fruit is proclaimed as the Globe's healthiest fruit (Guinness Book of Records, 2010) and the crop brings considerable net return per acre when compared to staple crops (FAO, 2005). In addition to its high nutritive values, avocados can also be used as shade trees, windbreaks, posts, and ornamentals (Albertin and Nair, 2004). Large plantations may play an important role in carbon storage and sequestration that mitigates environmental pollution (Kirby and Potvin, 2007).

Avocado's global production has now reached more than 3.8 million metric tons (FAOSTAT, 2010). These days, the crop is produced in several countries where

Ethiopia stands the 10th leading producer and 6th most important consumer in the world (FAOSTAT, 2010).

Avocado was first introduced to Ethiopia in 1938 by private orchardists in Hirna and Wondo-genet and production gradually spread into the countryside where the crop was adapted to different agro-ecologies (Edossa, 1997; Woyessa and Berhanu, 2010; Zekarias, 2010). Avocados are second in total volume of production, next to banana, in Ethiopia (Joosten, 2007). Annual avocado production in Ethiopia is 80,000 tons. The crop is now produced by more than half a million farmers countrywide who collectively farm more than 7,000 ha of land (CSA, 2008; FAOSTAT, 2010; Joosten, 2007).

Jimma Agricultural Research Center (JARC) pioneered the introduction of avocados to South-western Ethiopia (Edossa, 1997; Farm Africa and SOS Sahel, 2004; Mohamed et al., 2009; Woyessa and Berhanu, 2010; Zekarias, 2010). According to these sources, the center established the first avocado varietal orchard in 1969 with materials initially sourced from Wondo-genet and Debrezeit. Jimma is the 4th largest avocado producing zone of Ethiopia, after Wolayata, Sidama and Haditya zones. In Jimma zone, many households have relied on avocados as a major source of income (CSA, 2008). Avocados are the principal cash crop in South-western Ethiopia and large numbers of farming households rely on avocados for their livelihood (CSA, 2008; MoARD, 2008).

Despite relatively early establishment, the avocado industry in Ethiopia is in its infancy and has not yet utilized the immense potential of this crop. According to World Bank Group (2006), lack of concerted public support, scanty information, and lack of systematically documented knowledge that is readily accessible are the main constraints hampering the development of this sector. If these hurdles are not overcome, it is obvious that Ethiopia's capacity to produce avocado will not improve. In consideration of these facts, this work sought to identify impediments associated with the value chain of avocados in Jimma zone, South-western Ethiopia.

Objectives of the study

General objectives

1) To evaluate the role of avocado production in the coffee based farming system of South-western Ethiopia, Jimma zone.

Specific objectives

1) To characterize the avocado production system
2) To analyze determinants of market supply

METHODOLOGY

Description of the study area

Jimma zone is the most prominent avocado growing area in South-western Ethiopia, where coffee production prevails in Afromontaine forest remnants. The zone is located between 7° 13' and 8° 56' N latitude, and 35° 52' and 37° 37' E longitude. The zone is characterized by its humid tropical climate with heavy annual rainfall ranging from 1200 to 2000 mm with a temperature range of 25 to 30°C and a minimum temperature of 7°C. Largest part of the zone (62%) is mid-altitude (Jimma Zone Agriculture, 2010).

Sampling technique

This study relied on three stage random sampling techniques. In the first stage, three Woredas were selected randomly from the nine Woredas of the target zone. In the second stage, Kebeles were selected from these three Woredas based on population proportional to size (PPS) to have a total of eight sample Kebeles. In the third stage, similar approach was followed to select respondent households to have a total of 99 farmer respondents. Determination of sample size is resolved by means of Slovin's sampling formula with a 90% confidence level.

$$n = \frac{N}{1 + N (e)^2}$$

Where: n = Sample size for research use, N = total number of household head in eight avocado producing Kebeles, and e = margin of error at 10%.

Focus group discussions (FGD) and key informants

Purposive sampling was employed to collect data from knowledgeable people (elders, youth, and women farmers and responsible persons of different institutions) on avocado production in the three selected Woredas in Jimma zone, and at the terminal market at the capital city, Addis Ababa. Discussions were held to access community level information, so pertinent data was collected to satisfy the assumptions of the underlying analytical techniques (Haggablade and Gamser, 1991; Heisman, 1995). Thus, open FGD were held with three groups based on pre-determined checklists and a total of 27 key informants were interviewed from 12 different institutions. The data generated from discussions was combined with other relevant data collected for this project.

Methods of data collection

The exploratory survey was conducted in February, 2010. During this period, informal discussion was held with farmers, frontline extension personnel, subject-matter specialists, governmental and non-governmental offices to obtain prior-informed consents and to identify villages to participate in surveys. Techniques recommended by Hellion and Meijer (2006) were used, and field surveys were conducted between September and December, 2010. Exploration for pertinent information was continued through key informant and FGD, case studies, and secondary information sources such as official and unofficial reports, statistics, research papers, press clippings, websites, and journals.

Interview schedules were administered by trained researchers from JARC and pre-testing was done to eliminate irrelevant questions and to refine questions to ensure the intended metric was measured. Preliminary data from informal surveys was used to support formal surveys to better understand current situations and to capture insights into the actions and subsequent outcomes generated by participants. Data on agronomic practices, demographic issues, production trends, prices, costs, yields, endowments, employment, remuneration, and direction of trade and physical flows of produce were collected through semi-structured interview schedules, case studies, key informants, and FGD to satisfy pre-determined objectives set at number one and two above.

Method of data analysis

Raw data were coded before analysis in Statistical Package for Social Sciences (SPSS) Version 17 and Microsoft Excel 2010. The SPSS was run to generate tabulated reports, charts; plots of distributions and descriptive statistics. T-test; and Chi-square tests

The role of avocado production in coffee based farming systems of South Western Ethiopia...

121

Table 1. Demographic characteristics and access to services to the farming households.

Indicator	Mana (N = 36)		Goma (N = 35)		Seka-Chokorsa (N = 34)		Total		F-value
	Mean	STD	Mean	STD	Mean	STD	Mean	STD	
Household head age (years)	42.39	16.12	43.16	13.04	38.85	12.54	41.33	14.10	28.56***
Family size (No. of people)	6.22	2.66	6.6	1.5	6.85	2.53	6.54	2.36	27.09***
Experience (years)	11.42	5.22	8.08	2.74	11.21	5.267	10.46	4.89	20.87***
Access to infrastructures									x^2 value
Distance to road (km)	1.66	2.75	1.66	1.82	1.49	1.83	1.60	2.21	7.07***
Distance to DA office (km)	2.38	3.31	2.66	2.14	2.17	1.74	2.38	2.52	9.21***
Distance to market (km)	1.92	4.35	4.36	3.42	3.11	4.38	3.13	4.05	6.95***
Visit to demonstration (frequency in days)	19.17	24.11	15.32	16.60	1.19	2.87	17.25	20.36	20.74*

***, Significant at 1%; **, significant at 5%; *, significant at 10%. Source: Survey result (October, 2010).

were analyzed to identify mean differences among continuous and discrete variables, respectively.

RESULTS AND DISCUSSION

Socio-demographic characteristics of producers

Age of the household head

The mean age of household head was 41.33 years. The results further indicated smallest proportions of the respondents (48% of avocado producers) and are within a range of 18 to 30 years of age. The statistical test for homogeneity, which was run to compare means of continuous variable among Woredas was highly significant (P < 0.01) for age, family size and experience; depicting extreme variations among sample locations. Additionally, the distance to the nearest road and development agents' office, access to and use of land, and distance to market were significantly different between Woredas (P < 0.01) (Table 5). This result suggested that household outcomes were strongly influenced by access to resources and this had a positive effect on the marketable supply of avocado fruit.

Education

With an adult literacy rate of 73%, the study identified a good level of education. This result is twice that of the national average (35.5% literacy rate) and it has important implications for augmenting the volume of production and sales of avocado in the study areas (Table 1).

Indirect transfer of information was channeled through the implementation of programs for primary school children (45%). Consequently, this endeavor influenced parents' awareness of important issues. Consequently,

67.5% of respondents were indirect beneficiaries because they had at least one child in primary school that was exposed to this information.

The result is similar to that of Bezabih and Hadera (2007) who explained that a child's education can influence parents' decisions and direct attitude changes. This approach has the potential to amplify momentum in horticultural production and marketing in Eastern Ethiopia.

Family size

Survey results showed that an increase in family size was directly proportional to allotted productive labor sources for avocado production (Table 5). As a result, lower dependency ratios and larger family sizes positively affected the supply of avocados promoting better participation in markets (Wolday 1994). Bezabih and Hadera (2007) documented that different labor sources are employed in horticulture in Eastern Ethiopia and family laborers account for the majority of labor allotments.

Farming experience: Farming experience of more than 7 years was reported by 85% of respondents, which likely increased the probability of HHH to be better able to participate in production and marketing of avocados in study areas. This 7 years period is more than the minimum time required for plantings to bear at least one crop of avocado fruit. Further, research results from JARC (1995), demonstrated that time to first bearing can be reduced to 3 years through grafting when compared to the production of non-clonal avocados grown from seeds.

Dependency ratio

A dependency ratio of 0.80 is reported as a way of life in

Table 2. Average household size and dependency ratio.

Woredas	Non-working members	Working members	Dependency ratio (Mean)
Mana	180	207	0.87
Goma	125	187	0.67
Seka-chokorsa	170	202	0.84
Total	475	596	0.80

Source: Own computation (October, 2010).

the study area, which is better off to national average that is, 0.97 (CSA, 2008). The result indicated that, out of 100 working persons, 80 are economically inactive in the study area (Table 2), but more are unable to support income generation in nationwide. Thus, family labor endowments have positively affected participation in the chain, given labor-intensive nature of avocado on harvesting.

Access to all weather roads and distance to development agents

Availability and adequacy of roads was an important prerequisite to link producers with markets which also reduced transportation costs. The assessment on the continuum, measured in km, revealed 81 and 90% of respondents are within 3 and 2 km from all-weather roads and development agencies offices, respectively. Under these conditions, most households can access roads and offices within 30 min of walking. Paradoxically, most avocado farmers failed to use these access ways. Thus, weak and intermittent delivery of extension services was observed in study areas because of these under-utilized access ways.

Accessibility to markets

Accessibility to reach markets was measured in km. This distance study revealed that the transportation infrastructure in Jimma zone is generally satisfactory and comparatively close to nearby fruit markets. Ease of accessibility assisted farmers by lessening transport costs and facilitated market supply. Access to good roadways was associated with avocado production in study areas.

This result supports a World Bank (2004) finding which found that better road densities in the study area, 77 km/1000 km^2 was significantly better than the national average of 30 km/1000 km^2. Despite this finding, however, our data indicated road networks did not support the neighboring farming communities, since most farmers were accepting whatever lower prices offered them at the farm gate.

Conversely, some farmers explained that road access was conducive to selling a large proportion of avocado fruit at distant markets, but 44% of sales were made at the farm gate, followed by vending at nearby village markets (22%). However, despite these seeming inconsistencies regarding the importance of road access for selling avocados, the overall research results highlighted that access to close markets have assisted farmers to plant avocados in large numbers as they do not have to rely on costly road transport to move their produce to distant markets where they invariably sell at loss.

Access to extension services

Despite two institutionally assigned development agents to work-in production areas, extension services assisted little on avocado production and promotion of consumption. This agency failure was accompanied by a significant shortage of technical expertise by development agencies which ultimately resulted in an inability to access markets. The result further highlighted that a failure to extend knowledge failed to support households that participated in the avocado production chain. Thus, information exchange through informal routes has remained the most accessible channel for information (64%) followed by extension personnel (21%). But because of lack of facilities, instructional materials, and trained manpower, Farmer Training Centers (FTCs) remained dysfunctional to serve as knowledge promotion centers to study areas.

Belay (2003) and Sonko et al. (2005) had similar findings for extension efforts supporting cereal and fruit production. Lack of effective extension negatively affected fruit production and marketing because of weak linkages between stakeholders and affiliates.

Types of extension services required

Low frequency of extension visits and unfamiliarity of development agents with avocado production was reported by 81 and 75.3% of the respondents, respectively. These are two central failings to building capacity which ultimately reduces dissemination of information. For this reason, development agents are compelled to rely on informal knowledge sources which may be of poor quality or erroneous (Table 3).

Table 3. Rank of extension service acquisition on various themes.

Extension information is needed for	Rank					Total	Mean
	1st	2nd	3rd	4th	5th		
Disease management	32.60	24.20	13.70	11.60	8.40	90.50	18.06 (1st)
Planting material preparation	28.42	23.16	24.21	12.63	-	88.42	17.68 (2nd)
Post-harvest handling	12.60	5.30	17.90	13.70	20.0	69.50	13.90 (3rd)

Source: survey result (October, 2010).

Table 4. Major means of income for farming households.

Principal income sources	Best-bet income sources among tropical fruits	Rank
Coffee	Avocado	1
Fruit	Banana	2
Grain	Papaya	3
Other	Orange	4
Livestock	Pineapple	5

Source: survey result (October, 2010).

Similarly, 90% of respondents appreciated receiving extension services on disease management followed by planting material preparation (88.42%) and post-harvest handling (69.5%), respectively. But insects were not reported as important factor that needs intervention.

Access to and use of land

Average land holdings in the study area was 1.94 ha, which is two-fold more than the national average (that is, 0.8 ha) (CSA, 2008). These large land holdings are primarily for coffee production which provides an opportunity for avocado production which supports crop diversification in South-western Ethiopia.

Avocado fruit as a source of income

The annual revenue earned from avocados (81,000 Ethiopian Birr/ha) is better than that of coffee (18,000 Ethiopian Birr/ha) and many other crops. For instance, the average annual income of maize is (7,200 Ethiopian Birr/ha), whereas for sorghum revenues reach 5,400 ETB/ha. Avocado revenue is about 11 to 15 times larger than the latter crops, respectively. The high potential commercial returns revealed the significance of avocados in local farming systems which could lessen reliance on coffee as a cash crop through diversification with a high value fruit (Table 4).

This finding is supported by the FAO (2005) which confirmed that revenue from avocado in Ethiopia and Kenya is rapidly increasing. Weinberger and Lumpkin (2005) stated avocado has brought higher net return hectare^{-1} than other staples produced by farming households in Costa Rica, where it is an important over-storey providing shade for coffee production.

Cultural practices

About 55% of respondents reported that they intercrop avocados with maize, taro, ginger, chat, cabbage, and bananas. Gillard and Godefroy (1995) stated that intercropping of avocados with short cycled crops was very common in sub-Saharan Africa and an excellent way to utilize the empty space as crops established.

Role of livestock in avocado production

Donkeys and horses are used principally to transport avocado from the farm gate to accessible roads and markets. Manure produced by these animals is used as organic fertilizer study areas. However, their numbers are constrained by shortages of grazing areas and stored feed.

Average number of avocado trees owned by households

The average number of total avocado plants and bearing trees owned by individual farmer was 34.03 and 13.01, respectively. With standard deviation of 41.6, the range

Table 5. Avocado trees owned by growers.

Indicator	Mana (N = 36)		Goma (N = 25)		Seka-Chokorsa (N = 34)		Total		x^2 /t-value
	Mean	STD	Mean	STD	Mean	STD	Mean	STD	
Bearing tree number	15.53	32.29	2.76	3.84	17.88	57.02	13.01	39.65	3.20***
Nonbearing tree	33.33	44.61	15.53	32.29	3.83	8.30	23.05	32.67	6.88***
Number of died trees	3.83	8.30	.44	0.96	2.88	3.59	2.60	5.68	4.46***
Total trees	52.44	55.48	10.60	8.944	31.76	28.67	34.03	41.63	7.97***
Production (ton/ tree)	0.31	0.14	0.29	0.77	0.33	0.15	0.31	0.13	2.35***

Source: Own survey (October, 2010).

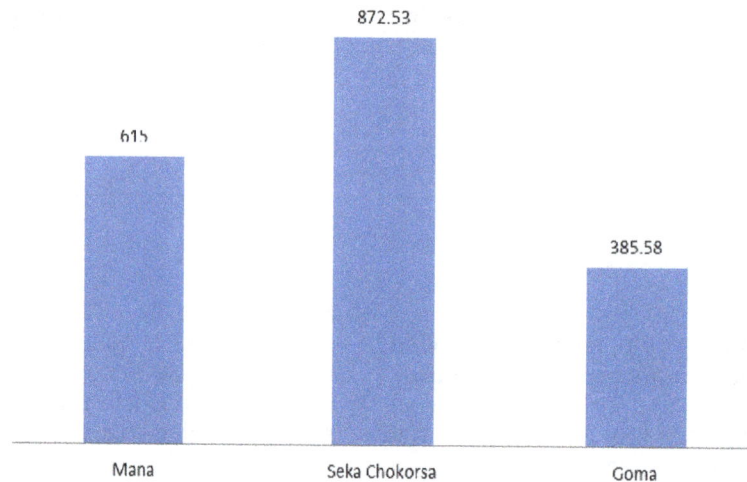

Figure 1. Volume of avocado channeled to markets in tons. Source: Survey result (October, 2010).

indicates that many juvenile avocado trees are not in the stage of fruit bearing, which in turn is indicating the prospect of increased fruit production with increased start of bearing of juvenile avocado trees.

Wasilwa et al. (2004) reported that Kenyan farmers own a greater number of fruit bearing trees and fewer juveniles, although the same total number of trees per farm is similar to that of Ethiopia, trees tend to be more mature on average (Table 5).

Production and productivity of avocados

Average productivity of 3.21 tons/ha is reported in the study area which is larger than the national average of 6.6 tons/ha (CSA, 2008). This productivity is within previously documented ranges of 1.56 to 7.80 ton/ha) (Woyessa and Berhanu, 2010; Zekarias, 2010).

Gillard and Godfroy (1995) reported average yields in Kenya being 33.2 tons/ha but lower in Coted'Ivore and Cameroon which produce less than 18 tons/ha. Edossa (1997), suggested that yield differences could results

from variation in productivity of cultivars, age of productive trees, and climate. The volume of avocado channeled to markets in tons is shown in Figure 1.

Total volume of production was 18721 of avocado fruit from the three study Woredas. Mana Woreda dominated by supplying 8725 ton (47%) followed by Seka-chokorsa and Goma Woredas with 61,500 (33%) and 38 45.8 ton (20%) of avocado supply, respectively. Avocado production from the study area is still increasing and 49% of this yield increase came from the commencement of fruiting from previously non-bearing trees.

Input utilization for avocado

Avocado production in Jimma zone is characterized by low inputs with farm yard manure (FYM) the major amendment made to soil to boost productivity. About 20% of respondents do not apply FYM and 60% applied FYM rates of 3.7, 9.4 and 18.7 ton/ha, respectively. The assessment further highlighted that, chemical inputs are not used for fertilization or pest treatment. Even though

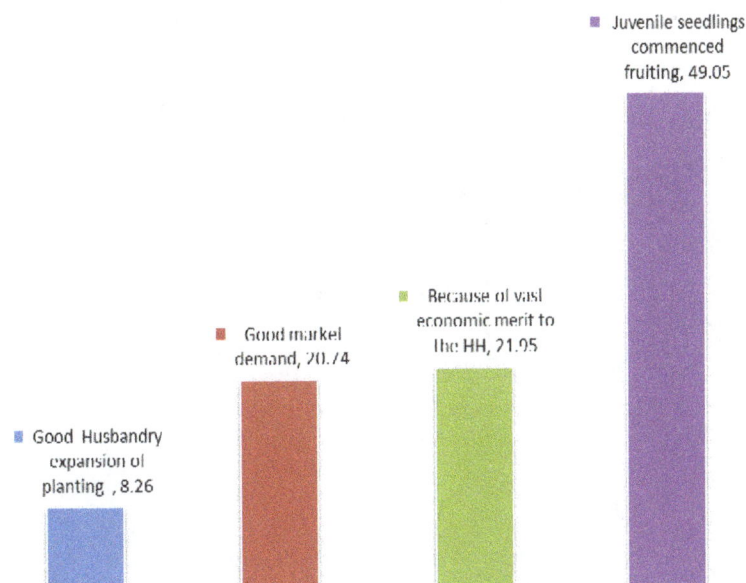

Figure 2. Reasons for boosted production in percentile. Source: Survey report (October, 2010).

this rate of application is drastically lower than the national recommendation (which is 29.84 ton/ha[1]), but it is greater than average rates of application [which is 0.55 ton/ ha in Ethiopia (Devi et al., 2007)]. In this study, the rate of FYM application was inversely proportional to increased number of avocado trees owned, as tree numbers increases, each tree got less FYM because of low availability.

Sources of avocado planting material

Avocado production in Jimma zone is exclusively based on distribution of mixed unimproved varietals that have lower than desired productivity. For this reason, 40% of respondents acquired planting materials from Woreda Agricultural offices followed by local planting materials purchased from unknown market sources (29%). The JARC and self-production by farmers were reported as the 3rd and 4th sources of avocado planting materials in the study areas (Figure 2). Local seed production is the major source of seedlings for distribution. However, seeds never breed true to parental lines and the use of non-clonal forms adds a lot of heterogeneity to production systems which can result in variable flowering, fruiting, and harvesting times, and varying fruit quality, size, and shape. All of which affects marketability.

Woyessa and Berhanu (2010), and Zekarais (2010) stated that JARC and Woreda Agricultural offices are the main sources of avocado seedlings in South-western Ethiopia. Gillard and Godfroy (1995) reported that natural propagation from seed was the prominent means of avocado dissemination for West and Central Africa.

Factors affecting availability of planting materials

Unavailability of planting materials and seedlings from unknown origin are the principal problems affecting avocado production in the study areas (Figure 3). Elfring et al. (2007) reported that producers complained about the unavailability of planting materials in terms of quantity and quality.

Fruit harvesting

Fruit harvesting usually commences at fruit drop, which is the principal maturity index used by farmers in the study areas. Around 76% of producers conduct harvesting subsequent to the commencement of fruit drop. This maturity index encouraged 23% of producers to leave fruit hanging on trees until better market prices appeared. Fruit harvesting is largely executed by child laborers who use picking hooks, shaking of trees, and knocking down fruit. However, the later practice has the potential to cause physical injury.

The research result supports findings by FAO (2005) which showed cuts, punctures and bruises to avocados increased ethylene production and hastened fruit softening and ultimately decay.

Sorting, grading, loading, and packaging: These functions are principally carried out at the farm gate and at primary procurement centers via the efforts of local collectors. Thus, fruit is sorted according to consignment needs of collectors where under-grades (that is, culls) such as shrunken, smaller sizes, with splits and

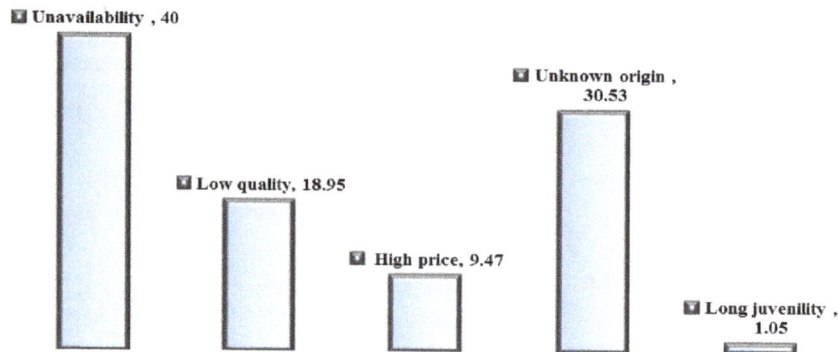

Figure 3. Problems encountered in using planting materials (percent). Source: Survey report (October, 2010).

Selling price of avocado across months (Birr/Kg)

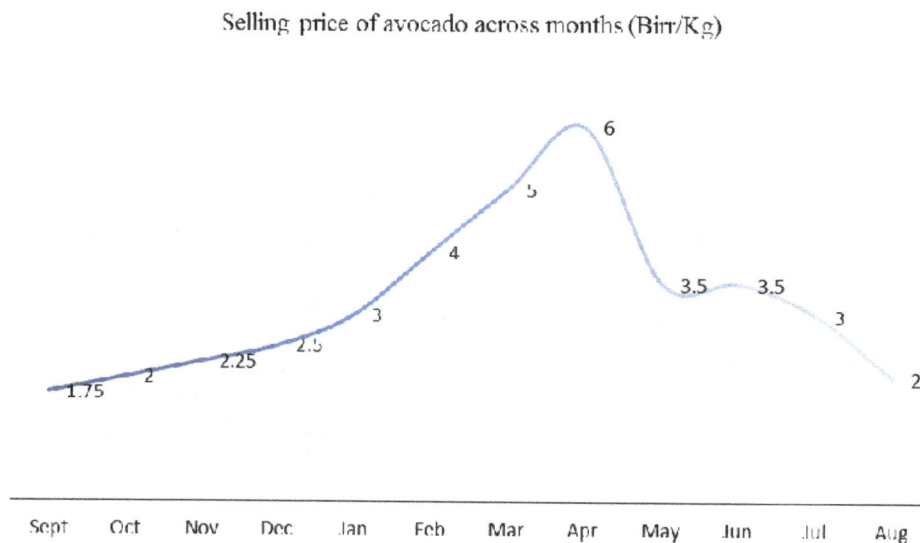

Sept   Oct   Nov   Dec   Jan   Feb   Mar   Apr   May   Jun   Jul   Aug

Figure 4. Purchase and selling prices, respectively. Source: survey result (October, 2010)

punctures are removed. But unsellable under-grades are not wasted as they are commonly consumed in farming households.

There is a shortage of standardized packaging materials for avocado fruit, and synthetic fiber sacks *"madaberiya"* are a popular packaging material to transport fruits from farm gate to primary procurement centers. Avocado packaging is an open sector for large private investment and introduction of modern technology and entry for investors.

Wiersinga and Jager (2009) stated that most available packing material in Ethiopia does not meet required standards for avocados. Consequently, exporters in Ethiopia import packing materials from the Netherlands and Israel. Efforts were recently launched by several new companies to produce fruit packing material in Ethiopia. However, the Ethiopian Commodity Exchange (ECX) is hesitant to implement quality control and market information services.

Perishablity

Around 73% of farmers have reported they are compelled to sell avocados at whatever price offered to them, while 15 and 9% of farmers have taken the risk to sell the fruit on another market day or to take it to another market, respectively in an attempt to get better prices (Figure 4).

Lack of effective post-harvest handling practices coupled with the short shelf-life of avocados has forced producers to sell at prevailing prices. Knowing this, wholesalers put pressure on producers to sell at low prices. Starting from production up to marketing, every farmer produces and sells on individual basis which affects their bargaining power during the sale of avocados. Local grower unions or cooperatives could

Table 6. Method of price setting and time of receive money after sale.

Price setting strategy	Average (%)
Negotiation with farmers	27.10
Set by demand and supply	38.90
Myself	34.03
Term of payment	
As soon as you sold	55.9
After some hours	11.8
On the other day after sale	32.4
Method of attracting suppliers	
Giving better price	58.8
By visiting them	23.5
Fair scaling /weighing	2.9
Giving pre-payment	5.9
Offering credit service	8.8

Source: Survey result (October, 2010).

greatly help producers get better prices for the fruit when they collectively bargain with fruit collectors.

Avocado root rot (*Phytophthera cinnamomi* Rands)

Incidence of this root disease is high, 95% of trees were infected in the study area which highlighted that almost all production areas are suffering from avocado root rot. This pathogen is the prevailing malady affecting avocado production in the study areas. Respondents indicated that there is no commonly accepted name for this disease; though some reported it as *"cholera"* or *"gogsa"*. For this reason, most farmers communicate by describing its symptoms, but canopy desiccation is the most common symptom.

Severity of avocado root rot

The assessment further signified a severity of 25, 18 and 16% of *P. cinnamomi* in Mana, Seka-chokorsa and Goma Woredas, respectively which indicated widespread presence. Thus, decline of avocado trees was detected in all the study areas.

Mohamed et al. (2009) reported that established avocado plots in the JARC research center were entirely devastated by avocado root rot disease; and survivors, about 30% of trees were drastically hampered by the fungus. Woyessa and Berhanu (2010), and Zekarias (2010) reported similar observations.

Avocado root rot versus boosting production

Despite the severity of avocado root rot, avocado production continued due to recent fruit setting by juvenile trees. Thus, 69% of respondents reported that, they will continue to farm avocados despite disease severity. Surprisingly, 82% of respondents revealed that they would plant new trees to replace those lost to avocado root rot. Most farmers consider avocados as a major means of financial insurance and a reliable relief crop when other crops fail. However, the result indicated that the practice of using resistant avocado rootstocks is absent indicating the need to start a root stock screening program to find rootstocks resistant/tolerant to avocado root rot.

Price of avocado

According to the pooled assessment, the avocado industry in the study area operated under an unregulated environment. Prices were exclusively determined by traders negotiating with farmers at time of procurement. For this reason, farm gate prices for the previous 5 years, was 8 birr/kg of fresh fruit, and has now dropped to 1.75 birr/kg except when producers are able to evade intermediaries and directly supply to wholesalers.

Over supply of fruit is the principal reason for price declines which affected 58% of farmers. However, 31% of respondents reported that, they were aware of price declines and did not know the market forces that cause prices to fall.

Price variation due to seasonality

Price variation from 1.75 to 8.00 birr/kg was reported during the major and off season production periods at wholesale markets in Jimma, respectively. According to 58%t of respondents, market prices are high for early arrivals but drop with an increase in supply, before increasing towards the end of the production cycle when supply gets scarce. June to early September are periods when prices are typically low and July to August is when prices are most depressed before increasing again in October.

Price setting and term of payment

Approximately, 34% of traders' reported they set avocado prices. A large proportion of traders (55.9%) earn their money instantly after transactions, while some (32.4%) receive their money on a different day after the sale (Table 6).

Place of sale

The research result indicated 59% of HHH sell avocados to retailers, while 18, 11 and 9% of them sold fruit to assemblers, individual consumers, and processors at

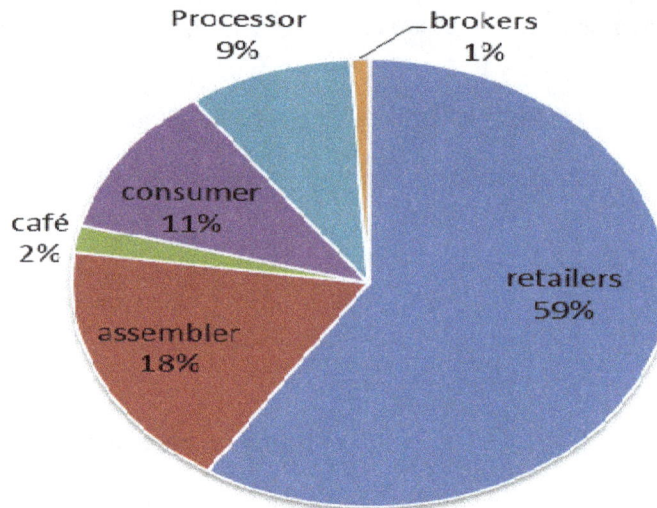

Figure 5. Actors role in avocado vale chain. Source: Survey report (October, 2010).

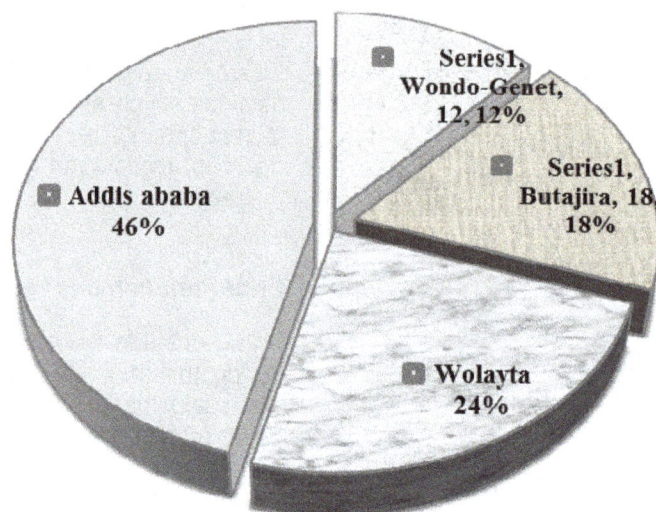

Figure 6. Preference of consumers to opt for their better variety. Source: Own survey (October, 2010)

local markets, respectively. But producers opted to sell their avocados to consumers due to their ability to pay better prices to farmers, followed by requests to sell to assemblers and cafés, respectively (Figure 5).

Spatial arbitrage

Failure to procure large avocado stockpiles for sale to meet local demand is the rationale behind spatial arbitrage in Jimma zone. Thus, 46, 24, 18 and 12% of consumers purchased avocado fruit from Addis Ababa, Wolayata, Butajira and Wondo-Genet, respectively

(Figure 6). Austin (2009) documented that the absence of a cold chain system in Rwanda led to importation of consignments of fruit from other locations to mitigate seasonality gaps in the supply of fresh fruit in local markets.

CONCLUSION AND RECOMMENDATIONS

Endowed with diverse range of agro-ecological zones, Ethiopia is one of the 10 major avocado producing countries of the world. But production is too traditional and poorly supported by scientific recommendations due to failures in institutional, social and economic factors to be nationally and internationally successful. Simultaneously, marketing activity is poorly linked along production channels (e.g., product sorting, grading, packing, transporting and marketing).

Although comparative rewards such as regional suitability for production, proximity to local markets and cheap provision of labor are opportunities for boosting production. However, declining prices due to oversupply, deadly fungal diseases, poor market integration, an absence of improved technologies, and provision of extension packages for growers are major setbacks that hamper production and marketing of avocados in Ethiopia. With existing prominent organic production, avocados are not yet organic certified in the study area.

Constraints hindering the development of avocados are found in all stages of the production chain. At the farm-level, lack of clean disease-free seedlings and grafted seedlings has compelled farmers to use inferior and low yielding varieties. Storage facilities are scarce all along the chain and absence of collective bargaining power has forced individual farmers to accept unfavorable deals.

Due to entire absence of improved varieties, avocado production is exclusively based on distribution of mixed materials. Consequently, the local seed system has come out as a best-bet arena and is now a common route for seedling dissemination in Jimma zone. However, there are many problems for this industry that relies exclusively on a crop with non-uniform production characteristics.

Even though most payments are made instantly, payment in small installments were sometimes reported in the study areas. Scaling deductions, quoting of low prices, and lack of market information were common market malpractices in the study areas. Deficiency in capital and credit availability was also reported as major setbacks that compelled farmers to sell their produce at whatever price was offered by traders who had loaned to the grower earlier. Absence of organized institutional support and a system of group marketing has positioned traders in a strong position to dictate pricing to the disadvantage of producers. Despite the closeness of four governmental and nine private commercial banks and five non-banking institutions, denial to formal credit was prevalent. Thus, an informal credit system was customary feature for avocado growers to deal with in the study areas.

A number of actions need to be undertaken in order to promote the development of avocados as a valuable crop. This includes, capacity building, technological applications, improved extension and outreach, and plant breeding activities.

Infrastructural development is a key to support the avocado agricultural sub-sector. In this arena, emphasis should be given to improved storage and transportation systems and offering credit and other services to improve effective production and marketing of the crop. Thus:

1) Efforts should be exerted to envelope avocado into an organized market through cooperatives which could further improve innovation and performance.

2) The Research-Extension-Farmers Advisory Council (REFLAC) should be strengthened to tackle constraints and to promote opportunities at local and regional levels.

3) Plausible post-harvest management and small-scale processing should be in place to strengthen and harmonize market chain development through effective and streamlined coordination among all workers involved in avocado production in Jimma zone. This could assist to eliminate duplication of efforts and promote greater production efficiencies.

4) The ECX should develop marketing services for avocados, similar to recently added commodities like sesame, haricot bean and maize. If done so, avocado prices will be stabilized and maintained at a premium.

5) Research should work towards generating improved production technologies e.g., variety and agronomic practices through cultivar diversification and practices mitigating avocado root rot. The disease management practice should include adoption of resistant root stocks in the production area.

6) Since high value crops are knowledge and technology intensive, the weak research and extension service should be built up and the necessary institutional supports should be put in place as quickly as possible to support this emerging industry. Thus, critical and continuous capacity building activities should be pursued to ensure success for avocado producers and marketers in Jimma zone.

7) Existing trees should be characterized in terms of growth habit, tree morphology, fruit quality, including post-harvest properties to assist in targeting markets and making selections that makes desirable varieties through cloning and grafting onto root rot resistant rootstocks

8) Establishing cottage industries for avocado processing should be entailed through micro enterprises to better manage surpluses when over production occurs.

9) Measures and incentive structures should be in place to encourage participation of stakeholder to solve the existing problems and promote the opportunities observed along the value chain of the industry.

10) Organize cooperatives to focus on avocado production starting with grass root structures that are developed up to the union level which empower them to compete with dominant groups by having their own sales outlets in major markets. Women should be particularly encouraged to join cooperatives and to participate in leadership positions.

11) Initiate a stakeholder co-ordination platform for future interventions on sub-sector development. Regular coordination meetings with regional stakeholders are recommended.

12) Training should be offered to dealers, nursery operators, growers, and small traders to help them understand the requirements for running a successful business and to better deal with shrewd operators in the

production and sales chain. Courses on basic business skills training, tax education, marketing, and demand analysis should be offered.

13) Impart skills to take participate in complementary activities such as nursery management, grafting services, and disease management.

14) Promote international standard good agricultural practices (GAP) through demonstration sites for growers to observe.

REFERENCES

Albertin A, Nair PKR (2004). Farmers' Perspectives on the Role of Shade Trees in Coffee Production Systems: An Assessment from the Nicoya Peninsula, Costa Rica, Human Ecol. 32:4.

Austin JE (2009). Study on Marketing, Post-Harvest and Trade Opportunities for Fruit and Vegetables in Rwanda, Final Report. 22 July 2009.

Belay K (2003). Agricultural extension in Ethiopia: the case of Participatory Demonstration and Training Extension System. Journal of Social Development in Africa 18(1):49-84.

Bezabih E, Hadera G (2007). Constraints and Opportunities of Horticulture Production and Marketing in Eastern Ethiopia, DCG Report. P. 46.

CSA (2008). the Federal Democratic Republic of Ethiopia, Central Statistical Agency, Agricultural Sample Survey, 2008, Volume I, Report On Area And Production Of Crops, (Private Peasant Holdings, Meher Season), Addis Ababa, June, 2008. Stat. Bull. P. 417.

Edossa E (1997). Selection of Avocado (Persea americana M.). Collection of Desirable Fruit Characteristics and Yield at Jimma, Proceedings of the 8th Annual Conference of the Crop Science Society of Ethiopia, Feb. 26-27, Addis Ababa. Ethiopia. pp. 26-35.

Elfring W, Yohannes A, Mulugeta T (2005).VALUE CHAINS IDENTIFICATION FOR INTERVENTION.

Devi R, Kumar A, Bishaw D (2007). Organic farming and sustainable development in Ethiopia. Scientific Research and Essay 2(6):199-203.

FAO (2005). Market Segmentation of Major Avocado Markets, Sugar and Beverages Group Raw Materials, Tropical and Horticultural Products Service Commodities And Trade Division. Food and Agriculture Organization of the United Nations.

FAOSTAT (2010). Preliminary 2009 Data for Selected Countries and Products http://faostat.fao.org/site/567/DesktopDefault.aspx?PageID=567#anc or

FARM Africa and SOS Sahel International (2004). Participatory Forest Management Program PFMP, Fruit Production in Agro-Forestry System in Bonga, http://www.pfmp-farmsos.org/Docs/fruitproduction_bonga.pdf Accessed on 13th, March, 2010.

Gillard JP, Godefroy J (1995). The Tropical Agriculturalist, Avocado, Macmillan Education Ltd. London.

Guinness Book of records (2010), Trippy Food, Holy Guacamole, Carpinteria. California.

Haggblade SJ, Gamser MS (1991). Field manual for subsector practitioners http://library.wur.nl/way/bestand/enclc/189236.pdf Accessed on 7th May, 2010.

Heisman G (1995). Research Method in Psychology, Fourth Edition. Houghton Mifflin Company. Boston. USA.

Jimma Agricultural Research Center (JARC), (2010). Center Profile, Jimma, Ethiopia.

Joosten F (2007). Development Strategy for Export Oriented Horticulture in Ethiopia http://library.wur.nl/way/bestanden/clc/1891396.pdf.

Kirby K. Potvin C (2007). Variation in Carbon Storage among Tree Species: Implications for the Management of Small-Scale Carbon Sink Project. For. Ecol. Manage. 246:208-221. www.elsevier.com/locate/foreco.

MOARD (2008). Draft Working Document to Establish National Agricultural Market Information Service in Ethiopia. June 2008.

Mohamed Y, Wodirad M, Eshetu A, Girma A, Dereje T, Temama H, Meki S (2009). In: Abrham Tadesse eds., Increasing Crop Production through Improved Plant Protection. Vol. II, Plant Protection Society of Ethiopia (PPSE), PPSE and EIAR, Addis Ababa. Ethiopia. P. 542.

Sonko S, Njue E, James M, Jager A (2005). Pro-Poor Horticulture in East Africa and South East Asia, the Horticultural Sector in Uganda, EAST AFRICA. January 2005.

Wasilwa, LA, Njuguna JK, Okoko EN, Watani GW (2004). Status of Avocado Production in Kenya. Kenya Agric. Res. Inst. Nairobi Kenya.

Weinberger K, Lumpkin TA (2005) Horticulture for Poverty Alleviation-The Unfunded Revolution. Shanhua, Taiwan: AVRDC - The World Vegetable Center, AVRDC Publication. Working Paper. 15(20):05-613

Wiersinga R, Jager A (2009). Business opportunities in the Ethiopian Fruit and Vegetable Sector, Wageningen University and Research Centre, Final version. February 2009.

Wolday A (1994). Food Grain Marketing Development in Ethiopia after Reform 1990. A Case Study of Alaba Siraro. The PhD Dissertation Presented to Verlag Koster University. Berlin P. 293.

World Bank (2004). Opportunities and Challenges for Developing High-Value Agricultural Exports in Ethiopia, Poverty Reduction and Economic Management, Country Department For Ethiopia. Africa Region.

World Bank Group (2006). Ethiopia: developing competitive value chain http://siteresources. worldbank.org/INTAFRSUMAFTPS/Resources/aftpsnote29F0610-17.pdf Accessed on 17th, December. 2009

Woyessa G, Berhanu T (2010). Trends of avocado (Persea americana M.) Production and Its Constraints in Mana Woreda, Jimma Zone: A Potential Crop for Coffee Diversification. Trends in Horticultural research, 2010 ISSN 1996-0735 / DOI: 10.3923/thr.2010.

Zekarias S (2010). Avocado Production and Marketing in South Western Ethiopia. Trends Agric. Econ. 3(4):190-206, 204. ISSN 1994-7933, 2010 Asian Network for Scientific Information.

20

Prevalence of apple scab and powdery mildew infecting apples in Uganda and effectiveness of available fungicides for their management

Arinaitwe Abel Byarugaba, Turyamureeba Gard and Imelda Night Kashaija

Kachwekano Zonal Agricultural Research and Development Institute, P. O. Box 421, Kabale, Uganda.

Apple scab caused by *Venturia inaequalis* (Cooke) wint, and apple powdery mildew, caused by *Podosphaera leucotricha* (Ell. and Ev.), are the most important diseases of apples in Uganda. Control of apple scab and powdery mildew require the application of fungicides in absence of resistant apple cultivars. This study was conducted to identify effective control fungicides and to document the status of these diseases in south western Uganda. The survey results of 2012 indicated that scab incidence was highest in districts of Kanungu (71.90%), followed by Kisoro (48.14%), Kabale (41.03%), Buhweju (29.23%), Mbarara (28.75%) and Rukungiri (17.80%). The severity of apple scab measured as percentage leaf area affected ranged from 20.12 to 76.19%. Powdery mildew incidence ranged from 30.00 to 70.00% with severity score ranging from 14.63 to 76.19%. Fungicides containing propineb (70% a.i), metalazyl (4% a.i) + macozeb (64% a.i), and bupirimate (25% a.i) were found to be effective at controlling apple scab with potential to reduce the disease severity by 61.11, 61.11 and 58.33%, respectively, while fungicides containing bupirimate (25% a.i), tebuconazole (43% a.i) and propineb (58% a.i) + cymoxanil (4.8% a.i) were more effective in the control of powdery mildew with potential to reduce the infection by 55.95, 39.12 and 20.84%, respectively.

Key words: Apple scab, powdery mildew and fungicides.

INTRODUCTION

Apple production is new agro enterprise that was established in Uganda since 2000 in South-western highlands mainly in Kabale district (ICRAF, 2003). 13 apple cultivars were introduced by NARO and ICRAF following farmers' demand for fruit trees. Two apple cultivars 'Anna and Golden Dorsett' were officially released in 2009 and are now widely grown in Uganda in the highlands of south western region as commercial varieties. The production of apples however is faced with a number of challenges among which diseases are one of the important constraints facing apple farmers. Apple scab and powdery mildew were identified as the most important diseases of apples in the country, with a potential to cause significant effects on the developing temperate fruit agro enterprise (Chemining et al., 2005). The apple scab pathogen *Venturia inaequalis* (Cooke) wint and the causal agent of powdery mildew (*Podosphaera leucotricha* (Ell. and Ev.) cause extensive crop losses in all apple production areas (Kerik, 2012; Brun et al., 2008). In the absence of proven diseases management technologies in Uganda, attempts to control these two diseases have been initiated through the use of fungicides and screening for tolerance among the introduced apple cultivars. However, no proper information is currently available regarding appropriate fungicides for the management of apple scab and powdery mildew. Therefore, this study was established to document the status of these diseases in the country and to evaluate the effectiveness of the available fungicides in the control of the diseases at Bungongi substation, in Kabale located at 1830 m a.s.l so that information on effective fungicides could be released and recommended for farmers to use.

MATERIALS AND METHODS

Prevalence of apple scab and powdery mildew in western district of Uganda

A survey was conducted in six district of the Western region of the country, where apples have been established since 2000. The survey districts consisted of Kabale, Kisoro, Kanungu, Rukungiri, Mbarara and Buhweju in western Uganda. Farmers with at least 5 year old apples were purposively selected and identified with help from the Agricultural extension officer. In each apple orchard, apple scab and powdery mildew disease incidence was estimated as percentage of apple tree leaves with visible symptoms. The disease severity was scored as percentage plant leaf area affected (PLAA) with scab according to a scale of 0 to 7 developed by Parisi et al. (1993), Croxall et al. (1952) 0 to 7 where 1 = 0% < percentage of scabbed leaf surface (sls)< 1%; 2 = 1%<sls<5 %; 3 = 5%<sls< 10%; 4 = 10< sls< 25%; 5 = 25%<sls<50%; 6 = 50%<sls<75%; 7 = 75%<sls while the severity of powdery mildew was evaluated visually on individual leaflets as percentage of infected area, using a 0 to 4 scale, where 0: no symptoms; 1: 1 to 5%; 2: 5.1 to 20%; 3: 20.1 to 40%; 4: 40.1 to 100% (Kim et al., 2004).

Screening fungicides to control powdery mildew and apple scab

Commercially available fungicides were evaluated in a nursery of apple rootstocks at Bugongi sub-station, for their efficacy in the control of apple scab and powdery mildew. The chemical fungicides were accessed from importers of Agricultural chemicals in Kampala. Among the fungicide evaluated, contact fungicides that inhibit fungal spore germination comprising of wettable suphur (80% a.i), Macozeb (80% a.i), and propineb (70% a.i), semi-systemic or systemic fungicides that retards spore movement and germination of sporangia that include propineb (58% a.i) + cymoxanil (4.8% a.i), Famoxadone (16.6% a.i) + cymoxanil (22.1% a.i) and Metalazyl (4% a.i) + Macozeb (64% a.i), Carbendazim (43% a.i), copper oxychloride (92% a.i), Tebuconazole (43% a.i) and Bupirimate (25% a.i) were used. For each fungicide, the industrial recommended concentration was applied to apple rootstocks grafted with "Anna and Golden Dorsett" that were infected with scab and powdery mildew.

Disease initiation for fungicide effectiveness in disease management

One-year-old apple seedlings varieties of 'Anna' and 'dorsett' grafted on rootstocks of "bidden fielder" cultivar were used in this experiment. One set of the experimental material was defoliated to 100%, while other was left intact. Both sets of seedlings were inoculated with spores of powdery mildew and apple scab, prepared from infected leaves as described. Infected leaves were collected and immersed in distilled water for four hours. The leaves

were then removed, the inoculum filtered, through a 0.2 mm sieve. The suspensions were adjusted with distilled water and standardized to 1×10^3 conidia per ml using a hemacytometer and used as a fine mist spray on the seedlings until run off. Inoculation was done in the evening to encourage spore germination and penetration. 48 h after innoculation, seedlings were subjected to treatments of contact and systemic fungicides. To prevent interplot interference due to fungicide drift; a split plot design was adopted, where fungicides constituted the main plot and apple cultivars, the sub-plots. The main plots (fungicides) were separated from each other with polythene sheeting. For each combination cultivar by fungicide, three replicates were used.

Fungicide effectiveness in scab and powdery mildew management

Apple scab and powdery mildew incidence and severity were scored on the first ten leaves following inoculation, from 2009 to 2011. Data on the number of leaves with powdery leaf lesions and scab lesion were counted and used to calculate disease incidence, as a proportion of disease-infected leaves. The mean number of powdery mildew lesions and scab lesions per leaf was considered as the measure of disease severity that was used to compute the percentage leaf area affected (PLAA). Apple scab PLAA was estimated using a scale of 0 to 7, where; 1 =: 0% < percentage of scabbed leaf surface (sls)< 1%; 2 = 1%<sls<5 %; 3 = 5%<sls< 10%; 4 = 10< sls< 25%; 5 = 25%<sls<50%; 6 = 50%<sls<75%; 7 = 75%<sls (Parisi et al., 1993). Powdery mildew percentage of affected area was estimated visually on individual leaflets using a 0 to 4 scale, where 0: no symptoms; 1: 1 to 5%; 2: 5.1 to 20%; 3: 20.1 to 40%; 4: 40.1 to 100% (Kim et al., 2004). Data were recorded 10 days after fungicide sprays, and, every week subsequently for a period of 8 weeks per season. These data generated on the intensity of powdery mildew and scab was used to compute the area under disease progress curve (rAUDPC) using the procedure of Campbell and Madden (1990).

$$AUDPC = \sum_{i=1}^{n}[(x_i + x_{i-1})/2][t_i - t_{i-1}]$$

Where; x_i= present disease severity, x_{i-1}= previous disease severity, $t_i - t_{i-1}$ = time difference between two consecutive disease severities

RESULTS AND DISCUSSION

Prevalence of apple scab and powdery mildew in Southwestern districts of Uganda

Apple scab and powdery mildew incidence and severity were significantly different (P=0.05). The results of the current work showed a high incidence of apple scab and powdery mildew in the majority of districts surveyed. Scab incidence was highest in Kanungu (71.90%), followed by Kisoro (48.14%), Kabale (41.03%), Buhweju (29.23%), Mbarara (28.75%) and Rukungiri (17.80%) (Table 1). The severity of apple scab was also high in all the districts surveyed, except for Rukungiri where the severity rates were of 20.12%, measured as percentage leaf area affected. Scab severity in Kanungu was 76.19%, in Kisoro, 45.00%, Kabale 44.29%, Buhweju 40.38%, Mbarara 33.45%, and Rukungiri, 20.12%. The most damaged plant part were the leaves, this is in

Table 1. Prevalence of apple scab in the two varieties grown in south western Uganda.

District	Cultivar Anna		Cultivar Golden Dorsett			
	Apple scab incidence (%)	Scab disease severity leaf area affected (PLAA) (%)	Apple scab incidence (%)	Scab disease severity leaf area affected (PLAA) (%)	Mean scab incidence	Mean scab disease severity leaf area affected (PLAA) (%)
Buhweju	23.33	41.67	34.29	39.29	29.23	40.38
Kabale	43.23	45.31	37.50	42.67	41.03	44.29
Kanungu	71.82	77.27	72.00	75.00	71.90	76.19
Kisoro	47.27	44.32	49.62	46.15	48.14	45.00
Mbarara	32.50	40.00	25.00	27.50	28.75	33.75
Rukugiri	18.24	19.12	17.50	20.83	17.80	20.12
Grand Mean	41.53	43.84	36.93	39.94	39.46	42.09
F.Pr					0.704	0.769
LSD (P=0.05)					12.78	11.89

agreement with work of Valiuskaite et al. (2009). Powdery mildew disease incidence was highest in Kanungu, at 70.00%, with severity score of 76.19%, followed by Kisoro district, with incidence of 38.00% and severity score 39.29%, in Kabale mildew incidence was 34.62% with severity of 41.47%, Buhweju 30.00% with 39.23% disease severity, while in Rukungiri incidence of mildew was at 12.93% with 14.63% disease severity.

The high incidence and severity of the diseases in most of the districts surveyed is attributed to the lack of proper management practices and lack of knowledge on the appropriate fungicides to be used for the management of these diseases. Developing cultivars resistant to both scab and powdery mildew and use of synthetic fungicides are important for scab and powdery mildew management. In this study it was noted that districts which experience lower temperatures and higher rainfall, scab was a major problem whereas powdery mildew was less important in areas where temperatures are higher and rainfall is lower, powdery mildew is a major problem, and

scab is less important. This is supported by report of (Blazek and Hlusickova, 2003). The study also showed no significant difference in infectivity of apple scab and powdery mildew between the two common apple varieties 'Anna' and 'Golden Dorsett' in all the district surveyed (Tables 1 and 2). Cultivars Anna and Golden Dorsett were equally affected by apple scab and powdery mildew in the surveyed district.

Effectiveness of the fungicides for control of powdery mildew and apple scab

The results of the study indicate a significant difference (P=0.05) in the effectiveness of fungicides to control powdery mildew for three seasons of the study. Fungicides containing bupirimate (25% a.i), Tebuconazole (43% a.i) and Propineb (58% a.i) + cymoxanil (4.8% a.i)), were more effective in reducing apple disease severity and controlled the development of powdery mildew on leaves and young shoots of apple of grafted apple seedlings. Fungicides containing

bupirimate (25% a.i) was able to reduce the severity of powdery mildew by 55%, followed by Tebuconazole (43% a.i) 30% and Propineb (58% a.i) + cymoxanil (4.8% a.i), 20%, Copper oxychloride (92% a.i) 19.54%, Macozeb (80% a.i) 19.45%, Famoxadone 22.1% (a.i)+ cymoxanil (30% a.i) 13.95%, Metalazyl (4% a.i) + Macozeb (64% a.i), 10%, Propineb (70% a.i) 7.79%, and Wettable suphur (80% a.i) 5.92% (Table 3). However, for fungicide Carbendazim (43% a.i), the intensity of powdery mildew was the same as that of the control (No spray) implying that there was no effect in controlling the disease development

In order to control apple scab, fungicides containing Propineb (70% a.i), Metalazyl (4% a.i) + Macozeb (64% a.i), and bupirimate (25% a.i) were more effective in controlling apple scab with potential to reduce apple scab severity by 61, 61 and 58%, respectively (Table 4). Defoliated apple seedlings exhibit lower disease severity (rAUDPC) for scab across all fungicide sprays, ranging from 0.5 to 5.8% for defoliated apples, whereas for intact plants, it ranged from 5 .1 to 8.5%. This

Table 2. Prevalence of powdery mildew in the two apple varieties grown in Western Uganda.

District	Cultivar Anna		Cultivar Golden Dorsett			
	Powdery mildew incidence (%)	Powdery mildew disease severity leaf area affected (PLAA) (%)	Powdery mildew incidence (%)	Powdery mildew disease severity leaf area affected (PLAA) (%)	Mean powdery mildew incidence	Mean powdery mildew severity leaf area affected (PLAA) (%)
Buhweju	26.67	41.67	32.86	37.14	30.00	39.23
Kabale	35.31	42.08	33.50	40.50	34.62	41.47
Kanungu	69.09	75.00	71.00	77.50	70.00	76.19
Kisoro	40.00	39.77	34.62	38.46	38.00	39.29
Mbarara	20.00	43.75	12.50	25.00	16.25	34.38
Rukugiri	20.00	22.35	7.92	9.17	12.93	14.63
Grand Mean	36.25	41.90	29.94	34.89	33.42	38.75
F.pr					0.369	0.357
LSD (P=0.05)					11.57	12.98

Table 3. Average rAUDPC for powdery mildew and percentage reduction in severity for three seasons, as influenced by different fungicides in apple crops.

Fungicide trade name	Active ingredient	Application rate	Disease severity for powdery mildew, measured as rAUDPC	% reduction in powdery mildew disease severity
Nimrod	Bupirimate 25%	15 g/100 L	19.57	55.95
Orius	Tebuconazole 43%	35 ml/100 L	27.05	39.12
Milraz	Propineb 58% and cymoxanil 4.8%	200 g/100 L	35.17	20.84
Cobox	Copper oxychloride 92%	500 g/100 L	35.75	19.54
Agrozeb	Macozeb 80%	200 g/100 L	35.79	19.45
Equatin Pro	Famoxadone 22.1% + cymoxanil 30%	50 g/ 100 L	38.23	13.95
Ridomil Gold	Metalazyl 4% +Macozeb 64%	250 g/100 L	39.98	10.02
Antracol	Propineb 70%	200 g/100 L	40.97	7.79
Thiovit	Wettable suphur 80%	200 g/100 L	41.8	5.92
Rodazim	Carbendazim 43%	30 g/100 L	47.35	-6.57
Control			44.43	
Grand Mean			37.58	
F.pr			0.036	
LSD (P=0.05)			13.77	

rAUDPC= relative area under disease progress curve, computed using the percentage of leaf area affected measured over a period of 8 weeks, for 2 consecutive experimental growing seasons.

Table 4. Average rAUDPC for apple scab and percentage reduction in the disease severity, for two seasons as influenced by different fungicides.

Fungicide trade name	Active ingredient	Application rate	Defoliated	Un defoliated	Grand mean	% reduction in scab severity
Antracol	Propineb 70%	200 g/100 L	0.5	5.1	2.8	61.11
Ridomil	Metalazyl 4% +Macozeb 64%	250 g/100 L	1	4.6	2.8	61.11
Nimrod	Bupirimate 25%	15 g/100 L	1.4	4.6	3.0	58.33
Equation Pro	Famoxadone 22.1% + cymoxanil 30%	50 g/ 100 L	1.2	4.9	3.1	56.94
Cobox	Copper oxychloride 92%	500 g/100 L	1	5.7	3.3	54.17
Milraz	Propineb 58% and cymoxanil 4.8%	200 g/100 L	2.6	5.2	3.9	45.83
Rodazim	Carbendazim 43%	30 g/100 L	2.5	5.2	3.9	45.83
Orius	Tebuconazole 43%	35 ml/100 L	0.8	9.0	4.9	31.94
Agrozeb	Macozeb 80%	200 g/100 L	2.1	9.1	5.6	22.22
Control			5.8	8.5	7.2	-
F.pr			<.001	<.001	0.001	
LSD (P=0.05)			0.98	0.98	2.19	

indicates that farmers should defoliate their apple orchards prior to fungicide spraying, as this cultural practice helps to reduce the disease innoculum, and subsequent disease severity. Therefore, it is recommended, for effective control of powdery mildew, that the plants are defoliated prior to spraying so that the product can reach newly formed leaves allowing an effective management of the spread of the disease. Based on the results of the current study, it is recommended that fungicide containing propineb (70% a.i), Metalazyl (4% a.i)+ Macozeb (64% a.i), and bupirimate (25% a.i) are used for effective management of apple scab, whereas Bupirimate (25%) and Propineb (58%) are recommended for the control of powdery mildew in Uganda. The order of the recommendations correspond to the effectiveness of the products in our experimental conditions.

Conclusion

The interchangeable use of fungicides containing propineb (70%), Metalazyl 4% + Macozeb 64%), and bupirimate (25%) promotes the control of apple scab to levels below those causing economic damages. The products of Bupirimate (25%), Tebuconazole (43%) and Propineb (58%) + cymoxanil (4.8%) are recommended for the management powdery mildew. Incorporating cultural methods, such as defoliation and use of resistant varieties, could also help to reduce the severity of the investigated diseases.

ACKNOWLEDGEMENTS

We acknowledge the Government of Uganda and National Agricultural Research Organization for providing funds to conduct the research.

REFERENCES

Blazek J, Hlusickova I (2003). Influence of climatic conditions on yield s and fruit performance of new apple cultivars from the Czech Republic. Acta. Hort. 622:443-448.

Brun L, Didelot F, Parisi L (2008). Effects of apple cultivar susceptibility to *Venturia inaequalis* on scab epidemics in apple orchards. Crop Prot. 27(6):1009-1101.

Campbell CL, Madden V (1990). Introduction to disease Epidemiology. John Wiley and sons, New York. P. 532.

Chemining WG, Mulagoli I, Mwonga S, Ndubi J, Tum J, Turyamureeba G (2005). Kabale apples: boom or burst? A study to develop strategies to exploit market opportunities for apple farmers in Kabale, Uganda. pp. 23-24.

Croxall HE, Gwynne DC, Jenkins EE (1952). The rapid assessment of apple scab on leaves. Plant Pathol. 1:39-41.

ICRAF (2003). Temperate fruits go tropical: Apples, peaches, pears and plums take to the hills of Uganda. ICRAF, Nairobi, Kenya.

Kerik DC (2012). Future fungicides for scab/mildew in the face of multiple fungicide resistance. Dept. of Plant Pathology and Plant-Microbe Biology. Cornell University, NYSAES, Geneva, NY 14456.

Kim JC, Choi GJ, Lee SW, Kim JS, Chung KY, Cho KY (2004). Screening extracts of *Achyranthes japonica* and *Rumex crispus* for activity against various plant pathogenic fungi and control of powdery mildew. Pest. Manag. Sci. 60:803-808. doi: 10.1002/ps.811.

Parisi L, Lespinasse Y, Guillaumes J, Kruger J (1993). A new race of *Venturia inaequalis* virulent to apples with resistance due to the Vf gene. Phytopathology 93:533-537.

Valiuskaite A, Raudonis L, Lanauskas J, Sasnauskas A, Survillene E (2009). Disease incidence on different cultivars of apple tree for organic growing. Agron. Res. 7 (1):536-541.

21

Biochemical and nutritional properties of baobab pulp from endemic species of Madagascar and the African mainland

Cissé Ibrahima[1], Montet Didier[2], Reynes Max[2], Danthu Pascal[3], Yao Benjamin[1] and Boulanger Renaud[2]

[1]INHP, BP1313 Yamoussoukro, Côte d'ivoire.
[2]CIRAD, UMR 95 Qualisud, TA B95/16, 34398 Montpellier cedex 5 – France.
[3]CIRAD, PD Forêts et Biodiversité, P. O. Box 853, Antananarivo, Madagascar and UR 105, Campus de Baillarguet, 34392 Montpellier Cedex 5, France.

The fruit of baobab (*Adansonia* sp.) is well known in Africa both for its medicinal properties and social uses. It is a very promising tropical fruit although it has been little investigated and exploited in Madagascar. One of the major challenges, in Africa, in the last years has been to establish the baobab as a commercial crop with an economic value. In order to know if Malagasy baobab fruits have the same potential, we proposed to study biochemical characteristics of its fruit pulp. To achieve this objective, five endemic baobab species from Madagascar and one from Côte d'Ivoire were studied. Contents in vitamin C, polyphenols, lipids, proteins and minerals were evaluated. The biochemical composition of the fruit pulp of Madagascar species was studied and compared to that of a Sudano-Sahelian species (*Adansonia digitata*). Results showed high variability in biochemical characteristics and mineral content between the five Malagasy species and the Sudano-Sahelian species. These data revealed that the composition and the interesting nutritional potential of the baobab pulp may be of high interest to Malagasy consumers, which would contribute to rank it as a commercial crop.

Key words: Malagasy baobab, *Adansonia*, biochemical composition, antioxidants activity, species discrimination.

INTRODUCTION

Baobab (*Adansonia*) is a big tree that grows principally in Africa and can live up to 1000 years. It originates from tropical Africa and has a wide distribution range. Baobab species are rustic trees, which are characteristic of Sahelian prairies and Sudano-Sahelian savannas (Wickens and Lowe, 2008) as well as the semiarid tropical zone of the western part of Madagascar. The genus *Adansonia* belongs to the Bombaceae family and the Malvales order. It comprises eight species, six of which are endemic to Madagascar, that is, *A. grandidieri* Baill., *A. madagascarensis* Baill., *A. perrieri* Capuron, *A.*

fony var. *rubrostipa* (Jum. & H. Perrier) H. Perrier, *A. suarezensis* H. and *A. za* Baill. *A. gregorii* F. Muell. is exclusively found in Northwestern Australia, whereas *A. digitata* L. is encountered in subtropical Africa where it plays key cultural roles in the beliefs of indigenous people (Kamatou et al., 2011; Sanchez et al., 2010).

The ovoid fruit, called monkey's bread, contains black seeds embedded in a white and chalky pulp. It is consumed as basic food in many regions of Central Africa. For example, the Haoussa ethnic group uses the baobab fruit as the main ingredient in a soup called *miyar*

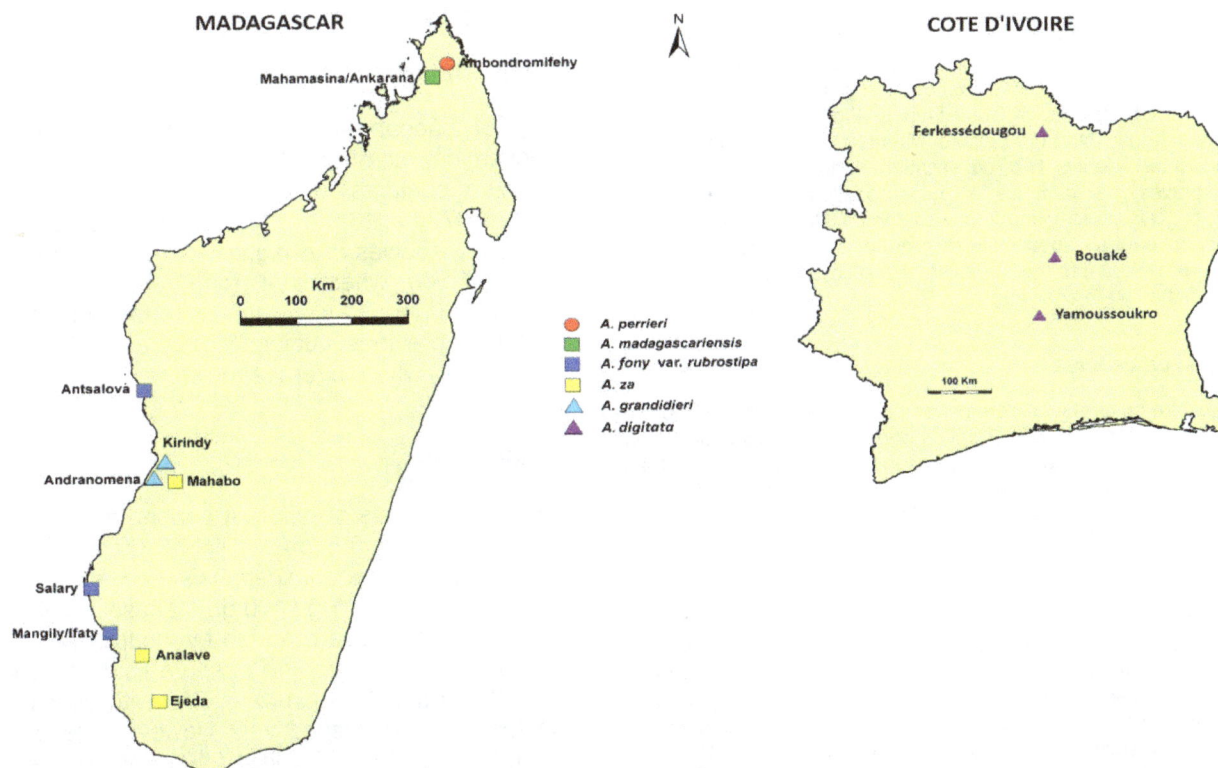

Figure 1. Zones of harvest of the six species of baobab; in Madagascar: ● *A. perrieri,* ■ *A. Madagascariensis,* ■ *A. fony var. rubrostipa,* ■ *A. Za,* △ *A. Grandidieri;* in Côte-d'Ivoire : ▲ *A digitata.*

kuka. The baobab fruit is also used daily in the diet of rural communities in West Africa (Assogbadjo et al., 2006; Codjia et al., 2001; Sidibé and Williams, 2002). The species contributes to rural incomes (Buchmann et al., 2010; Diop et al., 2005) and has various important medicinal and food uses (Kaboré et al., 2011, Assogbadjo et al., 2006; De Smedt et al., 2010a; 1997; Sena et al., 1998; Sidibé et al., 1996). The pulp is mainly consumed traditionally under different forms. It is used in the formulation and preparation of cereals and beverages.

Although, baobabs are widely known, the current scientific knowledge on the biochemistry and importance of its fruit in human nutrition is scarce. To date, most of the studies have concerned the species *A. digitata* in particular in relation to its botanical (Sena et al., 1998; Soloviev et al., 2004; De Smedt et al., 2010b), agronomical (Munthali et al., 2012; Obizoba and Amaechi, 1993; Codjia et al., 2003; De Caluwé et al., 2009; Wickens and Lowe, 2008) and biochemical characteristics (Parkouda et al., 2012; Chadaré et al., 2009; Diop et al., 2005; Assogbadjo et al., 2005 and 2012; Gebauer et al., 2002; Sidibé and Williams, 2002). Biochemical studies showed that especially the pulp of *A. digitata* is rich in dietary fibers (Chadaré et al., 2009; Cissé et al., 2008), carbohydrates (Solviev et al., 2004; Nour et al., 1980; Murray et al., 2001) and vitamin C (De

Caluwé et al., 2010; Chadaré et al., 2009; Sidibé et al., 1996), and that it could be used to produce beverages and nectars (Cissé et al., 2008; Ibiyemi et al., 1988; Obizoba and Anyika, 1994). The current knowledge on the biochemical properties of *A. digitata* is now well known however, these properties are lacking for Malagasy species. Studies have only been carried out on germination (Danthu et al., 1995; Razanameharizaka et al., 2006) and seed fat characterization (Gaydou et al., 1982; Ralaimanarivo et al., 1982).

Our objectives were to improve characterization of the biochemical composition of fruits of various sources, to assess fruit potential for development on a larger scale, and to identify an approach to preserve the pulp so that it may be used in the production of drinks and nectars. We hypothesized that the genetic diversity of endemic Malagasy species evolving in an island ecosystem might show variability in the fruit chemical composition compared to that of Sudano-Sahelian species (*A. digitata*). Therefore, the study aimed to characterize the pulp of Malagasy species and compare it with the African species (*A. digitata*).

MATERIALS AND METHODS

Mature fruits from five species of Malagasy baobabs (*A. za, A. perrieri, A. grandidieri, A. fony* var. *rubrostipa* and

A. madagascariensis) were collected in different geographic zones of Madagascar during the 2009-2010 harvests (Figure 1). Pulps were extracted, vacuum packed, and sent to CIRAD laboratory in Montpellier, France. Mature fruits of *A. digitata* were collected in three geographic zones of Côte d'Ivoire (Figure 1), specifically in Ferkéssédougou (North), Bouake (Center), and Yamoussoukro (South-Center) during two consecutive years (2009-2010). For every species, the samples of pulps resulted from a minimum mixture of six fruits resulting from 2 to 6 trees. Pulps from Malagasy and Ivorian baobab species were ground, sieved through a 0.08 mm sieve, packed into freeze-resistant plastic bags and stored at -80°C until they were analyzed.

Analytical procedures

The moisture content was determined by the gravimetric method at 104°C (Künsch et al., 1999). Total nitrogen content was determined with the Kjeldahl method using 6.25 as the conversion factor of total nitrogen to protein. Lipids were extracted with Avanti Soxtec and petroleum ether as solvent (Anon, 1990). The contents in ashes, carbohydrates, and dietary fibers were determined according to Van Soest et al. (1991).

Pulp nutritive mineral contents (Ca, Na, K, Mg, P) were determined by atomic adsorption spectrophotometry, whereas metals (Fe, Cu, Zn) were determined by flame photometry. The pulp pH was determined at 25°C after dilution at 1/10 (w/v) in deionized water using a pH meter (HANNA, pH 211 Microprocessor). The vitamin C content was determined by oxido-reduction of dichloro-2.6-phenolindophenol (2.6-DCPIP) after extraction with 100/80 (v/v) of metaphosphoric acid/acetic acid. Briefly, 3 g of pulp were homogenized in 30 ml of metaphosphoric acid solution for 30 min under gentle agitation. Samples were filtered using Watman filter paper, and 2 ml of the filtrate were titrated using 2.6-DCPIP until appearance of persistent pink coloration.

The antioxidant capacity was evaluated with the oxygen radical absorbance capacity (ORAC) method (Vaillant et al., 2005). Briefly, an amount of 0.5 g of pulp sample was mixed in 2 ml of a solution of acetone/water (50/50, v/v) for 60 min under gentle shaking at 1900 rpm, and then centrifuged at 14000 rpm at 10°C for 15 min. The supernatant was used in ORAC assays. Total polyphenolic contents were determined with the method described by Georgé et al. (2005). Polyphenols were extracted by mixing 0.3 g of pulp in 10 ml of ethanol/water (70/30 v/v). Extracts were agitated for 10 min using a rotating shaker (Heidolph Multi Reax, Germany) and filtered on Whatman Filter paper. Phenolic compounds were extracted in solid phase in order to reduce interferences associated with other substances such as reducing sugars, alcohols, tartric acid and other antioxidants (such as ascorbic acid) during the polyphenolic determination with the Folin-Ciocalteu reagent. Thus, 2 ml of sample were placed onto an OASIS cartridge column (OASIS simple extraction Product, Ireland), pre-wetted twice with 3 ml of ethanol and rinsed twice with 2 ml of water. The total polyphenolic content was determined by colorimetry using the Folin-Ciocalteu reagent. Gallic acid was used as external standard and the values were expressed as milligrams of polyphenolics per 100 g of sample.

Statistical analysis

In order to ensure reproducibility of the results, all the samples were collected from each species and each sample was analyzed in duplicate. Data were expressed as means, giving the relative standard deviations. The coefficients of variations between years were expressed by standard deviations. Two statistical analyses were performed with XI-STAT Prov. 7 (Addinsoft): an ANOVA for the year effect, and a principal component analysis (PCA) to detect differences that discriminate the baobab species.

RESULTS AND DISCUSSION

Water contents

Pulp of baobab is a dried pulp. Table 1 shows the biochemical composition of various baobab species. Moisture contents varied among the different species from 11.7 to 13.5% (12.5% on average). For the Malagasy species, the highest content was observed in *A. grandidieri,* whereas *A. fony* var. *rubrostipa* had the lowest. Our results were similar to those reported for *A. digitata* in previous studies (Murray et al., 2001; Soloviev et al., 2004) where average moisture contents were 12%.

Lipids

The lipid contents of different baobab species were very low, varying from 0.5 to 2.1 g lipid / 100 g dry matter (DM) (Table 1). *A. madagascariensis* had the highest lipid content with 2.1 g / 100 g. These values were comparable to those reported by Nour et al. (1980), and Lockett et al. (2000), who observe lipid contents as low as 0.21 and 0.41 g /100 g DM with gravimetric and Soxtec methods, respectively. However, these values are low compared to 15.5 g lipid / 100 DM reported by Glew et al. (1997) for *A. digitata*. This result is very surprising because the lipid content is higher than that found in baobab seeds, that is, 9 g lipid /100 g DM. The Soxtec method used in our study gave 0.94 g /100 g DM of lipid in *A. digitata*, that is, twice the values reported by Lockett et al. (2000) in the same species with the same method. Therefore, the observed variations may result from the analytical methods used, but also from the different baobab ecotypes and species studied.

Proteins

The protein content of the six baobab species varied from 2.5 to 6.3 g protein /100 g DM (Table 1). *A. fony* var. *rubrostipa* had the highest protein content (6.3 g/100 g DM), and *A. za* the lowest (2.5 g/100 g DM). Our results were very similar to those reported by Lockett et al. (2000), and Osman, (2004) who obtain 5.3 g/100 g DM in *A. digitata*. All investigated species, except *A. fony* var. *rubrostipa* (6.3%), had similar protein content. With the same analytical procedures and 6.25 as conversion factor, Chadaré et al. (2009) report a protein content comprised between 2.5 and 17 g/100 DM for *A. digitata*. Only Sena et al. (1998) and Obizoba and Amaechi (1993) report a very high protein contents, 17 and 19.1 g/100 g DM, respectively, but the values were not observed in all other studies realized on *A. digitata*.

Fibers

Fiber content of different baobab species varied from

Table 1. Nutritional composition of six baobab species from Madagascar and Côte d'Ivoire.

Macronutriments	Malagasy Species					African Species		References
	A. za [9]	A. perrieri [21]	A. grandidieri [14]	A. fony var. rubrostipa [14]	A. Madagascariensis [1]	A. digitata [9]	Literature g/100 g	
Moisture g/100 g	13.5	12.8	13.5	11.8	11.7	11.7	11.6	Lockett et al. (2000); Murray et al. (2001); Osman (2004); Soloviev et al. (2004)
Lipids g/100 g	0.5	1.2	1.7	1.6	2.1	0.5	0.2 – 15.5	Lockett et al. (2000); Murray et al. (2001); Osman (2004)
Proteins g/100 g	2.5	3.1	3.5	6.3	3.6	3.0	2.5 – 17	Lockett et al. (2000); Murray et al. (2001); Osman (2004)
Starch g/100 g	38.8	71.7	26.1	43.9	60.8	39.2		
Glucose g/100 g	5.3	2.9	8.9	3.8	2.9	7.9		
Fructose g/100 g	5.4	3.2	9.9	4.1	3.5	7.0		
Sucrose g/100 g	1.04	1.03	5.2	0.6	1.25	1.7		
Fibers g/100 g	25.78	17.20	25.09	27.89	25.82	25.25		
Ash g/100 g	5.3	7.8	6.1	6.9	7.1	5.2	4.9-6.4	Murray et al. (2001)
Acidity meq/100 g	161	132	112	142	95	102	>40	Diop et al. (2005)
Vit. C mg/100 g	138	70	60	92	76	67	300	Gebauer et al. (2002)
Polyphenols mg/100 g	1706	329	600	715	1126	1085	250	Cisse et al. (2008)
Antioxidant capacity µmoles TE/g	151	151	115	159	114	109	25	Besco et al. (2007)

[sample numbers]; TE: Trolox Equivalent.

17.2 to 27.9 g/100 g DM. Excepted A. perrieri with 17.2 g/100 g DM, all the species had fiber content around 26 g/100 g DM. Chadaré et al. (2009) note variability between 0.6 and 45.1 g/100 g DM. In fact, this variability could be explained by the method used. The average value observed by these authors is 13.7 g/100 g DM.

Carbohydrates

As in most fruits, in two studied species (A. perrieri and A madagascariensis), carbohydrates represented more than 60% of the dry matter and consisted of soluble sugars for half of it. In baobab fruits, among the soluble sugars, glucose was the least represented, but the content of reducing sugars (glucose + fructose) was greater than the sucrose content. Results showed variability in carbohydrate contents, especially for starch, as two species, A. perrieri and A. madagascariensis, had high values, 71.7 and 60.8 % DM, respectively, compared to the others, from 26.1 to 43.9 %. The highest sugar contents were found in A. grandidieri with 8.9% for glucose, 9.9% for fructose and 5.2% for sucrose, followed by A. digitata with 7.9% for glucose, 7.0% for fructose and 1.70% for sucrose (Table 1). We thus observed that the species with the highest starch contents had the lowest free sugar contents. The presence of sugar was similar to that mentioned by Soloviev et al. (2004), who found a total of

soluble sugar content ranging between 7.2 and 11.2 g/100 g DM in the pulp of baobab, whereas Nour et al. (1980) report a 23.2% total sugars and 19.9% reducing sugars. According to Murray et al. (2001), simple sugars represent about 35.6% of total carbohydrates. This explains the considerably sweet taste of the pulp. However, sweetness can vary depending on the species, maturity of the fruits, and environmental soil and climate.

Acidity

Very high acidity was observed in the pulp of all investigated species (Table 1). The highest value was 161 meq/100 g MS in *A. za*. Among the studied parameters, acidity and ascorbic acid contents showed the highest variations between samples. Pedoclimatic conditions and storage conditions of the pulp were among factors that might explain such variations.

Ash

Ash contents were 7.8 g/100 g DM in *A. perrieri* (the highest value) and 5.3 g/100 g DM in *A. za* (the lowest value). These results were similar to those of Murray et al. (2001), and Lockett et al. (2000) who report a content of 5.1 to 5.7 g/100 gDM, respectively, whereas a very low value of 2.4 g/100 g DM is reported in Obizoba and Amaechi (1993). The incidence of soil and climatic conditions, and ripeness stage at harvest were factors that could explain these variations.

Vitamin C

Table 1 shows the vitamin C content of the six baobab species, which was comprised between 60 and 138 mg/100 g DM. *A. za* had the highest vitamin C content, whereas that of *A. grandidieri* was twice as low. These data clearly showed the high variability of vitamin C content between species that are endemic to Madagascar. This variability existed also between regions in Sudano-Sahelian species *A. digitata* (Data not shown). Indeed, besides the variability between species observed in our data, Scheuring et al. (1999) also reports a high variability between trees of the same species. Our study showed that the vitamin C content of baobab pulp (60-138 mg/100 g) was similar to that of kiwi fruits (98-180 mg/100 g), higher than that of oranges (37-92 mg/100 g) and papayas (62 mg/100 g) (Rodrigues et al., 2001), but lower than that of jujubes (500 mg/100 g). Nonetheless, our values for the genus *Adansonia* were lower than those of 300 mg vitamin C /100 g DM in *A. digitata* reported by Gebauer et al. (2002).

Polyphenols

As with vitamin C and proteins, the polyphenolic content of the six baobab species showed high variability among species (Table 1). *A. perrieri* had the lowest polyphenolic content (329 mg/100 g), whereas *A. za* had the highest (1706 mg/100 g). Table 1 shows that *A. za*, *A. digitata* and *A. madagascariensis* can be considered as the ones producing fruits with high polyphenolic contents. Lamien-Meda et al. (2008) also report high polyphenol content in *A. digitata,* in relation to 13 other fruits of Burkina Faso.

Antioxidant capacity

The ORAC method was used to evaluate the antioxidant

capacity of the pulp samples collected from Malagasy baobab species and the African mainland species, *A. digitata*. The antioxidant capacities of the fruit pulp from the six species were comprised between 109 and 159 µmol TE/g. All Malagasy species had higher antioxidant capacity than *A. digitata*. Consumption of foods rich in antioxidant is highly recommended as a factor contributing to the prevention and reduction of human cell death caused by oxidation, and thus reducing the incidence of some diseases. The baobab high antioxidant capacity could contribute to increase its economic value. Its antioxidant capacity was higher than many widely consumed fruits and vegetables. Indeed *A. digitata* had 108 µmol TE/g DM antioxidant capacity. For comparison, oranges have 100 µmoles TE/mg, kiwis 340 µmoles TE/mg (Vertuani et al., 2002), strawberries 15 µmoles TE/mg, lentils 81 µmol TE/g, grapes 87 µmol TE/g, blackberries 72 µmol TE/g and tomatoes 67 µmol TE/g, and these products are well known for their high antioxidant capacity (Lam et al., 2005).

Minerals and metals

The mineral composition of juices and fruits are among the criteria that guide consumers' choice. Among the minerals analyzed (Table 2) in this study, the K content was higher than 1528 mg/100 g DM in all six species indicating that K was the most predominant mineral, and *A. za* had the highest K content (3054 mg/100 g). The baobab fruit thus ranks among species with a very high source of K. The content of P and Na varied from 57 to 116 mg/100 g DM and 2.3 to 43.4 mg/100 g DM, respectively. Ca and Mg are major minerals in human nutrition, and their levels were comprised between 313 and 658 mg/100 g DM for Ca, and 176 and 255 mg/100 g DM for Mg. Therefore, the baobab pulp might be an important source of Ca, higher than that of milk. The Mg content was also high and similar to those found in almonds and hazelnuts (USDA, 2011a). Na and Fe are also important in biological systems, mainly as electrolytes and as heme for Fe in blood cells; their levels ranged between 0 and 40 mg/100 g DM, and 49 and 202 ppm, respectively. Arnold et al. (1985), and Osman (2004) reported a high variation in the Fe content of *A. Digitata*, 1.1 and 10.4 mg/100 g DM, respectively.

The high variability in mineral and metal contents in the baobab pulp has been largely highlighted (Nour et al., 1980; Arnold et al., 1985; Sena et al., 1998; Osman, 2004). It may be associated, at least in part, with the soil type and origin of samples.

Potential impact on daily intake

For 4-to-8-year old children, the recommended daily intake (RDI) is 0.025 g/100 g vitamin C, 0.8 g/100 g Ca,

Table 2. Mineral composition of pulp from six baobab species from Madagascar and Côte d'Ivoire.

Minerals (mg/100 g)	Malagasy species					African species		References
	A. za [9]	A. perrieri [20]	A. grandidieri [14]	A. fony var. rubrostipa [13]	A. madagascariensis [1]	A. digitata [3]	Literature mg/100 g	
P	57	116	72	94	80	80	106	Obizoba and Amaechi (1993); Saka and Msonthi,(1994); Glew et al. (1997); Sena et al. (1998)
K	3054	2728	2221	2735	2252	1528	1794	Saka and Msonthi (1994); Sena et al. (1998); Osman (2004)
Ca	464	658	356	313	372	345	302	Sena et al. (1998); Lockett et al. (2000); Osman (2004)
Mg	255	224	176	198	225	199	195	Glew et al. (1997); Sena et al. (1998); Lockett et al. (2000); Osman (2004)
Na	9.0	5.9	8.9	43.4	5.7	2.3	14.8	Glew et al. (1997); Sena et al. (1998); Osman (2004)
Cu	0.7	0.3	0.8	0.5	0.9	1.5	0.9	Obizoba and Amaechi (1993); Sena et al (1998); Lockett et al. (2000); Osman (2004)
Fe	10.5	9.2	16.6	10.6	4.9	10.0	4.3	Arnold et al. (1985); Obizoba and Amaechi (1993);
Mn	1.4	0.70	1.4	2.2	0.6	2.1	0.7	Saka and Msonthi (1994); Glew et al. (1997); Sena et al. (1998); Lockett et al. (2000); Osman (2004)

[sample numbers].

0.01 g/100 g Fe, and 19 g/100 g proteins (USDA 2011b). The same authors recommend 0.085 g/100 g, 1g/100 g, 0.027 g/100 g, and 71 g/100 g of vitamin C, Ca, Fe, and proteins, respectively, for 19-to-30-year-old pregnant women. Thus, consumption of 40 g of pulp of A. digitata (the richest species in vitamin C) and A. za (the poorest) by a child aged 4-8 years covers 220 and 96% vitamin C RDI, respectively. For pregnant women aged 19-30 years, the consumption of 40 g of baobab pulp covers from 30 to 60% of RDI depending on the species (Table 3). The calcium, protein and iron values found in our species (Table 3) were similar to those observed by Chadaré et al. (2009) for A. Digitata. The highest values reported by these authors show that the consumption of 40 g of baobab pulp covers the following RDIs: 41.5% of Fe, 25.4% of zinc, and 35% of calcium for children aged 4 to 8 years. The consumption of 40 g of pulp by a pregnant woman covers from 84 to 141% of RDI of vitamin C, considering the lowest and highest vitamin C contents of the pulp reported by Chadaré et al. (2009). The baobab pulp is undoubtedly a valuable source of vitamin C.

Discrimination of baobab species

A principal component analysis (PCA) was performed on all biochemical data measured on the six species. The first three main axes explained 40.18, 22.46 and 16.44%, respectively, of the total variance. Figure 2 shows the circles of correlation of plans 1-2 and 1-3 and species representation on these plans. A. za is clearly separated from the others, A. digitata and A. grandidieri form a second group, which is distinct from other species. A third group consists of three Malagasy species, A. fony var. rubrostipa, A. perrieri and A. madagascariensis.

The species of this group are clearly separated in the graphical representation of plan 1-3. Glucose, fructose, sucrose, Ca and K variables were positively correlated, and starch and

Table 3. Recommended daily intake (RDI) for children (aged 4-8 years) and pregnant women (aged 19-30 years).

Macronutriments	Subject	Need/day	A.za (40 g)	A. perrieri (40 g)	A. grandidieri (40 g)	A. fony var. rubrostipa (40 g)	A. madagascariensis (40 g)	A. digitata (40 g)
			(%) Supply	(%) Supply	(%) Supply	(%) Supply	(%) Supply	(%) Supply
Vitamin C	Children 4-8 years	0.025	220	113	96	148	122	107
	Women 19-30 years	0.085	65	33	28	43	36	32
Ca	Children 4-8 years	0.8	23	33	16	16	19	17
	Women 19-30 years	1	18	26	13	12	15	14
Proteins	Children 4-8 years	19	5	7	7	13	8	6
	Women 19-30 years	71	1	2	2	4	2	2
Fe	Children 4-8 years	0.01	42	37	81	42	20	40
	Women 19-30 years	0.027	16	14	30	16	7	15

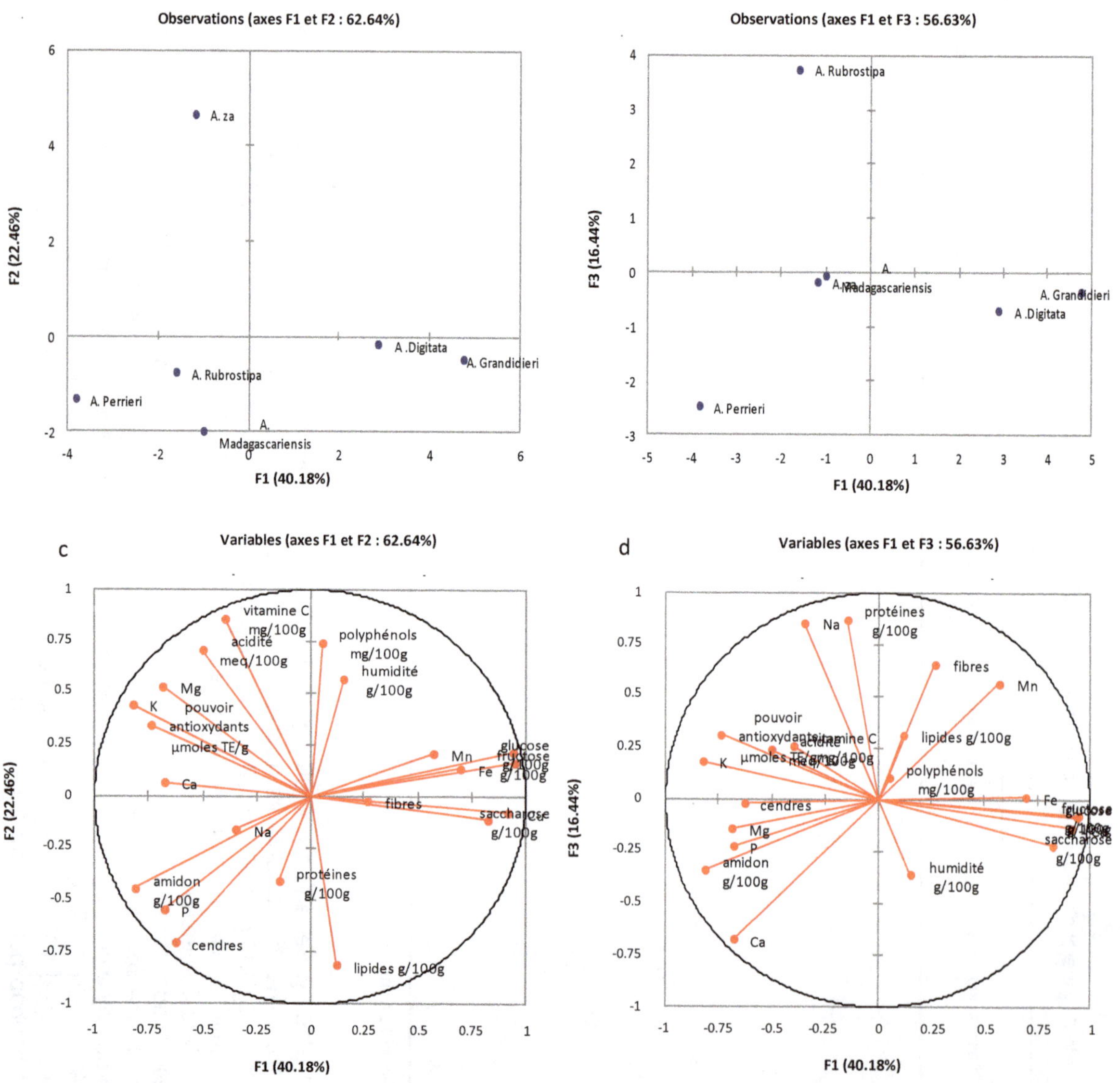

Figure 2. Biplot of baobab species loadings (a and b) and variable scores (c and d) on the first three axes of the PCA.

antioxidant capacity negatively (Figure 2). Vitamin C and polyphenol variables were positively correlated to axis 2, and lipids and ash negatively. Variables proteins and Na were positively correlated to axis 3, and Ca was negatively correlated. *A. za* was characterized by a high content of vitamin C and a low fat content. *A. digitata* and *A. grandidieri* were characterized by low levels of sugar (glucose, fructose, sucrose) and Ca. *A. fony* var. *rubrostipa*, *A. perrieri*, and *A. madagascariensis* were characterized by higher starch, ash and fat contents than those in other species (plan 1-2). The protein content and Na were highest (plan 1-3) in *A. fony* var. *rubrostipa*, whereas *A. perrieri* was characterized by its Ca content.

Conclusion

This study showed a high variation in the biochemical, mineral and nutritional characteristics of five endemic species of Madagascar and one African Sudano-Sahelian species, *A. digitata*. All investigated species showed higher vitamin C and mineral contents, and stronger antioxidant capacity than commonly consumed fruits. Because of its biochemical and nutritional characteristics, the baobab pulp is quantitatively and qualitatively nutritive, and suitable for marketing on a large scale. To increase the added-value of the baobab product, efficient processing methods to preserve pulp quality during storage or transformation are necessary. To date, despite the baobab fruit nutritional importance, the lack of knowledge on pulp preservation causes loss. Future research should focus on increasing the pulp storage time while preserving its nutritive and sensorial value, and ensure its stability.

ACKNOWLEDGEMENTS

The authors are grateful to the CIRAD technicians who provided assistance. They also thank Dr. B. Fofana (Agriculture and Agri-Food Canada) for his comments and suggestions during the preparation of the manuscript. This research received a financial support from the French Government and I. Cisse received a fellowship to conduct this study from the Government of Côte d'Ivoire.

REFERENCES

Anon (1990). Fruits and fruits products, in: HelrichK. (Ed.), Official methods of analysis of the association of official analytical chemists (AOAC), Arlington, USA, pp. 910-928.

Arnold T, Well M, Wehmeyer A (1985). Khoisan food plants: taxa with potential for economic exploitation. In:Wickens, J. R., Goodin, and Field. D. V. London Ed. Plants for Arid Lands. Allen and Unwin, pp. 69-86.

Assogbadjo A, Chadaré F, Kakaï R, Fandohan B, Baidu-Forason J (2012). Variation in biochemical composition of baobab (*Adansonia digitata*) pulp, leaves and seeds in relation to soil types and tree provenances. Agric. Ecosyst. Environ. 157:94-99.

Assogbadjo A, De Caluwé B, Sinsin J, Codjia T, Van Damme P (2006). Indigenous Knowledge of Rural People and Importance of Baobab Tree (*Adansonia digitata* L.) in Benin. in Z. F. Ertug, ed., Proceedings of the Fourth International Congress of Ethnobotany (ICEB 2005). Yeditepe University, Istanbul, 21–26 August 2005. pp. 39-47.

Assogbadjo A, Sinsin J, Van Damme P (2005). Caractères morphologiques et production des capsules de baobab (*Adansonia digitata* L.) au Benin. Fruits 60:327-340.

Besco E, Bracioli E, Vertuani S, Ziosi P, Brazzo F, Bruni R, Sacchetti G, Manfredini S (2007). The use of photochemiluminescence for the measurement of the integral antioxidant capacity of baobab products. Food Chem. 102:1352-1356.

Buchmann C, Prehsler S, Hartl A, Vogl C (2010). The importance of baobab (*Adansonia digitata* L.) in rural West African subsistence - Suggestion of a cautionary approach to international market export of baobab fruits. Ecol. Food Nutr. 49:145-172.

Chadaré F, Linnemann A, Hounhouigan J, Nout M, Van Boekel M (2009). Baobab food products: a review on their composition and nutritional value. Crit. Rev. Food Sci. 49:254-274.

Cisse M, Sakho M, Dornier M, Diop C, Reynes M, Sock O (2008). Caractérisation du fruit du baobab et étude de sa transformation en nectar. Fruits 64:19-34.

Codjia J, Fonton-Kiki B, Assogbadjo A, Ekué R (2001). Le Baobab (*Adansonia digitata*), une Espèce à Usage Multiple au Bénin. Cent. Int. d'Ecodéveloppement Intégré. (CECODI), Cotonou, Bénin, West Africa, P. 45.

Codjia J, Assogbadjo A, Ekué M (2003). Diversité et valorisation au niveau local des ressources forestières alimentaires végétales du Bénin. Cah. Agric. 12:321-331.

De Caluwé E, De Smedt S, Assogbadjio A, Samson R, Sinsin B, Van Damme P (2009). Ethnic differences in use value and use patterns of baobab (*Adansonia digitata* L.) in Northern Benin. Afr. J. Econ. 47(3):433-440.

Danthu P, Roussel J, Gaye A. and El Mazzoudi E. (1995). Baobab (Adansonia digitata L.) seed pretreatments for germination improvement. Seed Science and Technology 23: 469-475.

De Caluwé E, Halamova K, Van damme P (2010). A review of traditional use, phytochemistry and pharmacology. Afrika focus 1, 23:11-51.

De Smedt S, Simbo D, Van Camp J, De Meulenaer B, Potters G, Samson R (2010a). Opportunities for Domestication of the African Baobab Tree (*Adansonia digitata* L.) in Mali Tropentag, September 14-16, Zurich, Germany "World Food System - A Contribution from Europe.

De Smedt S, Alaerts K, Kouyaté A, Van damme P, Potters G, Samson R (2010b). Phenotypic variation of baobab (*Adansonia digitata* L.) fruit traits in Mali. Agroforest. Syst. 82:87-97.

Diop A, Sakho M, Dornier M, Cisse M, Reynes M (2005). Le baobab africain (*Adansonia digitata* L.) : principales caractéristiques et utilisations. Fruits 61:55-69.

Gaydou E, Bianchini JP, Ralaimanarivo A (1982). Cyclopropenoid Fatty Acids in Malagasy baobab *Adansonia grandidieri* (Bombacacae) seed oil. Fett. Wiss. Technol. 12:468-472.

Gebauer J, El Siddig K, Ebert G (2002). Baobab (*Adansonia digitata* L.): a review on a multipurpose tree with promising future in the Sudan. Gartenbauwissenscha 67:155-160.

Georgé S, Brat P, Alter P, Amiot M (2005). Rapid determination of polyphenols and vitamin C in plant-derived products. J. Agric. Food Chem. 53:1370-1373.

Glew R, Van der Jagr D, Laeken C, Griveui L, Smith G, Pastuszyn A., Millson M (1997). Amino acid, fatty acid, and mineral composition of 24 indigenous plants of Burkina Faso. J. Food Comp. Anal. 10:205-217.

Ibiyemi S, Abiodun A, Akanji S (1988). *Adansonia digitata, Bombax* and *Parkia filicoideae* Welw: Fruit Pulp for the Soft Drink Industry. Food chem. 28:111-116.

Kaboré D, Sawadogo-Lingani H, Diawara B, Compaoré C, Dicko M, Jakobsen M (2011). A review of baobab (*Adansonia digitata*) products: Effect of processing techniques, medicinal properties and uses. Afr. J. Food Sci. 5:833-844.

Kamatou G, Vermaak I, Viljoen A (2011). An updated review of

Adansonia digitata: A commercially important African tree. S. Afr. J. Bot. 77:908-919.

Künsch U, Shärer H, Patrian B, Hurter J, Conedera M, Sassella A, Jermini M, Jelmini G (1999). Quality assessment of chestnut fruits. Act. Hortic. 494:119-128.

Lam H, Proctor A, Howard L, Cho M (2005). Rapid fruit extracts antioxidant capacity determination by Fourier transform infrared spectroscopy. J. Food Sci. 70:545-549.

Lamien-Meda A, Lamien C, Compaoré M, Meda R, Kiendrebeogo M, Zeba B, Millogo J, Nacoulm O (2008). Polyphenol Content and Antioxidant Activity of Fourteen Wild Edible Fruits from Burkina Faso. Molecules 13:581-594.

Lockett C, Calvert C, Grivetti L (2000). Energy and micronutrient composition of dietary and medicinal wild plants consumed during drought. Study of rural Fulani, Northeastern Nigeria. Int. J. Food Sci. Nutri. 51:195-208.

Munthali C, Chirwa P, Akinnifesi F (2012). Phenotypic variation in fruit and seed morphology of Adansonia digitata L. (baobab) in five selected wild populations in Malawi. Agroforest. Syst. 85:279-290.

Murray S, Schoeninger M, Bunn H, Pickering T, Marien J (2001). Nutritional composition of some wild plant foods and honey used by Hadza foragers of Tanzania. J. Food Comp. Anal. 13:1-11.

Nour A, Magboul B, Kheiri N (1980). Chemical composition of baobab fruit (Adansonia digitata). Trop. Sci. 22:383-388.

Obizoba I, Amaechi N (1993). The effect of processing methods on the chemical composition of baobab (Adansonia digitata L.) pulp and seed. Ecol. Food Nutri. 29:199-205.

Obizoba I, Anyika J (1994). Nutritive value of baobab milk (gubdi) and mixtures of baobab (Adansonia digitata L.) and hungry rice, acha (digitaria exilis) flours. Plant Food Hum. Nutr. 46:157-165.

Osman M (2004). Chemical and nutrient analysis of baobab (Adansonia digitata) fruit and seed protein solubility. Plant Food Hum. Nutr. 59(1):29-33.

Parkouda C, Sanou H, Tougiani A, Korbo A, Nielsen D, Tano-Debrah K, Ræbild A, Diawara B, Jensen J (2012). Variability of baobab (Adansonia Digitata L.) fruits' physical characteristics and nutrient content in the West African Sahel. Agroforest. Syst. 85:455-463.

Ralaimanarivo A, Gaydou E, Bianchini JP (1982). Fatty acid composition of seed oils from six Adansonia with particular reference to cyclopropane and cyclopropane acids. Lipides 17:1-10.

Razanameharizaka J, Grouzis M, Ravelomanana D, Danthu P (2006). Seed storage, behaviour and seed germination in African and Malagasy baobabs (Adansonia species). Seed Sci. Res. 16:83-88.

Rodrigues R, De Menezes H, Cabral L, Dornier M, Reynes M (2001). An amazonian fruit with a high potential as natural source of vitamin C: the camu-camu (Myrciaria dubia). Fruits 56:345-354.

Saka J, Msonthi J (1994). Nutritional value of edible fruits of indigenous wild trees in Malawi. Forest Ecol. Manage. 64:245-248.

Sanchez A, Osborne P, Haq N (2010). Identifying the global potential for baobab tree cultivation using ecological niche modelling. Agroforest. Syst. 80:191-201.

Scheuring J, Sidibé M, Frigg M (1999). Malian agronomic research identifies local baobab tree as source of vitamin A and vitamin C. Sight Life Newslett. 1:21-24.

Sena L, Vanderjagt C, Rivera A, Tsin I, Muhamadu O, Mahamadou M, Millson A, Pastuszyn A, Glew R (1998). Analysis of Nutritional Components of Eight Famine Foods of the Republic of Niger. Plant Food Hum. Nutr. 52:17-30.

Sidibé M, Scheuring J, Tembely D, Sidibé M, Hofman P, Frigg M (1996) Baobab homegrown vitamin C for Africa. Agroforestry Today 8:13-15.

Sidibé M, Williams J (2002). Baobab – Adansonia digitata. Fruits for the future 4, International Centre for Underutilised Crops, Southampton, UK, P. 105.

Soloviev P, Niang, T, Gaye A, Totte A (2004). Variabilité des caractères physico-chimiques des fruits de trois espèces ligneuses de cueillette, récoltés au Sénégal: Adansonia digitata, Balanites aegypriaca et Tamarildus indica. Fruits 59:109-119.

USDA (2011a). National Nutrient Database for Standard Reference. Available at http://ndb.nal.usda.gov/ndb/foods/list, accessed 09.03.2012.

USDA (2011b). Food and nutrition information center. Available at http://www.iom.edu/Activities/Nutrition/SummaryDRIs/~/media/Files/Activity%20Files/Nutrition/DRIs/5_Summary%20Table%20Tables%201-4.pdf, accessed 09.03.2012.

Vaillant F, Perez A, Davila I, Dornier M, Reynes M (2005). Colorant and antioxidant properties of red-purple pitahaya (Hylocereus sp.). Fruits 60:3-12.

Van Soest P, Robertson J, Lewis B (1991). Methods for dietary fiber, detergent fiber, and non-starch polysaccharides in relation to animal nutrition. J. Dairy Sci. 74:3583-3597.

Vertuani S, Braccioli E, Bulwni V, Manfredini S (2002). Antioxidant capacity of Adansonia digitata fruit pulp and leaves. Acta Phytotherapeutica 5:2-7.

Wickens G, Lowe P (2008). The baobabs: Pachycauls of Africa, Madagascar and Australia. Dordrecht: Springer, London, P. 500.

22

Growth and expansion of strawberry fruit (*Fragaria* x *ananassa* Duch.) under water stress conditions

Modise D. M.

School of Agriculture and Life Sciences, University of South Africa (UNISA), P/Bag X6, Florida 1710, South Africa.

This investigation was carried out to find out if a constant number of strawberry fruit in the plants would have differing rates of growth and expansion when subjected to different levels of water stress at specific growth stages. Soil water stress treatments were imposed at flowering (flo) and at fruiting (fru), by withholding water until the available soil water were 0.40 to 0.45% v/v for the normal stress treatment (normal), 0.35 to 0.40% v/v mild stress (ms) and 0.25 to 0.35% v/v for severe stress (ss). The ms fru, ss fru and ss flo treatments showed significantly lower fruit weights than other treatments while fruit firmness was significantly increased by ms fru and ss fru treatments in the primary, secondary fruit and tertiary fruit. The total soluble solids (TSS) were not affected significantly by the water stress treatments. Osmotic adjustment may be attributed to the ability of the water stressed strawberry fruit to grow and expand post anthesis. This research provides an understanding of the effects of water deficits on fruit quality when other factors such as fruit number and fruit positioning on the inflorescence are similar in all experimental units. Strawberry producers may consider reduced crop loading to ameliorate reduced fruit size, when faced with water deficit irrigation regimes.

Key words: Crop load, fruit size, fruit weight, total soluble solids (TSS), water stress.

INTRODUCTION

Plants carrying a heavy crop load have a lower turgor potential when compared to those that have a light load thus are likely to show reduced fruit growth and fruit size (McFadyen et al., 1996). The premise is that reduction in fruit size, diameter and weight can be counteracted by the reduction in crop load of water stressed strawberries. Naor et al. (1997, 1999) and Mpelasoka et al. (2001) suggested that there would be increased levels of assimilate availability through increased photosynthesis (Pn) and subsequently increased fruit turgor potential (Ψ_{fp}) and fruit growth due to reduced crop load. Pomper and Breen (1995, 1997) suggested that osmotic adjustment may enable fruit expansion to take place due to solute accumulation in the apoplast of strawberry fruit.

Dwyer et al. (1987) remarked that since the fruit is a major sink in plants, water stress imposition even if mild could reduce fruit yield significantly. Although a lot of research work has been done on the effect of crop load on water relations in fruit, none have addressed the possibility of reducing crop load to counteract the reduction in fruit size and weight in water stressed fruit of strawberries. There are no reports in the literature of any studies that have been conducted on strawberry fruit expansion under deficit irrigation where fruit load is reduced to single trusses. The objective of this experiment was therefore to determine the rate of fruit expansion under deficit watering in primary, secondary and tertiary fruit, when the crop load had been reduced to a few fruit in the strawberry plant. It is postulated that water stress effects can be mitigated by reducing crop load in strawberry without negative implications on fruit quality as determined by Naor et al. (2008) and Lopez et al. (2010) in studies on apples and peach trees respectively.

The use of deficit irrigation strategies have been applied previously to conserve water and to control vegetative growth of plants where water shortages are increasingly becoming a problem in peach (Chalmers et al., 1981), pear (Chalmers et al., 1986), grapevines (Matthews et al., 1987) and on apple (Ebel et al., 1995) thus increasing farmers profits (Fereres and Soriano, 2007; Geerts and Raes, 2009) and increasing the quality of fruit produced (Trought and Naylor, 1988).

MATERIALS AND METHODS

Plant material and treatments

This study was conducted at the University of Nottingham in the United Kingdom. Bare rooted strawberry seedlings cv. Elsanta were grown in 13 cm pots containing Levington M2 compost. The plants were established in a polytunnel and transferred to the glasshouse three weeks later when roots had fully developed. Plants only received natural sunlight in the glasshouse. Upon flowering (on the 5th week), the primary, secondary and tertiary flowers were tagged using different colour tags for ease of identification. The rest of the flowers were removed and plants were not allowed to develop any more flowers. Only one truss per plant (primary truss) was allowed to grow and develop fruit therefore the primary, secondary and tertiary fruit were all located on the primary truss. The experimental unit did not encompass a full crop load as a control as this would have produced an undesired dimension to this study. The aim was not to compare heavy with light loads but to measure performance under various water regime treatments. To achieve good fruit set and to ensure normal fruit development, flowers were hand pollinated at anthesis every 2 to 3 days between 10H00-12H00 h, with a soft squirrel brush.

Three levels of soil water stress were imposed (normal, mild and severe) by withholding water for different periods until the available soil water was 0.40 to 0.45% v/v for the normal stress treatment, 0.35 to 0.40% v/v mild stress (ms) and 0.25 to 0.35% v/v for severe stress (ss) as measured using a Theta probe – Soil Moisture Sensor (Type ML2X, Delta-T Devices Ltd, Cambridge). The plants were re–watered to achieve field capacity after each stress period. The stresses were applied at flowering (flo), that is, when at least 80% of the flowers had opened, and at fruiting (fru), at the green fruit stage that is, 10 days after anthesis.

The plants were arranged in three randomised blocks. Each block consisted of five plots with each plot being representative of a stress treatment. Each plot had 6 replicate plants with guard plants at the end of the back row and guard rows at the sides of each plot. A row was thus randomly allocated a water stress treatment and this was replicated over the other two additional blocks. Therefore there were 30 plants per block giving a total of 90 experimental plants for all 3 blocks.

Fruit measurements

After ripening, the primary, secondary and tertiary fruit were weighed, analysed for texture and measured for firmness, total soluble solids (TSS), fruit length and diameter.

Fruit weight

Fruit were weighed when ripe using an electronic balance (PJ Precisa Junior 500 C,Precisa Balances Ltd, Bucks, UK).

Texture

The freshly picked strawberries were analysed for firmness using a Stevens – LFRA Texture Analyser (Stevens, Coventry, UK) with a penetration probe of 13.6 mm diameter applied to the longitudinal axis of the fruit. The maximum force required for the probe to penetrate the fruit by 6 mm at a speed of 1.0 mm $^{s-1}$ was recorded. Fruit from each plant was individually weighed and used for firmness determination. The same fruit was also used for measurement of total soluble solids.

Total soluble solids (TSS)

TSS measurements were also taken from the same fruit used for texture analysis. A Delta refractometer (Bellingham and Stanley Ltd, Kent, UK) was used for measuring the TSS. Juice was squeezed from the fruit by hand at the distal end to release about 0.01 ml juice onto the lens of the refractometer.

Fruit length

Measurements were taken every other day in the morning longitudinally on primary and secondary fruit using an electronic vernier calipers (Mitutuyo (UK) Ltd.

Diameter

Measurements were taken every other day in the morning on the equatorial axis on primary and secondary fruit also using the electronic vernier calipers.

Plant measurements

Canopy height

Measurements of canopy height were taken at both the fruiting and the ripening stage using a ruler. The canopy heights measured from the soil surface were recorded.

Number of leaves

Leaves were counted manually during the fruiting stage and at the ripening stage of the fruit.

Plant fresh and dry weights

At the end of the experiment, plant biomass (whole canopies) were cut at the soil surface and weighed before placing in pre-weighed paper bags for drying in the oven.Dry weights were determined after 48 h of drying in a 70°C oven. All plant, fruit growth and quality measurements were taken on 6 plants per treatment/block, each plant had 3 fruit (primary, secondary and tertiary) giving a total of 18 fruit per treatment per block.

Data analysis

Data collected were subjected to analysis of variance (ANOVA)

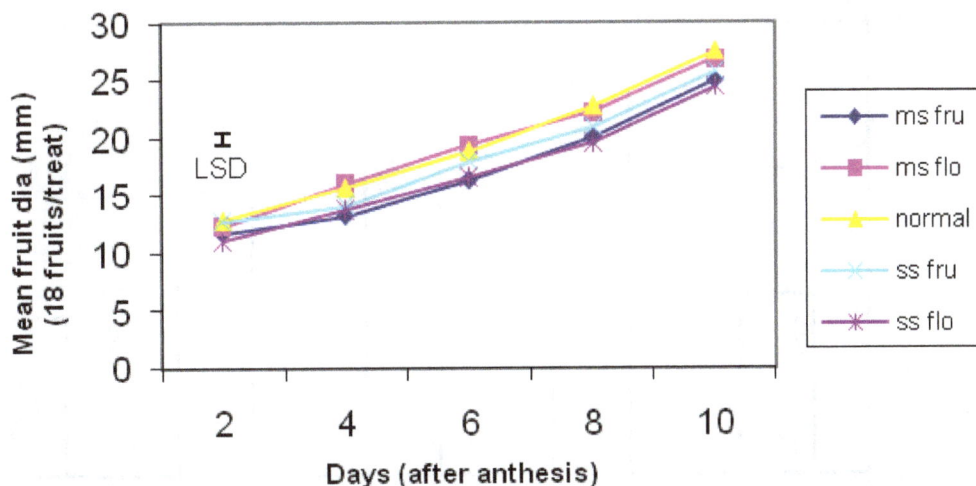

Figure 1. Effect of treatments on primary fruit length per 2 day interval diameter increase per 2 day interval.

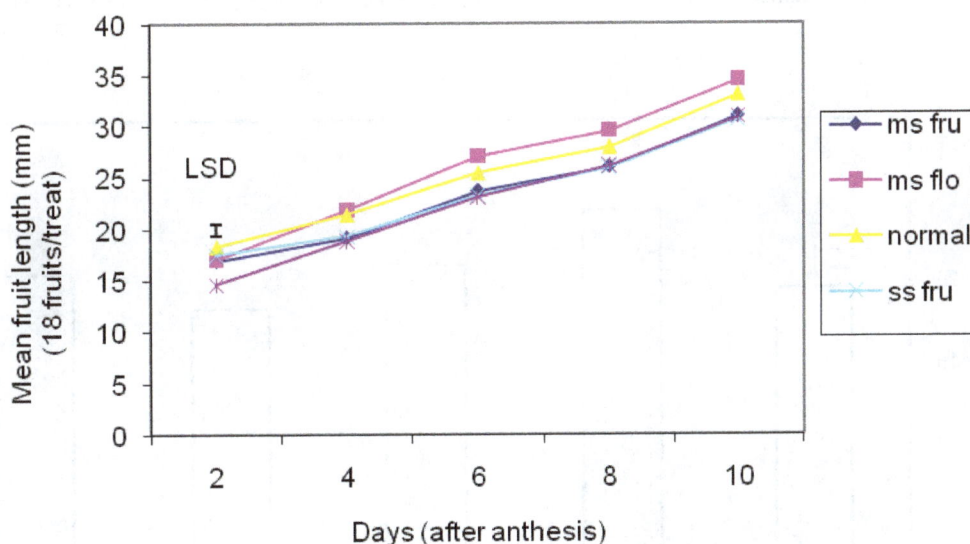

Figure 2. Effect of treatments on primary fruit length per 2 day interval diameter increase per 2 day interval.

using Genstat (Rothamstead) and the results were considered to be significant P < 0.01 level of probability. Means were separated using the Least Significant Difference (LSD).

RESULTS

Primary fruit

There were differences in the diameter and in the length of primary fruit among the treatments (Figures 1 and 2). Fruit from ms fru, ss flo and ss fru treatments were significantly smaller than those from normal and ms flo treatments. On average, every 2 days there was an increase of between 3.3 and 3.65 mm in the diameter of primary fruit up to the 10th day when fruit became ripe. The largest mean diameter increases were between 2nd and 6th day and between 8th and 10th day after anthesis. The mean length of primary fruit increased by between 3.29 and 4.36 mm per day until ripening on the 10th day after anthesis (Figure 2). The difference between mean length due to ss fru and normal treatments when computed over overall mean diameter increase for all treatments every 2 days (3.84 mm) was 14%. The largest mean length increasing trend was between 2 and 4 days and 8 to the 10th day after anthesis (R^2 = 0.95) and were shown by ms flo, ss flo and normal treatments (Figure 2).

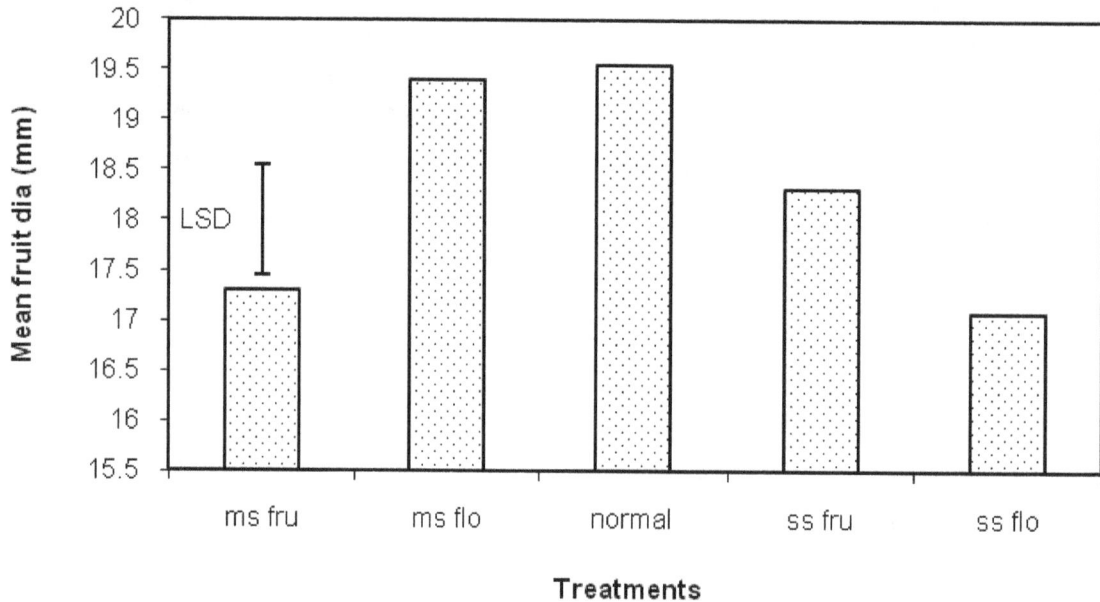

Figure 3. Effect of treatments on final diameter of primary fruit at ripening stage (that is, 10 days after anthesis).

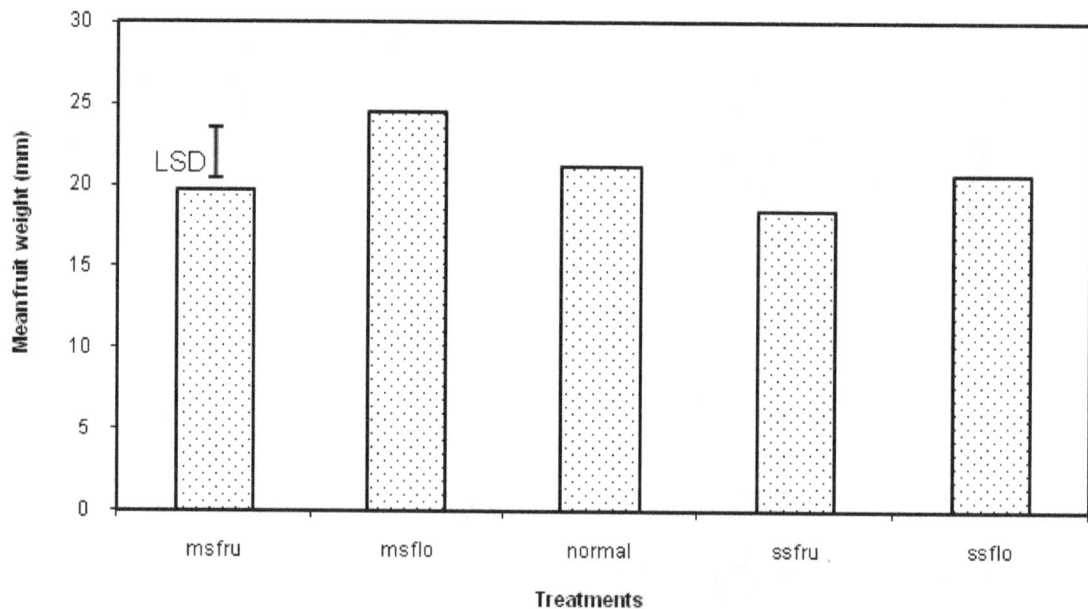

Figure 4. Effect of treatments on weight of primary fruit at harvest at ripening stage (that is, 10 days after anthesis).

At the end of the experiment the ms fru, ss fru and ss flo treatments resulted in significantly smaller fruit in terms of diameter (Figure 3) compared to normal treatment and lower weight for ms fru and ss fru (Figure 4). The ms flo and normal treatments were not significantly different from each other. TSS was plotted against mean fruit weight (of all primary fruit) using CurveExpert 1.3 (Zen University). There was a linear decrease in TSS of

primary fruit with increased fruit weight, the correlation coefficient was 0.78 (Figure 5).

Secondary fruit

The secondary fruit showed a different trend to the primary fruit. The diameter was not significantly affected

S = 1.64191099
r = 0.78526916

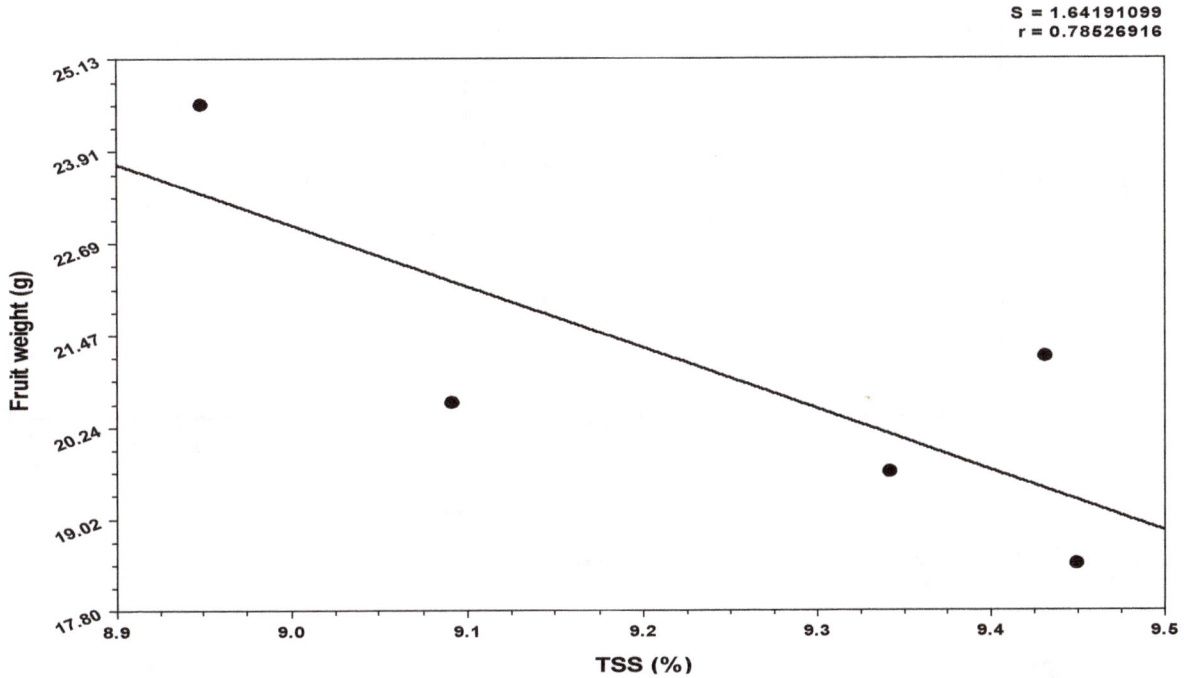

Figure 5. Relationship of TSS with fruit weight of primary fruit (mean of 6 fruit).

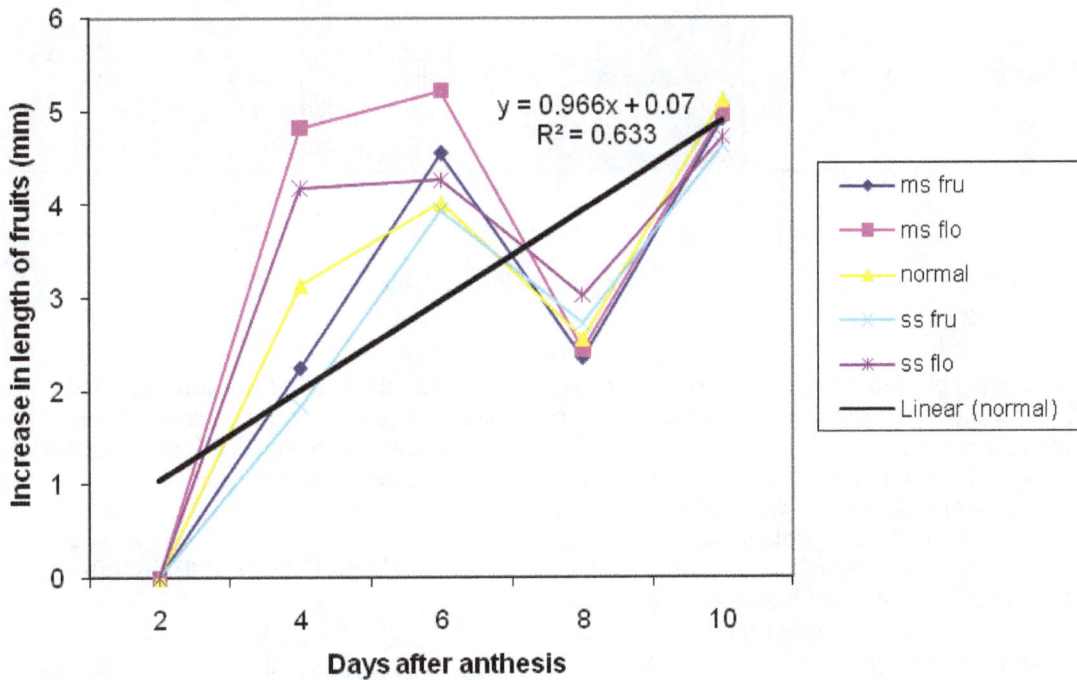

$y = 0.966x + 0.07$
$R^2 = 0.633$

Figure 6. Mean increase in secondary fruit diameter during growth and development until ripening, due to treatments.

by the treatments but there was a significant difference in the mean diameter of the fruit (Figure 6). The ss fru and ms fru treatments caused significantly smaller fruit compared to other treatments. There were no interactions between treatments and days but there was a date effect. The largest mean increase of the diameter thus was between the 2nd and 4th day and also between the 8th and 10th (Figure 6).

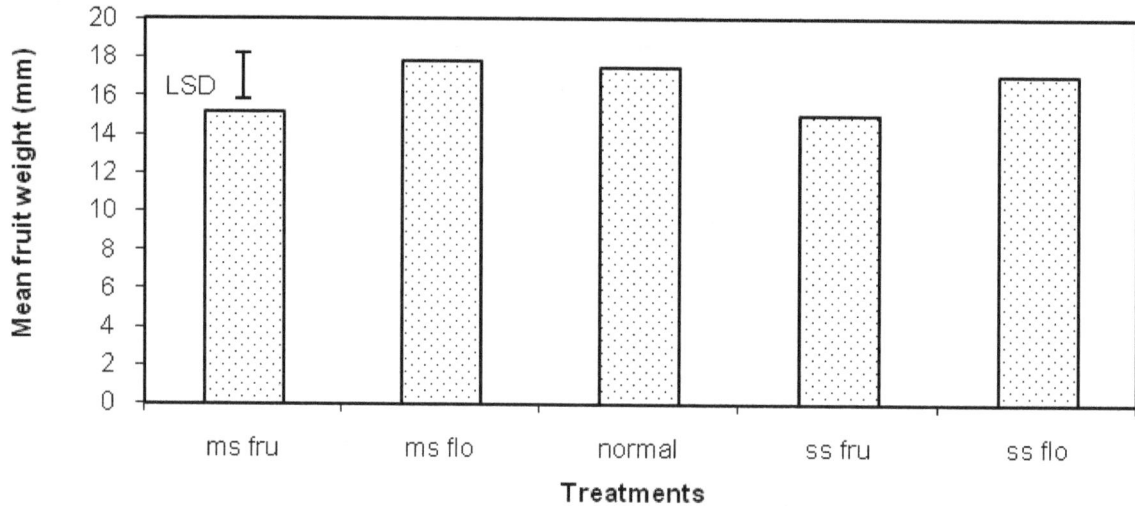

Figure 7. Effect of treatments on secondary fruit weight at harvest.

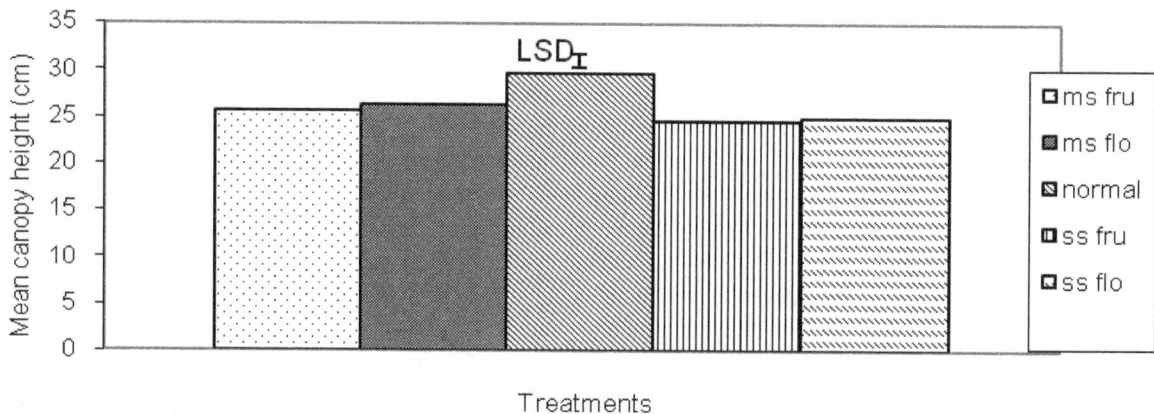

Figure 8. Effect of water stress on canopy height during the fruiting stage.

The mean length of secondary fruit increased by between 3.19 and 3.81 mm per 2 day interval until the fruit were ripening by the 10th day. There was a 7% difference between the mean length of ss fru (3.23 mm) and normal (3.49 mm) treatments when calculated over overall mean increase of all treatments (3.49 mm). There was a date effect and largest mean increases in length of fruit were between 2nd and 4th day followed by a decline and also between the 8th and 10th day (Figure 7).TSS and texture were not affected significantly by the treatments (results not shown). Similarly to primary fruit, there was a decreasing a trend in TSS to increasing fruit weight when these parameters were plotted against each other.

Tertiary fruit

As in the primary and secondary fruit, the ms fru, ss fru and ss flo treatments showed significantly lower fruit

weights than other treatments. TSS was not affected significantly by the water stress treatments. A relationship between TSS and fruit weight could not be established in tertiary fruit.

Measurement of plant parameters

Canopy height

The canopy height of the normal treatment was significantly higher than all other treatments during the fruiting stage of growth (Figure 8) but ss fru, ms fru and ss flo had significantly lower canopies than the other treatments during ripening stage of the fruit.

Plant fresh and dry weights

The plant fresh weights were significantly lower in ss fru

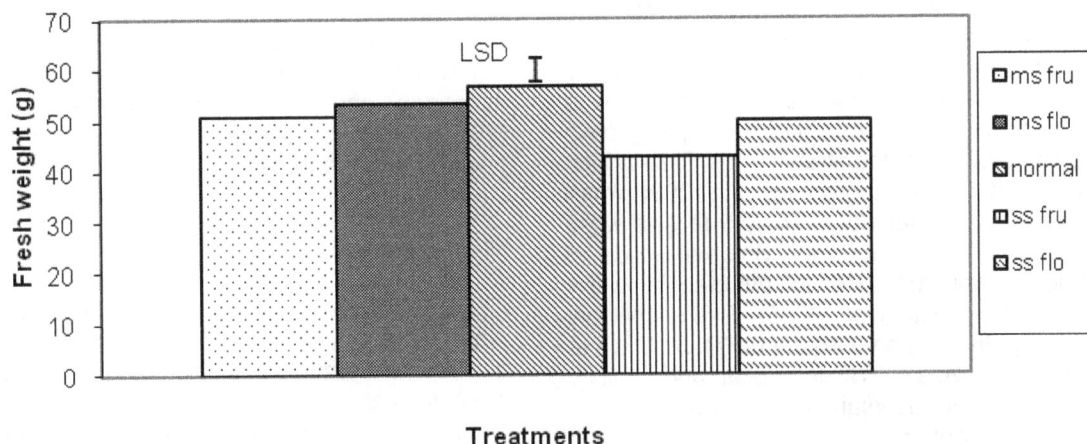

Figure 9. Effect of water stress on fresh weight of the plants.

and ss flo when compared to the rest of the treatments (Figure 9) and so were the dry weights.

Leaf area

Leaf area followed a similar trend to the fresh and dry weights. It was significantly lower in ss fru, ms fru and ss flo also had lower leaf areas though not statistically significant.

DISCUSSION

Primary fruit

The ss fru and ms fru treatments generally resulted in smaller primary fruit in terms of diameter, length and weight, the difference in mean increase between the ss fru and normal treatment was 6% for diameter and 14% for length. The ss fru treatment was selected over other treatments for comparison with the normal treatment because it showed the smallest increases in diameter and length in the measurements in every 2 days. These differences, though statistically significant are not very substantial considering that Mpelasoka et al. (2001) found about 4% reduction in the fruit growth rate of 'Braeburn' apples due to deficit irrigation when the crop load was light and 13% on full crop load. Caspari et al. (1994) proposed that water deficit might inhibit growth by 17% when it was practiced on 'Hosui' Asian pears.

The largest mean increases in the diameter and length of the primary fruit were between 2 and 4 days and 8 to 10 days after anthesis respectively and the treatments that caused this increase were ms flo and ss flo. This finding supports the theory of Caspari et al. (1994) that a short term water deficit that is, applied at anthesis (in this study ms flo and ss flo) when shoots are growing

vigorously, does not affect fruit growth, thus the similar trend between ms flo and ss flo. Water deficits applied at the fruit growth stages (ms fru and ss fru) will inhibit fruit growth (Caspari et al., 1993). These investigators were not able to explain the mechanism of the effects of short-term water deficits. It can be presumed though that there are changes in the hydraulic lift in the soil-plant continuum leading to partitioning of hydraulic conductance into the soil, root and stem components. This is one of the plant adaptations that act to buffer plants against damaging effects of water deficits (Richards and Caldwell, 1987). The results indicate that the 2nd to 4th day after anthesis were periods of rapid fruit growth followed by a lag phase of the 4th to the 6th day and then finally the rapid fruit growth phase leading to fruit ripening. Johnson and Handley (2000) acknowledged that there are rapid periods of growth after bloom followed by a lag phase that precedes the rapid fruit growth prior to harvesting. This type of growth is often referred to as double sigmoidal growth because it is characterized by two periods of rapid growth with an intervening slow phase of growth (Perkins-Veazie, 1995). Schwab and Raab (2004) observed that the growth of fruit fits a single sigmoidal curve or is biphasic depending on the cultivar.

There were no significant increases in the diameter and length of the primary fruit at 10 days after anthesis as the fruit had reached the pink-red colour stages and were beginning to ripen. The maturation period was relatively short (about 13 days after green fruit stage). Perkins-Veazie (1995) reported comprehensively on the disagreements among researchers on the pattern of berry growth. Knee et al. (1977) and Cheng and Breen (1992) have reported a stoppage of cell division between 7 and 15 days after anthesis. They further noted that rapid cell expansion follows cell division resulting in a sharp rise in fruit cell volume 10 days after anthesis. The large diversity in fruit growth patterns is influenced by the number of receptacle cells per achene, which in turn is a factor of environmental conditions, genetic variation and cultivar type (Cheng and Breen, 1992). The temperatures were high during the experimental period since it was summertime, and this may have aided the rapid growth and development process of the strawberries. This notion was ably explained by Warrington et al. (1999) when affirming that growth and maturity of apple fruit is affected by early season temperatures. Total Soluble Solids of primary and secondary fruit were not affected by water stress treatments (Reynolds et al., 2005). This is contrary to conclusions by Irving and Drost (1987) that water stress increased the levels of TSS in apples, and findings by Mpelasoka and Behboudian (2002) that TSS is increased by deficit irrigation.

Secondary fruit

Unlike the primary fruit, the mean diameter of secondary fruit was not significantly affected by water stress treatments. The mean length and mean fruit weight however, followed a similar trend to the primary fruit; they were significantly lower in the ms fru and ss fru treatments. The differences between the mean length of the ss fru and normal treatments were once more not substantial at 7% although ss fru and ms fru treatments resulted in smaller fruit in length and weight. Primary fruit were slightly larger than secondary fruit in mean diameter and mean length. The differences in the sizes of the fruit was caused by the fact that primary fruit flowers were the first to bloom and were therefore on a superior position on the inflorescence. Fruit size generally declines with fruit placed on inferior positions such as secondary, tertiary and quaternary (Moore et al., 1970). This concept could not be fully established in this research because the blooming of the primary and secondary flowers was almost simultaneous in most cases. TSS and texture of primary fruit was not affected significantly by the various water stress treatments.

Tertiary fruit

The weight of tertiary fruit was significantly low in ms fru, ssfru and surprisingly also in ss flo. The fruit subjected to ms fru and ss fru were smaller than those of other treatments. It can be deduced that water stress reduces fruit size in agreement with Mpelasoka and Behboudian (2002) and Lopez et al. (2010). Total Soluble Solids were not affected significantly by water stress in the tertiary fruit as were the primary and secondary fruit. The terminal (primary) fruit on the peduncle are stronger sinks of substrates unloaded by phloem, followed by secondary fruit and tertiary fruit in order of position on the inflorescence. The fruit weight therefore declines with inferior blossom position (Janick and Eggert, 1968). In this study, the inflorescence developed at the same time

and the hierachial order of flower development was not obvious, thus resulting in differences that were not statistically significant between the weight of primary, secondary and tertiary fruit. Nonetheless, Kassai et al. (2002) found no correlation between berry growth and position within a truss.

Plant measurements

At the fruiting stage that is, at the beginning of anthesis, plant canopy height was significantly lower in all treatments compared to normal treatment. At the fruit ripening stage though, ss fru showed significantly lower canopy height than all treatments followed by ms fru and ss flo. At fruiting, the shoot growth is more rapid therefore water stress would not significantly reduce canopy height, but during the ripening stages particularly the lag phase, shoot growth declines and vegetative growth may be reduced. Predictably, the ss fru treatment also demonstrated lower levels of fresh weight, dry weight and leaf area.

Conclusion

Although water stress can cause reductions in fruit diameter and length, the reduction is on average less than 10% for combined means of primary and secondary fruit when compared to plants that were not water stressed. This study suggests that imposing water stress (mild and/or severe at fruiting results in lower fruit size and weight. Primary trusses generally have larger fruit as the fruit are the first to flower as they are at a superior position. The loss in fruit size was expected to give a compensatory gain by the enhancement of fruit quality e.g. significantly increased sugars in fruit but that was not the case in the present experiment. Even though the strawberry plants were subjected to various water stress treatment levels, evidence presented here shows that fruit were still growing and expanding at a rate of between 3 and 4 mm every 2 days up to 10 days post anthesis. Thus, in countering water stress, reduced crop load may be applied but caution should be exercised not to lose quality of fruit in terms of size.

REFERENCES

Caspari HW, Behboudian MH, Chalmers DJ, Renquist AR (1993). The pattern of seasonal water use of Asian pears as determined by lysimeters and the heat-pulse technique. J. Am. Soc. Hort. Sci. 118:562-569.

Caspari HW, Behboudian MH, Chalmers DJ (1994). Water use, growth and fruit yield of 'Hosui' Asian pears under deficit irrigation. J. Am. Soc. Hort. Sci. 119:383-388.

Chalmers DJ, Mitchel PD, van Heek L (1981). Control of peach tree growth and productivity by regulated water supply, tree density and summer pruning. J. Am. Soc. Hort. Sci. 106:307-312.

Chalmers DJ, Burge G, Jerie PH, Mitchell PD (1986). The mechanism of regulation of 'Bartlett' pear fruit and vegetative growth by irrigation withholding and regulated deficit irrigation. J. Am. Soc. Hort. Sci. 111:904-907.

Cheng GW, Breen P (1992). Cell count and size relation to fruit size among strawberry cultivars. J. Am. Soc. Hort. Sci. 117:946-950.

Dwyer LM, Stewart DW, Houwing L, Balchin D (1987). Response of strawberries to irrigation scheduling. HortScience 22:42-44.

Ebel RC, Proesting EL, Evans RG (1995). Deficit irrigation to control vegetative growth in apple and monitoring fruit growth to schedule irrigation. HortScience 30:1229-1232.

Fereres E, Soriano MA (2007). Deficit irrigation for reducing agricultural use. J. Exp. Bot. 58 (2):147-159.

Geerts S, Raes D (2009). Deficit irrigation as an on-farm strategy to maximise crop water productivity in dry areas. Agric. Water Manag. 96:1275-1284.

Irving DE, Drost JH (1987). Effects of water deficit on vegetative growth, fruit growth and fruit quality in Cox's Orange Pippin apple. J. Hort. Sci. 62:427-432.

Janick J, Eggert DE (1968). Factors affecting fruit size in the strawberry. Proc. Am. Soc. Hort. Sci. 93:311-316.

Johnson RS, Handley DF (2000). Using water stress to control vegetative growth and productivity of temperate fruit trees. HortScience 35:1048-1050.

Kassai T, Mosoni P, Patyi R, Denés F (2002). Investigation of the dynamics of fruit growth in two strawberry varieties. In: Hietaranta T, Linna MM, Palonen P, Parikka P (eds.). Proc. Fourth International Strawberry Symposium, Finland.

Knee M, Sargent JA, Osborne DJ (1977). Cell wall metabolism in developing strawberry fruit. J. Exp. Bot. 8:377-396.

Lopez G, Behboudian MH, Vallverdu X, Mata M, Girona J, Marsal J (2010). Mitigation of severe water stress by fruit thinning in 'O' Henry' peach: Implications for fruit quality. Scientia Horticulturae 125(3):294-300.

Matthews MA, Anderson MM and Schultz HR (1987). Phenological and growth responses to early and late season water deficits in Cabernet Franc. Vitis 26:147-160

McFadyen LM, Hutton RJ, Barlow EWR (1996). Effects of crop load on fruit water relations and fruit growth in peach. J. Hort. Sci. 71:469-480.

Moore JN, Brown GR, Brown ED (1970). Comparison of factors influencing fruit size in large-fruited and small-fruited clones of strawberry. J. Am. Soc. Hort. Sci. 95:827-831.

Mpelasoka BS, Behboudian MH, Mills TM (2001). Water relations, photosynthesis, growth, yield and fruit size of 'Braeburn' apple: Responses to deficit irrigation and to crop load. J. Hort. Sci. Biotechnol. 76:150-156.

Mpelasoka BS, Behboudian MH (2002). Production of aroma volatiles in response to deficit irrigation and to crop load in relation to fruit maturity for 'Braeburn' apple. Postharvest Biol. Technol. 24:1-11.

Naor A, Naschitz S, Peres M, Gal Y (2008). Response of apple fruit size to water stress and crop load. Tree Physiol. 28:1255-1261.

Naor A, Klein I, Doron I, Gal Y, Z. Ben-David Z (1997). The effect of irrigation and crop load on stem water potential and apple fruit size. J. Hort. Sci. 72:765-771.

Naor A, Klein I, Hupert H, Grinblat Y, Peres M, Kaufman A (1999). Water stress and crop level interactions in relation to nectarine yield, fruit size distribution and water potentials. J. Am. Soc. Hort. Sci. 124(2):189-193.

Perkins-Veazie P (1995). Growth and ripening of strawberry fruit. Hort. Rev. 17:267-297.

Pomper KW, Breen PJ (1995). Levels of apoplastic solutes in developing strawberry fruit. J. Exp. Bot. 46 (288):743-752.

Pomper KW, Breen PJ (1997). Expansion ans osmotic adjustment of strawberry fruit during water stress. J. Am. Soc. Hort. Sci. 122(2):183-189.

Richards JH, Caldwell MM (1987). Hydraulic lift: Substantial nocturnal water transport between soil layers by Artemisia tridentata roots. Oecologia 73:486-489.

Reynolds GA, Parchomchuk P, Berard R, Naylor A, Hogue E (2005). Gewurztraminer grapevines respond to length of water stress duration. Int. J. Fruit Sci. 5(4):75-94.

Schwab W, Raab T (2004). Developmental changes during strawberry fruit ripening and physic-chemical changes during postharvest storage, p. 341-369. In: Dris R, Jain SM (eds). Production Practices and Quality Assessment of Food Crops, Vol 3, "Quality Handling and Evaluation, Kluwer Academic Publishers.

Trought MCT, Naylor AO (1988). Irrigation responses in a cool climate. In: Smart RE, Thornton RJ, Rodriguez SB, Young JE (Eds): Proc. 2nd In. Symp. Cool Climate Viticulture and Oenology: 156-160, New Zealand Society for Oenology and Viticulture, Auckland.

Warrington IJ, Fulton TA, Halligan EA, de Silva HN (1999). Apple fruit growth and maturity are affected by early season temperatures. J. Amer. Hort. Sci. 124:468-477.

23

Sensory analysis of banana blended shrikhand

Rita Narayanan and Jyothi Lingam

Department of Dairy Science, Madras Veterinary College, Chennai, TamilNadu, India.

Shrikhand is a popular Indian dessert prepared by fermentation of milk. It has a semi-soft consistency and is sweetish sour in taste. Fresh Curd (dahi) prepared was partially strained through a cloth to remove the whey and produce a solid mass called chakka. Chakka was finely mixed with sugar and flavouring agents, to give a sweetish-sour taste. Shrikhand was prepared from dahi with a constant level of sugar (40%) and supplementing with banana pulp at 10% (T_1), 20% (T_2) and 30% (T_3). T_0 served as control with no supplementation, sensory analysis showed a significant difference in different sensory attributes of T_2 sample with the rest of the treatments. T_2 (20%) supplementation of banana pulp to shrikhand was much preferred. Total solids of T_2 were 59.96 ± 0.35. Storage of 20% supplemented shrikhand showed no significant difference in sensory attributes up to 14 days.

Key words: Fruit shrikhand, banana shrikhand, value based shrikhand.

INTRODUCTION

The increasing demand from consumers for dairy products with 'functional' properties is a key factor driving value sales growth in developed markets. This has led to the promotion of added-value products such as probiotic and other functional yoghurts, reduced-fat and enriched milk products, fermented dairy drinks, and organic cheese (Rudrello, 2004). Present day consumers prefer foods that promote good health and prevent diseases. Furthermore, these foods must fit into current lifestyles providing convenience of use, good flavour and an acceptable price value ratio. Such foods constitute current and future waves in the evolution of the food development cycle (Chandan, 1999).

Since time immemorial, a significant proportion of milk has been used in India for preparing a wide variety of dairy delicacies, an unending array of sweets and other specialties from different regions of the country. In the process, the basic limitation of milk and its perishable nature has been tastefully overcome. It's processing aims is to extend the shelf-life of milk, while converting it into mouth-watering tit-bits. Thus, diverse methods to prepare as well as preserve milk products have been developed. An estimated 50 to 55% of the milk produced in India is converted into a variety of traditional milk products, using processes such as coagulation (heat and/or acid), desiccation and fermentation.

In Indian households, the life of milk is extended from 12 to 24 h by repeated boiling. It is preserved by souring with the aid of lactic cultures, which imparts an acid taste, particularly refreshing in hot climate. Dairy products are likely to remain as important dietary components because of their nutritional value, flavor, and texture. There will be a continuous demand for traditional, high quality dairy products, despite increasing competition from non-dairy based products (Rathore et al., 2007).

Shrikhand is a popular Indian dessert prepared by the fermentation of milk. It has a semi-soft consistency and a sweet and sour taste. Shrikhand originated in Persia

Table 1. Sensory evaluation of shrikhand blended with different levels of banana pulp[@].

Treatment	Sensory parameter				
	Colour and appearance	Body and texture	Sweetness	Flavour	Overall acceptability
T_0	7.50 ± 0.22^a	7.00 ± 0.01^a	8.00 ± 0.03^{NS}	7.83 ± 0.16^a	8.00 ± 0.01^a
T_1	7.16 ± 0.16^a	7.50 ± 0.22^a	8.00 ± 0.01^{NS}	7.66 ± 0.21^a	8.00 ± 0.02^a
T_2	8.66 ± 0.21^b	8.50 ± 0.22^b	8.00 ± 0.01^{NS}	8.66 ± 0.21^b	8.66 ± 0.21^b
T_3	7.50 ± 0.22^a	7.33 ± 0.21^a	8.00 ± 0.02^{NS}	7.50 ± 0.22^a	8.16 ± 0.16^a

Values with different superscript differ significantly (P < 0.01); [@] values are averages of 6 trials; [NS]Non significant.

using Frasi-shir (milk) and khand (sugar), and was later brought to the shores of Gujarat by the Parsi Zohrastrian settlers.

Fresh curd (dahi) is partially strained through a cloth to remove the whey and produce a solid mass called chakka. Chakka is finely mixed with sugar and flavouring agents, giving a sweetish-sour taste. Typically shrikhand constitutes 39.0% moisture and 61.0% of total solids of which 10.0% is fat, 11.5% is proteins, 78.0% is carbohydrates and 0.5% is ash, on a dry matter basis with a pH of about 4.2 to 4.4 (Boghra and Mathur, 2000; Kulkarni et al., 2006).

Some workers have attempted to improve the sensory and nutritive characteristics of Shrikhand by adding fruit pulp. Nigam et al. (2009) have studied the effect of papaya pulp on the quality characteristics of Shrikhand.

The present investigation was undertaken to explore the possibility of the use of banana pulp in shrikhand to produce a novel fermented milk product and assess its sensory attributes during its storage.

MATERIALS AND METHODS

Shrikhand was prepared by adopting the method of Sunil et al. (2011).

Preparation of shrikhand

Fresh milk was procured from the dairy plant at the Madras Veterinary College, Chennai, Tamil Nadu, India with 3.5% fat and 8.5% Solid not fat (SNF). Dahi was procured from local market and used as culture. Milk was boiled and then cooled down at 28 to 30°C and inoculated with dahi at the rate of 1.5% and incubated at 30 to 32°C for 10 to 12 h until a firm coagulum was formed. Coagulum was then crushed and transferred to a double muslin cloth and hung for expulsion of whey for 8 to 10 h in refrigerated conditions (4 ± 1°C). The semi solid chakka obtained after drainage of whey was used as the base for shrikhand. The level of sugar was adjusted at 40%. The sugar was powdered and kneaded uniformly with the chakka. Shrikhand was prepared by supplementing different levels of banana pulp viz. 10, 20, and 30% to chakka. Shrikhand prepared without banana pulp served as control and was compared with the treatments.

The sensory evaluation of the product was carried for attributes, namely colour and appearance, flavour, body and texture, sweetness and the overall acceptability of fresh shrikhand and samples stored up to 14 days by a panel of trained members based on a 9-point hedonic scale, wherein 9 denoted "extremely desirable" and 1 denoted extremely undesirable. The product was cooled to 4 ± 1°C coded and served cold to the panelists.

The scores for qualitative data such as colour and appearance, flavour, body and texture, sweetness, and the overall acceptability given by different judges were tabulated. The total solid content of the different treatment samples was determined by the method described in IS 2802 (Part II), 1964 by ISI and compared. The data thus obtained was analyzed as per one way ANOVA by Snedecor and Cochran (1994).

Storage studies

On the basis of various sensory parameters, shrikhand containing 20% banana pulp was selected as optimum. The optimum product was further packed in polystyrene cups and stored under refrigerated conditions at 4 ± 1°C for a period of two weeks.

RESULTS AND DISCUSSION

Sensory attributes

The mean values of various sensory parameters of shrikhand containing 0, 10, 20 and 30% of banana pulp are presented in Table 1.

Sensory evaluation of shrikhand

Colour and appearance

It is revealed from Table 1 that, the colour and appearance of shrikhand was significantly (P < 0.01) affected due to blending of banana pulp. The average score for colour and appearance attributes of shrikhand in different treatments and control viz. T_0, T_1, T_2, and T_3 were 7.50, 7.16, 8.66, and 7.50 respectively. The average score for colour and appearance attributes of shrikhand was highest in 20% T_2 (8.66) and lowest in control T_0 (7.50). Sunil et al. (2011) observed that, there was a decline in the trend in appearance score with increase in apple pulp, though the decline was not significant. Gavane et al. (2010) reported that, the scores for colour and appearance were highest with 2% supplementation of custard apple pulp to shrikhand.

Table 2. Total Solids content of the shrikhand blended with Banana pulp at different levels[@].

Treatment	Total solids (%)
T_0	57.98 ± 0.23^a
T_1	58.40 ± 0.18^a
T_2	59.96 ± 0.35^b
T_3	60.41 ± 0.71^b

Values with different superscript differ significantly ($P < 0.01$); [@] values is average of 6 trials.

Table 3. Sensory evaluation of shrikhand blended with 20% of banana pulp@ at different storage period in days.

Tr	Sensory parameters											
	Body and texture			Sweetness			Flavour			Overall acceptability		
	0	7th	14th	0	7th	14th	0	7th	14th	0	7th	14th
T_2	$8.50\pm$ 0.22^{NS}	$8.50\pm$ 0.22^{NS}	$8.50\pm$ 0.22^{NS}	$8.00\pm$ 0.00^{NS}	$8.01\pm$ 0.00^{NS}	8.02 $\pm0.0^{NS}$	$8.66\pm$ 0.21^{NS}	$8.66\pm$ 0.21^{NS}	$8.66\pm$ 0.21^{NS}	$8.66\pm$ 0.00^{NS}	$8.66\pm$ 0.00^{NS}	$8.66\pm$ 0.00^{NS}

[@]Values are average of 6 trials; [NS]Non significant.

Body and texture

It was observed that the body and texture of shrikhand was significantly ($P<0.01$) affected due to blending of banana pulp at 20 percent level. (T_2). The score for body and texture of shrikhand prepared under each treatment ranged from 7.00 (T_0) to 8.50 (T_2). The highest score for body and texture of shrikhand was recorded for T_2 (8.50). Gavane et al. (2010) reported that, blending of a maximum of 2% of custard apple pulp had a positive appeal on the body and texture of shrikhand.

Sweetness

It was noticed that, there was no significant difference in sweetness in all the treatment samples indicating that the sugar blended was equal in all treatments.

Flavour

It was found that, the mean score for flavour of shrikhand were 7.83, 7.66, 8.66, and 7.50 in different treatments viz. T_0, T_1, T_2 and T_3, respectively. The flavour of shrikhand was significantly ($P<0.01$) affected due to blending of banana pulp at 20% level. The highest score (8.66) was recorded for shrikhand blended with 20% banana pulp (T_2), the lowest being recorded for T_0 (7.83). Sunil et al. (2011) reported that, the scores for flavour showed a significantly increasing trend with increasing level of apple pulp supplementation.

Overall acceptability

The scores for overall acceptability was highest in T_2 (8.66) and lowest in T_0 (8.00) and it was significantly ($P < 0.01$) affected due to blending of banana pulp at 20% level. Sunil et al. (2011) reported that, 20% supplementation of apple pulp had a higher overall acceptability score than the control, 10 and 30% supplementation levels.

It was noticed that, the total solids content of shrikhand was significantly ($P < 0.01$) affected due to blending of banana pulp at different levels. The mean total solids content of shrikhand in different treatments viz. T_0, T_1, T_2 and T_3, were 57.98, 58.40, 59.96 and 60.41% respectively (Table 2). The mean total solids content of shrikhand in T_3 was highest (60%.) and lowest in T_0 (57.98%). This is in corroboration with Patel and Abd-El-Salem (1986) who reported the total solids content in plain shrikhand as 57.6%. Gavane et al. (2010) reported that, the mean total solids content of shrikhand bended with custard apple at various levels of custard apple pulp were higher than plain shrikhand. Shinde (1994) reported the average total solids content of plain shrikhand as 59.4%.

Table 3 showed that, there was no significant effect on all the sensory parameters with 20% banana pulp blended Shrikhand during storage period of 14 days after which the samples deteriorated.

REFERENCES

Boghra VR, Mathur ON (2000). Physico-chemical status of major milk constituents and minerals at various stages of shrikhand preparation. J. Food Sci. Technol. 37:111-115.

Chandan RC (1999). Enhancing market value of milk by Adding Cultures. J. Dairy Sci. 82:2245-2256.

Gavane PM, Zinjarde RM, Rokde SN (2010). Studies on preparation of shrikhand blended with custard apple pulp- A new fermented milk product. Indian J. Dairy Sci. 63(1):11-15.

IS 2802 (1964). Ice cream, Bureau of Indian Standards for Dairy producs and equipment.

Kulkarni C, Belsare N, Lele A (2006). Studies on shrikhand rheology. J. Food Eng. 74:169-177.

Nigam N, Rashmi S, Upadhayay PK (2009). Incorporation of Chakka by papaya pulp in the manufacture of shrikhand . J. Dairying Foods. 28(2):115-118.

Patel SS, Abd-El-Salem MH (1986). Shrikhand – as Indian analogue of Western quarg. Cult. Dairy Prod. J. 21:6.

Rathore R, Middha S, Dunkwal V (2007). Microbial safety while handling milk products. Proceedings of the Souvenir, International Conference on Traditional Dairy Products Nov. 14-17, NDRI, pp. 90-95.

Rudrello F (2004). Health trends shape innovation for dairy products (online). Euromonitor international archive; Oct 5.

Shinde RK (1994). Comparative study of fat and SNF Losses during the manufacture of shrikhand by different methods. M.Sc. Thesis submitted to Dr. B.S. K.K.V. Dapoli.

Snedecor G, Cochran WG (1994). Statistical Methods VIII ed. Oxford and IBH Publishing Co., New Delhi.

Sunil K, Bhat ZF, Pavan K (2011). Effect of Apple Pulp and Celosia argentea on the quality characteristics of shrikhand. Am. J. Food Technol. 6:817-826.

24

The competitiveness of the Saudi Arabian date palm: An analytical study

Mohammad Samir El- Habba and Fahad Al- Mulhim

Economics of Date Palm Chair", King Faisal University, Alahssa, Saudi Arabia.

The Kingdom's date production has not kept pace with export activities, where the demand of foreigners as workers or visitors in KSA is increased from year to other, so as the percentage of exports is about 6.8% of the domestic production, which is equivalent to 8.7% of the global exports, moreover, the average price of export reached 1065 US$/ton. The low export price of Saudi dates is attributed to the lack of focus on the production of high quality varieties of dates. Only 6.8% of the production is marketed externally, this indicates that the Kingdom's exports to the world markets is very low although the Kingdom is the second largest producer of dates worldwide producing 14.1% of world production in 2010. The outcome of excess production of dates, low consumption and weak export activities and processing of Saudi dates, resulted in a large surplus of dates of about 400 thousand tons in 2010, and it is expected to exceed 600 thousand tons by the year 2022. This is considered as a waste of water resources as well as the financial resources of the Kingdom. To evaluate the level of competitiveness of dates in the Gulf Cooperation Council (GCC) countries, the study used 3 main measures, that is, revealed comparative advantage (RCA), revealed trade advantage (RTA) and The trade entropy index (TEI), for the period 2000-2009. All the results showed that The KSA had revealed comparative advantage for export of dates.

Key words: Gulf cooperation council (GCC), revealed comparative advantage, revealed trade advantage, trade entropy index.

INTRODUCTION

Kingdom of Saudi Arabia (KSA) is a major dates producing country and is ranked the second in the world as per FAO statistics 2010 in terms of quantities produced (14.4%), where the per capita consumption of dates is the highest in the world (36 kg/year). Dates occupies a special place in the economic structure of Saudi agriculture with respect to production, consumption and marketing due to the Kingdom's supportin order to increase production while improving quality. The area planted with palm trees had increased by 152% during the period from 1997 to 2009, and the production increased by 153% during the same period. The total

planted area with date palm trees in the Kingdom during 2009 was about 162 thousand hectares, while the number of palm trees had reached nearly 23 million trees; with more than about 400 varieties. The best of these varieties being Khalas, Sukkary, Helwah, Ajwah, Ruthana, Segae, Barhi and Rushodia. Palm trees are grown in the various regions of the Kingdom, which are characterized by the diversity of climate; the most important palm growing regions are Riyadh, Qassim, Eastern Province and Medina. Despite the increase in the area planted, the productivity per hectare has declined in recent years, due to the fact that a large number of newly planted palms did not enter the production phase yet. This is likely to increase the production of the Kingdom of dates significantly in the coming years.

Dates production in Saudi Arabia had reached about 1.078 million tons in 2010, which is equivalent to 14.4% of world production, and is ranked second in the world in the production after Egypt, which produced about 1.353 million tons in the same year.

This research consists of 6 sections viz: Introduction, justification and objectives of the study followed by the literature cited, and methodology. Then results and discussion and conclusion

Justifications and objectives of the study

Competitiveness is now crucial to developing countries. Competitiveness is a process of change. It involves, private sector initiative, government initiative and effective dialogue between the two. A competitive advantage is an advantage over competitors gained by offering consumers greater value, either by means of lower prices or by providing greater benefits and service that justifies higher prices.

The Kingdom's date production has not kept pace with export activities, as the percentage of exports is about 6.8% of the domestic production, which is equivalent to 8.7% of the global exports, and the average price of export reached 1065 US$/ton. The low export price of Saudi dates is attributed to the lack of focus on the production of high quality varieties of dates.

The major problem of the production of dates in the KSA is related to dates processing, where only about 6.1% of the production is processed, using unspecialized equipment in the field of date processing sector. Only 6.8% of the production is marketed externally, this indicates that the Kingdom's exports to the world's markets are very low compared to its level of production.

The outcome of excess production of dates, low consumption and weak export activities and processing of Saudi dates, resulted in a large surplus of dates at about 400 thousand tons in 2010, and it is expected to exceed 600 thousand tons by the year 2022. This is considered as a waste of water resources as well as the financial resources of the state.

This research aims at assessing the competitiveness of the Saudi Arabian dates in the world with special emphasis on the Gulf States (GCC).

LITERATURE REVIEW

Al-Abbad et al. (2011) conducted a study on the economic feasibility of date palm cultivation in the Al-Hassa oasis of the Kingdom and estimated the average annual yield of dates to be 48.0 kg per palm with a selling price estimated at SR 4.00 per kg. The net income from date palm cultivation in the oasis was found to be

SR5800.00 / ha (SR 38.67 / palm). Significant number of farmers (23%) sells their produce in the farm itself, of which 57% is to known customers indicating sizeable "farmer-consumer" loyalty. Date palm farmers of Al-Hassa were also found to be quality conscious, as they adopt diverse measures (pre topost harvest) to ensure quality production of dates. SWOT analysis indicated spiritual attachment to the land by the farmers as a strength of the system, however; bureaucratic hurdles to obtain subsidies and lack of exploitation of facilities by traditional farmers, is a major threat to date farming in the oasis. There also exists a good possibility to develop logistics that support marketing of dates, especially through agricultural cooperatives, besides further enhancing exploitation of state subsidies for date palm cultivation.

A country market study of Saudi Arabia examining the trade flows between South Africa (SA) and Saudi Arabia was conducted by Rensburg and Letswalo, 2010. The major objective of the study was to identify agricultural products that have the potential to be exported to Saudi Arabia. A trade potential index (TPI) was drawn up in this study and it was noted that there are opportunities for deepening trade with Saudi Arabia. The "trade chilling" analysis was also conducted.

Liu and Pascal (2004) studied the marketing potential of date palm fruits in the European market. The purpose of the study was to evaluate the potential of various date varieties (including "non-traditional" ones) in the EU market. The study found that there was room for increased imports of Deglet Nour dates (or other varieties with similar taste and texture) provided high standards of quality (including low infestation rate), packaging and traceability could be met. However, prices were not expected to increase substantially from their present level. Medjool has attracted major interest in the United Kingdom and France and fetched high prices. It appeared to have good market prospects but some logistical constraints due to the low supply volume and retailer hesitations still needed to be solved. Conversely, the potential for Hayani and Bahri seemed limited to a small ethnic market.

MATERIALS AND METHODS

Secondary data was used in achieving the study objectives. Thedata covers quantities and values of dates traded from Saudi Arabia during the last decade between the GCC, as collected from local, national and international sources.

The following Modules were used to evaluate the competitiveness of the dates in the selected countries: (Türkekul et al., 2007).

Revealed comparative advantage (RCA)

A country's comparative advantage is determined by its relative factor scarcity. However, it is well known that measuring comparative advantage and testing the Hecksher-Ohlin (H-O) theory have some difficulties (Balassa, 1989) since relative prices under autarky are not observable. Given this fact, Balassa (1965) proposed that it may not be necessary to include all constituents effecting country's comparative advantage. Instead, he suggested that comparative advantage is revealed by observed trade patterns, and in line with the theory, one needs pre-trade relative prices which are not observable. Thus, inferring comparative advantage from observed data is named "revealed" comparative advantage (RCA). In practice, this is a commonly accepted method to analyzing trade data.

Revealed Comparative Advantage can be written as:

$$RCA1 = (Xij / Xit) / (Xwj / Xwt) = (Xij / Xwj) / (Xit / Xwt)$$

Where: X_{ij} = Country i's export of goods j, X_i = Country i's exports of all goods ,X_{wk} = World exports of good k ,X_w = World exports of all goods, t = is a set of commodities, n = is a set of countries. If RCA1 > 1, Comparative Advantage revealed., RCA1 < 1, No Comparative Advantage revealed.

One problem with the basic Balassa index is that it is not symmetrically distributed around the neutral value 1.0, ranging from 0 to 1 for comparative disadvantage and indefinitely upward from 1.0 for comparative advantage products. This problem is easily corrected by taking natural logarithms of the ratios with the index defined as follows: [Poramacom and Nongnooch (2002)].

$$RCA2 = \ln (Xij / Xit) / \ln (Xwj / Xwt)$$

The revised index is now symmetric around 0. This form is particularly useful for econometric studies.

Relative trade advantage (RTA)

Vollrath (1991) offered mainly 3 alternative ways of measurement of a country's RCA. These alternative specifications of RCA are called the relative trade advantage (RTA), the logarithm of the relative export advantage (ln RXA), and the revealed competitiveness (RC). In this study, for the sake of being systematic, we call them as RTA1, RTA2, and RTA3 respectively. It is clear that the advantage of presenting latter two indices (that is RTA2 and RTA3) is that they become symmetric through the origin. Positive values of Vollrath's 3 alternative measures of revealed comparative advantage reveal a comparative/competitive advantage whereas negative values indicate comparative /competitive disadvantage. This measure is the relative trade advantage (RTA), which accounts for imports as well as exports. It is calculated as the difference between relative export advantage (RXA, and its counterpart, relative import advantage (RMA): Thomas (1991).

$$RTA1 = RXA - RMP$$

Revealed Relative Comparative Advantage Export Index is defined as a country's export share relative to all other countries export of the specific product category), which equates to the Balassa index,

$$RXA = RCA (B)$$

Revealed relative import penetration index is defined as a country's import share relative to all other countries imports of the specific product category (Suleiman, 2011).

$$RMA = (Mij / Mit) / (Mnj / Mnt)$$

Where, M represents the imports.

Vollrath's second RCA measure is the logarithm of the relative export advantage (here as VRC2):

$$RTA2 = \ln RXA = \ln RCA1$$

The third measure of Vollrath is the revealed competitiveness (RC) (here as VRC3), expressed as:

$$RTA3 = RTA2 - \ln RMA$$

If RTA > 0 the goods have certain competitive advantages; RTA < 0 the goods has not competitive advantages.

The trade entropy index (TEI)

The third measure used to evaluate the competitiveness of trade in this research is the Trade entropy index (TEI). It is used in trade analysis for measuring the concentration or dispersion of trade. These trade flows can be either in terms of imports or exports (Arzolnal, 2003). The higher the index the more dispersed is the export (import) pattern of that country. The validity of the index derives from weighting each component of share (bij) by its relevance ln(1/bij). That means if the value of bij for a country is very high, it will be scaled down by the ln (1/bij) term and the maximum value is achieved when all shares are equal.

In this research the first equation (that is, the export equation) will be used

$$Ixi = aijln(1/aij) \text{ with } 0 < aij < 1 \text{ and } \Sigma \ aij = 1$$
$$Imi = b \ ln(1/b_{ij}) \text{ with } 0 < bij < 1 \text{ and } \Sigma bij = 1$$

where: Ixi: Entropy index of export. Imi: Entropy index of import. aij: Export share of country i to country j. bij: Import share of country i from country j.

In the formula each entity (share of a commodity) is weighted by its relevance and very high export (or import) share is weighted with correspondingly low weights and the very low ones are weighted with higher weights, consequently the higher values (in sum) are obtained for approximately equally distributed shares. In brief, the higher the index the more dispersed is the export (or import) pattern of that country.

RESULTS AND DISCUSSION

Date palm trade in the Kingdom of Saudi Arabia

The Central Department of Statistics and Information (CDSI) in the KSA publish annual trade statistic for all the commodities traded in the Kingdom (CDSI, 2011). The date palm fruit traded data was grouped according to country groups, the GCC, Non GCC Arab, Non-Arab Islamic Countries, Non-Arab Non-Islamic Asian Countries, Non-Arab Non-Islamic African Countries, Australia and Pacific Islands, North America Countries, EU countries, Western Europe Countries, and Other Countries.

Table 1 shows that about half the date exports from the KSA were mainly directed to Non-GCC Arab Countries, while about 23% was directed to Islamic Non-Arab Countries and about 12% were directed to the GCC.

Figure 1 shows that the annual trade of dates to the GCC during 2001 to 2010 fluctuated from year to year, it increased from 2.8 thousand tons in 2001 to 8.4 thousand tons in 2003 then dropped to about 1.7

Table 1. Quantity and Price of Exported Dates From the KSA to the World Markets During 2004 to 2011.

Year		The GCC	Other Arab Countries	Islamic Non-Arab Countries	Asian Non-Arab Non-Islamic Countries	African Non-Arab Non-Islamic Countries	Australia and Pacific Islands	North America Countries	EU Countries	West Europe Countries	Other Countries
						Country group					
2004	Q	5753	6980	337						128	54
	P	3.60	1.58	2.53						5.11	3.61
2005	Q	5707	9860	396						129	91
	P	4.29	1.49	3.18						4.36	4.05
2006	Q	2525	3139	131							133
	P	4.83	2.07	5.18							2.07
2007	Q	4031	6983	3105	788	478	12	73	380	98	8
	P	5.63	1.50	1.34	2.02	0.95	13.33	3.38	8.29	3.50	7.13
2008	Q	9209	9135	3716	1215	267	42	19	400		2
	P	8.32	1.86	1.91	1.74	5.39	11.45	18.58	3.82		1.50
2010	Q	2639	8866	4748	1837	521		140	383	25	
	P	7.91	2.49	2.93	4.46	5.15		5.67	4.95	10.92	
2011	Q	2242	8835	4243	1719	462		151	823		19
	P	10.70	2.49	3.30	2.33	6.49		6.62	8.51		
Share in 2011	%	12.12	47.77	22.94	9.29	2.50		0.82	4.45		0.10

Q: Tons; P: RS/kg.

thousand ton in 2005. The highest exports were in 2008 (About 16.2 thousand tons) and the lowest were in 2009 (about 705 tons).

Date palm price analysis in the Kingdome of Saudi Arabia

The Ministry of Agriculture in the KSA publishes annual average wholesale prices of the different varieties of dates in the main wholesale markets in the main producing regions (MOA, 2011).

The average wholesale price for dates in the kingdom in 2010 was SR 8.75/kg for all the varieties. The highest wholesale prices were for Sukkary (RS 13.74 /kg), Naboot Saif (RS 13.37 /kg), Khlass (RS 13.17 /kg) and Seqe'e (RS 12.54/kg). The lowest prices were for Nabtet

Rashed (RS 3.53 /kg) and Al-Helweh (RS 5.50 /kg) (Figure 2).

It is worth mentioning here that the average export prices for fresh, dried and stuffed dates in the year 2010 were RS 4.55 /kg, RS 4.12 /kg, and RS 4.15 /kg respectively, with a weighted average of RS 4.14 /kg for the three types of dates. From this, it is noticed that the local wholesale prices of dates is almost double the export prices in the

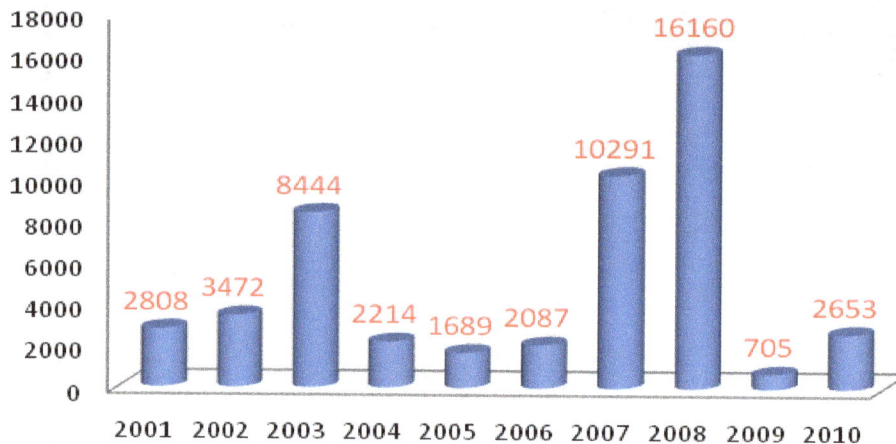

Figure 1. Date Palm Exports to the GCC during 2001-2010

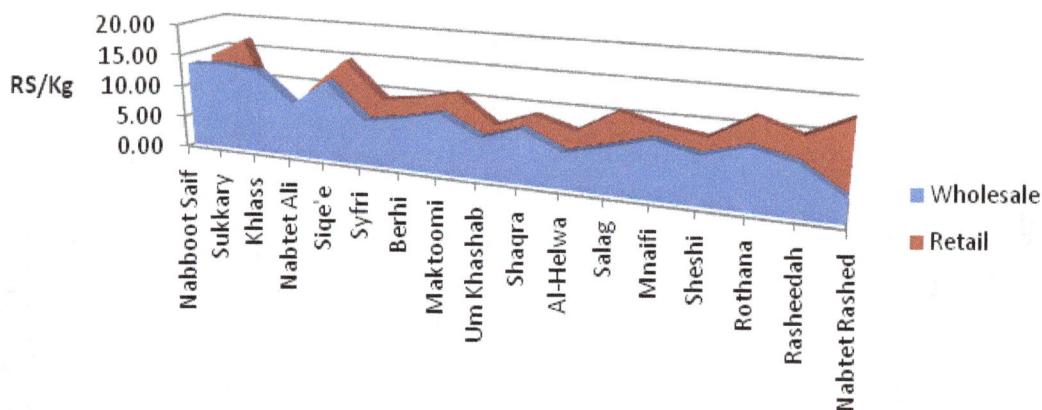

Figure 2. Average Wholesale and Retail Prices for the Different Varieties in the KSA in 2010.

same year. On the other hand, the highest price of sukkary was in Al-Qassim (RS 23.6 /kg) and in Hael for NabbootSaif (RS 35 /kg). The average wholesale prices of all the varieties were RS 15.74 /kg in Hael, RS 13.05 /kg in Najran, RS 12.91 /kg in Aseer, and 11.44 /kg in Al-Baha. Moreover, Figure 2 shows that the marketing margins were relatively high in all dates varieties (RS 2.2 /kg) which comprises 34% of the average wholesale prices compared to the services rendered by the middlemen in the dates markets.

Competitiveness measures for the Date Palm in Saudi Arabia

Revealed comparative advantage

Balassa revealed comparative advantage indices: Two RCA indices were evaluated here, that is, RCA1 and RCA2. The first index (RCA1) was evaluated for the 6 GCC, while the second was evaluated for the KSA only.

Tables 2, 3 and Figure 3 show that Bahrain and Kuwait have revealed comparative disadvantage during the studied period, while the rest of the GCC have revealed comparative advantage in Dates trade. Moreover, UAE show a stronger revealed comparative advantage during the period 2000 to 2009.

The revised Revealed Comparative Advantage index (RCA2) shows that KSA and the UAE still have higher level of comparative advantage in dates, since the UAE have a stronger RCA than the KSA. The rest of the GCC countries have comparative disadvantage during the studied period (Figure 4).

Relative trade advantage

The first relative trade index (RTA1), calculated as (RXA-RMA) emphasizes the results found in calculating of the RCA for the GCC. The UAE have the strongest comparative advantage in these countries then comes the KSA (Figure 5). According to this indicator, it was

Table 2. The Revealed Comparative Advantage Indices for Date in the GCC

Country	2000	2001	2002	2003	2004	2005	2006	2007	2008	2009	Average
KSA	27.7	68.0	50.2	30.8	47.6	28.6	34.4	16.7	24.1	1.1	32.9
UAE	106.9	122.6	109.6	91.1	39.0	86.4	24.4	69.0	73.9	94.0	81.7
Bahrain	0.11	0.00	0.13	0.00	0.09	0.03	0.00	0.03	0.00	0.08	0.0
Kuwait	0.5	0.0	0.4	0.9	0.1	1.8	0.4	0.2	1.3	0.6	0.6
Oman	0.0	30.5	9.9	8.1	18.5	8.7	14.5	19.2	10.9	13.9	13.4
Qatar	3.0	0.0	1.6	4.3	8.9	3.2	5.8	8.0	3.5	30.6	6.9

Table 3. Revised RCA2 for the Dates in the GCC

Year	2000	2001	2002	2003	2004	2005	2006	2007	2008	2009
KSA	1.14	1.22	1.19	1.15	1.20	1.15	1.16	1.10	1.13	0.79
UAE	1.30	1.31	1.30	1.28	1.18	1.27	1.13	1.25	1.26	1.28
Bahrain	0.25	0.00	0.35	0.00	0.20	0.10	0.00	0.15	0.00	0.29
Kuwait	0.51	0.00	0.44	0.52	0.00	0.66	0.42	0.36	0.57	0.57
Oman	0.00	1.12	0.99	0.98	1.04	0.97	1.01	1.07	1.01	1.03
Qatar	0.63	0.00	0.44	0.61	0.65	0.59	0.56	0.68	0.66	0.97

Give Average for RCA2 in this table as in Table 3.

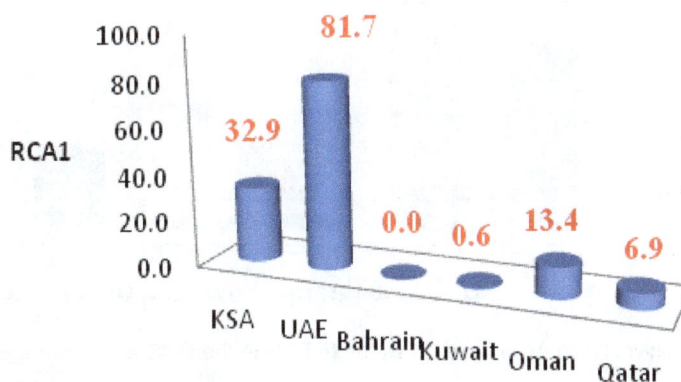

Figure 3. Revealed Comparative Advantage (RCA1) for the GCC Dates as an Average of 2000-2009

found that Oman has also revealed comparative advantage in dates trade.Moreover, the RTA2, calculated as (ln RXA = lnRCA1) values followed the same patterns as RTA. The RTA3, which is calculated as the difference between the RTA2 and the natural logarithm of RMA, showed that the KSA have higher Comparative advantage than the UAE (Figure 6).

The trade entropy index (TEI)

Table 5 presents the absolute trade entropy indices (TEI) calculated for dates exports and imports in KSA with respect to the world. The values of these indices were small, which means that the share of trade for dates in

the world market is small, that is, export (or import) pattern of that country are less dispersed. Moreover, it is noticed here that the Export Entropy Index fluctuated during the period, but taking a decreasing pattern. The Import Entropy Index, on the other hand showed very low values but more constant.

Conclusions

Although the KSA is considered the second date producing country, but its exports of this product are weak. The Competitiveness Analysis shows that the KSA had Revealed Comparative Advantage in dates trade, but second to the UAE.

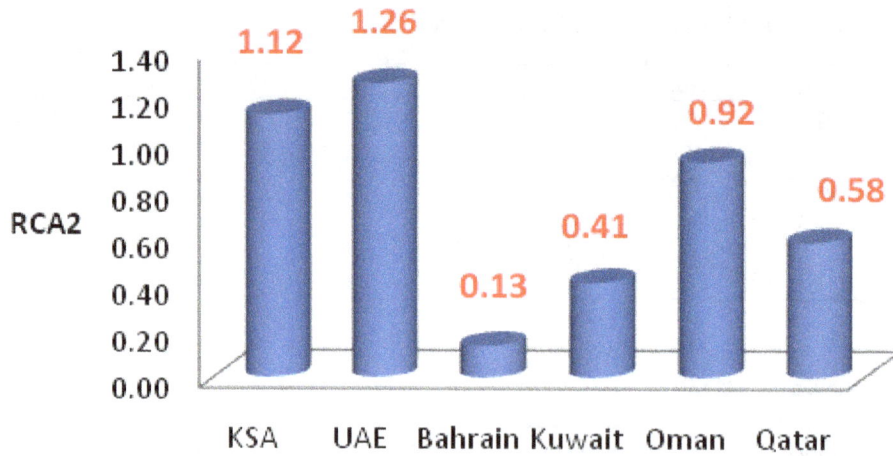

Figure 4. Revealed Comparative Advantage (RCA2) for the GCC Dates as an Average of 2000-2009.

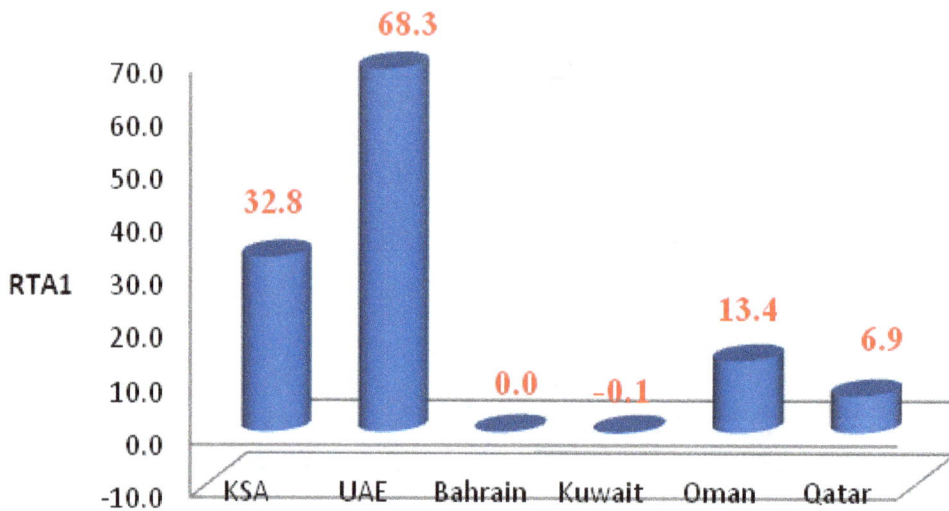

Figure 5. Relative Trade Advantage for Dates (RTA1) in the GCC as an average of the period 2000-2009

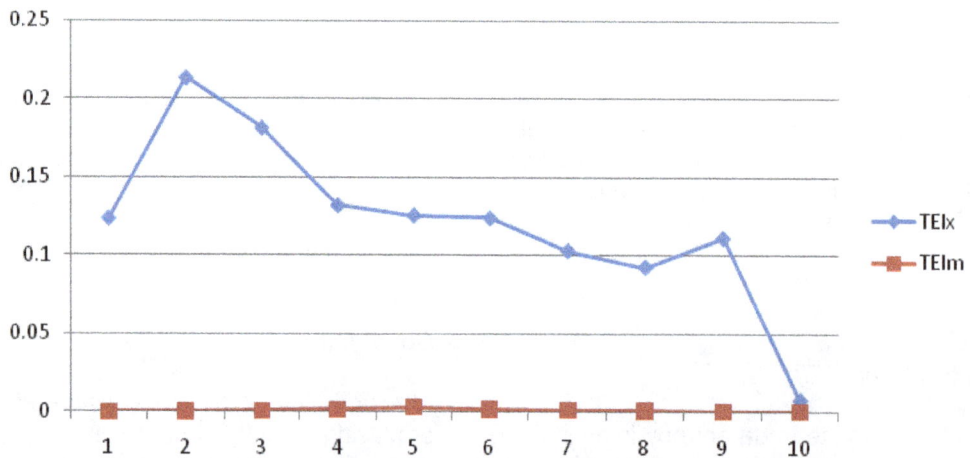

Figure 6. Trade Entropy Index for Exports and Imports of Dates in the KSA and the World

To enhance the level of competitiveness of KSA dates trade it is necessary to increase the promotion campaigns in the external markets after implementing the export markets standards and requirements.

ABBREVIATIONS

CMS, Constant market share; **DRC,** domestic resources cost; **KSA**, Kingdom of Saudi Arabia, **RCA,** revealed comparative advantage; **RTA,** revealed trade advantage; **SCB,** social cost benefit; **TEI,** trade entropy index **TPI,** trade potential index

REFERENCES

Al-Abbad A, Al-Jamal M, Al-Elaiw Z, Al-Shreed F, Belaifa H (2011), A Study on the Economic Feasibility of Date Palm Cultivation in the Al-Hassa Oasis of Saudi Arabia. J. Develop. Agric. Econ. 3(9):463-468.

Arzolnal G (2003). A Study into Competitiveness Indicators. Sabansi University, Turkey. 2003.

Balassa, Bela (1965). Trade Liberalization and Revealed Comparative Advantage," Manchester School of Economic and Social Studies, 33:99-123.

Balassa B (1989)."Revealed" comparative advantage revisited', in: B. Balassa (ed.), Comparative Advantage, Trade Policy and Economic Development, New York University Press, New York, pp.63–79.

Central Department of Statistics and Information (CDSI). (2011). Foreign Trade. Ministry of Economy & Planning. KSA.

Ministry of Agriculture (MOA) (2011). Average Wholesale Prices for the Most Important Agricultural Commodities (Local and Imported) in the Markets in the Main Areas in the KSA in 2010. P. 12.

Liu, Pascal (2004). The marketing potential of date palm fruits in the European market. Food and Agricultural Organization for the United States (FAO).

Poramacom, Nongnooch (2002), Revealed Comparative Advantage (RCA) and Constant Market Share Model (CMS) on Thai Natural Rubber. Kasetsart Journal (Soc. Sci) 23:54-60.

Rensburg G, Letswalo J (2010). Country Market Study: Saudi Arabia.

Suleiman H (2011). Commodity Chain Analysis and Exports of Dates in Jordan. An MSc Thesis in the University of Jordan, Jordan.

Türkekul, BC, Günden C. Abay B, Miran A (2007). Market Share Analysis of Virgin Olive Oil Producer Countries with special respect to Competitiveness. Paper prepared for presentation at the I Mediterranean Conference of Agro-Food Social Scientists. 103 rd EAAE Seminar 'Adding Value to the Agro-Food Supply Chain in the Future Euro-Mediterranean Space'. Barcelona, Spain, April 23rd - 25th.

Volrath T (1991). A theoretical evaluation of alternative trade intensity measures of revealed comparative advantage, "Review of World Economics (WeltwirtschaftlichesArchiv), Springer 127(2):265-280.

25

Simple and multiple linear regressions for harvest prediction of Prata type bananas

Bruno Vinícius Castro Guimarães[1], Sérgio Luiz Rodrigues Donato[2], Victor Martins Maia[3], Ignacio Aspiazú[3], Maria Geralda Vilela Rodrigues[4] and Pedro Ricardo Rocha Marques[2]

[1]Federal Institute of Education, Science and Technology of the Amazon, IFAM, Campus São Gabriel da Cachoeira, BR 307, km 03, Estrada do Aeroporto, Cachoeirinha, ZIP Code 69750-000, São Gabriel da Cachoeira, AM, Brazil.
[2]Federal Institute of Education, Science and Technology of Bahia – Campus Guanambi, P. O. Box 09, Ceraima District, ZIP Code 46.430-000, Guanambi, BA, Brazil.
[3]Montes Claros State University– Campus Janaúba, Center for Exact Sciences and Technology, Department of Agricultural Sciences, 2.630 Reinaldo Viana Av., PO Box 91, Bico da Pedra, ZIP Code 39.440-000, Janaúba, MG, Brazil.
[4]Agricultural Research Company of Minas Gerais / Regional Unit Epamig Northern Minas Gerais, ZIP Code 39525-000, Nova Porteirinha, MG, Brazil.

This study aimed to fit regression models for harvest prediction in Prata type bananas. The experiment consisted of plants and bunches of bananas carried out in Guanambi, BA, with genotypes Dwarf Prata (AAB) and BRS Platina (AAAB), planted at a spacing of 3.0 × 2.5 m, with irrigation. Measurements of vegetative and yield characteristics were sampled at random. Models of simple and multiple linear regression were estimated, considering as independent variables with highest correlation coefficients with the masses of the bunch (MB) and hands (MH). The simple linear regression models allow prediction of the masses of the bunch and hands, according to the number of hands (NB), with better precision, for both genotypes and at least 120 days prior to harvest. For the' Dwarf Prata', the equations were: $MCA = -10.05 + 3.08NH; r^2 = 0.99$, e, $MPE = -12.38 + 3.55NH; r^2 = 0.99$. For the 'BRS Platina' equations were: $MCA = -1.37 + 3.02NH; r^2 = 0.97$ e $MPE = -2.74 + 2.88NP; r^2 = 0.97$. The determination coefficients for the adjusted models ensure consistency of the regressions for the estimation of Prata type banana production.

Key words: Estimated harvest, Dwarf Prata (AAB), BRS Platina (AAAB) genotypes, the regression models.

INTRODUCTION

The banana (*Musa* spp.) is the largest herbaceous monocot grown in the world. Over the centuries, the crop has expanded and is now grown in over 120 countries, highlighting bananas as the most consumed fruit in the world (Cordeiro and Moreira, 2006). It is considered an important food, because of its chemical composition and

vitamins and minerals content, especially potassium. It also constitutes an important element in the diet, not only by the high nutritional value, but also by the low cost (IBRAF, 2005).

The fruit industry is among the main generators of income, employment and rural development of all world agribusiness (BRASIL/MAPA, 2007). Brazil is the fifth largest producer of bananas, with 7,329 million tons produced in 503,354 ha, resulting in average yield of 14.5 t ha^{-1} (FAO, 2011a), and the per capita consumption is 29.10 kg yr^{-1} (FAO, 2011b). Production and consumption in Brazil show very peculiar characteristics, prevailing in most Brazilian regions the dessert varieties, such as AAB bananas, Prata type, 'Dwarf Prata', 'Pacovan' and 'Common Prata', which represent about 80%. Because of its economic importance, the banana crop is a matter of growing interest of researchers worldwide (Dantas and Soares Filho, 2006).

In these studies, typically, the researcher is interested in the identification and selection of superior genotypes that meet desirable characteristics such as appropriate size, major pests and diseases resistance and adaptation to different ecosystems. Thus, the experimental analysis addresses the biometric characteristics of plants (plant height, leaf number, pseudostem perimeter) and bunch (bunch weight, number of hands, number of fruits per bunch, fruits length and diameter) (Silva et al., 2000). These variables are quantitative, easy to measure, and may be under polygenic control, being under environmental influence, having direct and indirect economic importance (Ortiz, 1997), and most of them show significant correlation between themselves (Donato et al., 2006). This way, the analysis of plant behavior and expression is very interesting to those engaged in research in crop production, being of great application and essential to proper planning of agricultural activities. Thus, several statistical studies, linear and nonlinear models, have been developed with the purpose of obtaining future information and describe plant growth over time (Hernández et al., 2007; Maia et al., 2009).

In this context, several authors make use of predictive models to reduce or circumvent the interference of biotic or abiotic environment in the expression of the variable of economic interest (Savin et al., 2007; Zhang et al., 2007). In addition, simulation models are strategic tools for estimating the duration of stages of plant development, choosing the time of planting, predicting an abnormal production and thus use these data in genetic improvement programs (Roberto et al., 2005; Stenzel et al., 2006; Bíscaro, 2007), and are, therefore, important for breeders and producers. From this perspective, there is the use of mathematical models also to forecast the harvest in several crops (Streck et al., 2007; Scarpari and Beauclair, 2009; Wyzykowski, 2009). However, mathematical modeling in the estimation of banana production is limited and has low expression in the literature. In this respect, Jaramillo (1982) and Meyer (1975) describe studies of this nature in Cavendish type cultivars in Costa Rica and Soares et al., (2013), in Brazil, adjusted models for harvest prediction of the cultivar Tropical, type Maçã. Hence, it becomes clear the importance of identifying the variables that explain part of the variation in productivity through mathematical modeling to estimate harvest, based on agronomic traits measured throughout the crop cycle. In this case, the mathematical model, if adjusted in a functional form, allows those involved with the banana crop, whether researcher and/or producer, planning and organization of operations of pre and post-harvest. In this context, this study has the objective of adjusting simple and multiple linear regression models to predict harvest in bananas type Prata, 'Dwarf Prata' and 'BRS Platina'.

MATERIALS AND METHODS

The experiment was established in a Red-Yellow Latosol (Hapludox), medium texture, hypoxerophilous caatinga phase, flat to moderate topography. The experimental area is located at the Federal Institute of Bahia, Campus Guanambi, BA, 14° 13' 30" S, 42° 46' 53" W, at an altitude of 545 m, average annual precipitation of 660 mm and average temperature of 26°C. The local climate is type Aw, according to Köppen"s classification. Micropropagated plantlets were used, planted in a spacing of 3.0 x 2.5 m and submitted to the system of fixed conventional sprinkler irrigation with micro-sprinklers. Installation and cultivation followed the recommendations for the crop, and fertilizers were applied based on analysis of soil and leaves. The two evaluated genotypes were: Dwarf Prata, triploid (AAB), susceptible to yellow and black Sigatoka and Panama disease, and the hybrid BRS Platina, tetraploid (AAAB), resistant to yellow Sigatoka and Panama disease, derived from the cross between 'Dwarf Prata' (AAB) and M53 diploid (AA), formerly known in prerelease as PA42-44.

Measurements were made at the time of harvest. Each plant, the basic unit, was considered as a replicate. Therefore, to assess vegetative and yield characteristics, the plants were sampled at random into the two genotypes with different numbers of replicates, 98 for 'Dwarf Prata' and 96 for 'BRS Platina'. Those measurements consisted of phenotypic vegetative descriptors plant height, pseudostem perimeter at ground level, 30 and 100 cm in height, number of live leaves at harvest. Measurements of bunch yield were also carried out: Bunch weight, number of hands and fruits per bunch, hands mass, stalk mass, length and diameter. Hands yield was also assessed: number of total fruits and fruits per bunch, and mass, internal and external length and diameter of the central fruit on the external and internal rows. The measured values were obtained according to the methodological proposal contained in catalogs of standard morphological descriptors for banana (IPGRI, 1996). For each evaluated genotype, 'Dwarf Prata' and 'BRS Platina', phenotypic correlations were estimated regarding the associations between the masses of bunch and hands with the evaluated yield and vegetative characteristics, based on the Pearson correlation (Pimentel-Gomes, 2000). The correlations were tested by Student's t test at 1% probability. Data from observations of individual replicates of each genotype were used for establishing associations between characteristics. From the estimates of the correlations between all measured variables with the masses of the bunch and hands, the significant associations and with highest values were considered to proceed the regression analysis with the subsequent choice of the best-fitted model.

To evaluate the importance of the variables related to yield and vegetative characteristics and their influence in the masses of the bunch and hands, estimations were made of multiple linear regression equations using the variable selection procedure called backward elimination (Ribeiro Júnior, 2001), in SAEG software (Statistical Analysis System), version 9.1, Federal University of Viçosa (SAEG, 2007). This way, the statistical model with k independent variables can be determined with the following multiple regression equation: $Yi = \beta 0 + \beta 1X1i + \beta 2X2i + ... + \beta kXki + \varepsilon i$. In the model, Yi refers to the response variable: mass of bunch or hands, as a function of the regressive variables of bunch yield (number and average mass of hands; mass, length and diameter of stalk and number of fruits). It also refers to hands yield (mass and number of fruits of the hand; mass and external length of the external and internal rows fruit; internal length of the external and internal rows fruit; diameter of the external and internal rows fruit ($Xi; ... ; Xk$). The error associated to the i-esime observation is εi, assumed as normal and independently distributed; constant $\beta0$ is inherent to the model and $\beta1... \beta k$ model coefficients. To elucidate the relationship with each variable, given that it was significant in the regression analysis with the final bunch of the mass, Pearson parametric corrrelations were estimated (Pimentel-Gomes, 2000).

In the present work, the statistical procedures made to estimate the prediction equations of the values of masses of bunch and hands for Prata type bananas, 'Dwarf Prata'and 'BRS Platina', only the variables with significant correlation coefficients and with highest values were selected. Significance of the regression coefficients by the "t" test, at a 1% probability level, behavior of the biological phenomenon, determination coefficient (r^2) and the significance of the F test for the regression analysis of variance were considered for the prediction equations adjusted, in each particular case. For that, the backward elimination procedure in the software SAEG (SAEG 2007) was used, according to Ribeiro Jr. (2001). In these cases, the data from individual observations of the replicates for each genotype, 'Dwarf Prata'and 'BRS Platina', were used separately.

Simple regression models also were fitted between masses of bunch and hands for each evaluated genotype, 'Dwarf Prata' and 'BRS Platina', with the yield and vegetative characteristics. For that, regressions were fitted from means of the replicates, considering as independent variable, for each genotype, the pseudostem perimeter measured at ground level and the number of hands, because these showed highest correlation coefficients with the masses of bunch and hands. Another reason for using those variables in the models is because they are easy to determinate, by direct counting (number of hands) and by simple measure with a measuring tape (pseudstem perimeter). They are also non-destructive and can be obtained at the stage of flowering, well before the harvest of the bunch of 'Dwarf Prata' and 'BRS Platina', about 120 to 150 days (Donato et al., 2009). This enables an efficient planning associated with the technique of marking the bunch by age as a criterion for the harvest time (Lichtemberg et al., 2008; Soto Ballestero, 2008), allows to predict the timing and amount of harvest. The number of repetitions for each case was variable, with the stratification of the number of hands on classes of observations, ranging from 9 to 13 bunches for the 'Dwarf Prata' and seven to 12 bunches for hybrid 'BRS Platina'. Similar were the proceedings for the pseudostem perimeter measured at ground level, with stratification according to the number of replicates of each class and using the average of the repetitions in the procedure for simple and multiple linear regression in the software SAEG (2007).

RESULTS AND DISCUSSION

The association between agronomic traits in banana is crucial for estimating the production of the bunch, and can be evaluated by means of phenotypic, genetic and environmental correlations (Rocha, 2010). Thus, it should be added to the study of harvest prediction of the correlation analysis, in order to determine which variables influence, to a greater or lesser degree, the production. Correlations between the masses of bunch and hands and the vegetative characteristics plant height, pseudostem perimeter at ground level, 30 and 100 cm in

height and number of live leaves at harvest, for the 'Dwarf Prata' and 'BRS Platina' bananas, were significant and positive for all variables. This directly indicates the variation of the variables yield and masses of bunch and hands, with the vegetative variables analyzed in this study (Table 1). However, the association between the characters of yield, masses of the bunch and hands, and pseudostem perimeter measured at ground level expressed more strongly in relation to other variables (Table 1). Different studies relating vegetative and reproductive characteristics in banana reported a significant correlation between the perimeter of the pseudostem and the production of the bunch (Lima Neto et al., 2003; Arantes et al., 2010). In addition, Siqueira (1984) found from clones of banana 'Prata', that among the characters related to vegetative development, the pseudostem perimeter was the most positively correlated with the characters of production, which suggests it as an effective variable to compose the harvest prediction model.

Measurements on vegetative descriptors are widely used in practice, because they are easy to determine and require simple tools. Additionally, they can be measured at flowering, about 120 to 150 days before harvesting Prata type bananas (Donato et al., 2009), which subsidizes a efficient harvest programming. Thus, given the high correlation usually found between the yield factors and the perimeter of the pseudostem at ground level, combined with ease of measurement and its non-destructive character, this variable was tested as a component of the equation to predict the masses of bunch and hands for 'Dwarf Prata' and 'BRS Platina' bananas. In this way, simple linear regression models were fitted between these variables from the means of the repetitions, taking as the independent variable the pseudostem perimeter measured at ground level (Table 2). The linear regression models were significant and positive for the variable pseudostem perimeter at ground level, both for the bunch massand for the mass of hands, and with appropriate determination coefficient (r^2) for both genotypes (Table 2). This result suggests the possibility of planning the harvest and its implications, such as the logistics of harvest and postharvest, as well as marketing, transportation and climatization, with considerable reliability and with three to four months in advance (at flowering), by means of an easily measured indicator.

The model estimates that for every centimeter of increase in the pseudostem perimeter of 'Dwarf Prata' banana, the masses of bunch and hands increase, respectively, 327 and 289 g. In this order, the determination coefficients were adjusted in 0.76 and 0.77. Thus, the r^2 shown in Table 2 represents the fit of the data to predict the harvest, as a function of the pseudostem perimeter at ground level for both genotypes. Still, according to Table 2, it can be suggested for hybrid BRS Platina that the coefficients of the models estimate that, for each centimeter of increase

Table 1. Correlation coefficients between the masses of the bunch and hands, in association with the vegetative characteristics in type Prata bananas 'Dwarf Prata' and 'BRS Platina', Guanambi, BA, 2009.

Vegetative characteristics	Genotypes		Genotypes	
	Dwarf Prata	BRS Platina	Dwarf Prata	BRS Platina
	*Mass of bunch***		*Mass of hands***	
Plant height	0.63	0.49	0.63	0.47
Pseudostem perimeter at ground level	0.75	0.75	0.75	0.73
Pseudostem perimeter at 30 cm from ground	0.69	0.64	0.68	0.63
Pseudostem perimeter at 100 cm from ground	0.68	0.74	0.67	0.72
Number of leaves on the harvest	0.35	0.52	0.35	0.52

** $P<0.01$.

Table 2. Prediction models for the masses of bunch and hands on type Prata bananas, 'Dwarf Prata' and 'BRS Platina', as a function of the pseudostem perimeter at ground level, Guanambi, BA, 2009.

Genotype	[a]Estimate ($\hat{y}$)	Simple linear regression equation**	(r²)
Dwarf Prata	MB	$\hat{Y} = -10.12 + 0.33\ PPGL$	0.76
Dwarf Prata	MH	$\hat{Y} = -8.74 + 0.29\ PPGL$	0.77
BRS Platina	MB	$\hat{Y} = -14.31 + 0.36\ PPGL$	0.77
BRS Platina	MH	$\hat{Y} = -13.45 + 0.32\ PPGL$	0.76

[a]Yield estimate ($\hat{y}$): *MB* = mass of bunch; *MH* = mass of hands; [b]Independent variable (X): *PPGL* = Pseudostem perimeter at ground level. ** $P<0.01$.

in pseudostem perimeter, of the bunch mass and the mass of hands increase, respectively, 355 and 324 g. Subsequently, the determination coefficients were adjusted in 0.77 and 0.76. In this sense, harvest estimates for the masses of bunch and hands indicated 77 and 76%, respectively, of reliability on the final production determination.

Although the pseudostem perimeter in this study was measured at harvest time, this would present the same dimensions if it were measured at flowering, as the characteristic remains constant after that. After flowering, the banana ceases emission of roots and leaves, beginning senescence of these organs and culminating with the maturing of the bunch (Soto Ballestero, 2008; Robinson and Galán, 2010). Additionally, the prediction equations were composed by variables of easy measurement and simple application in practice. Therefore, these data allow the use of non-destructive methods to determine production and productivity. However, Soares et al. (2012) in a study of banana cv. Tropical for harvest prediction by means of multiple linear regression observed that the vegetative variables plant height, pseudostem perimeter and number of live leaves at flowering had low response in the equation fit with r² of 0.13. For the author, this value may have been due to the little influence that each variable has on the final mass or due to the reduced number of variables that compose this model, which denotes that a number of other factors not considered in the study may influence the mass of the

bunch. However, the r²value found was low, although the coefficient of variation was also relatively low (CV = 16%). This is justifiable, because the associations between the bunch mass and the other characters in banana may vary between genotypes and cycles and even between hybrids and their respective parents (Donato et al., 2006; Arantes et al., 2010).

The simple linear regression equations involving masses of bunch and hands were estimated as a function of the means of the repetitions of the independent variable pseudostem perimeter measured at ground level (Table 2). The developed models express magnitudes consistent with the correlation studies and with determination coefficients values which may be used. Regarding the correlation of the plant descriptors, similar results were found by Donato et al. (2006) with respect to the character pseudostem perimeter and its relationship with the bunch mass to other genotypes, ST12-31, Grand Naine, PV42-85 and Nanicão. However, that study did not aim to establish equations for predicting the mass of the bunch, but only the relationship between the characters.

Statistical procedures for multiple linear regression were used to estimate the equation for predicting the values of the masses of bunch and hands for the 'Dwarf Prata' and 'BRS Platina' bananas. To this end, only variables that showed significant correlation coefficients and with highest values in association with the masses of bunch and hands for each genotype were selected. The

Table 3. Correlation coefficients between the masses of the bunch and hands, in association with the yield characteristics in type Prata bananas 'Dwarf Prata' and 'BRS Platina', Guanambi, BA, 2009.

Yield characteristics	Genotypes			
	Dwarf Prata	BRS Platina	Dwarf Prata	BRS Platina
	Mass of bunch**		Mass of hands**	
Number of hands	0.70	-	0.68	-
Number of fruits	0.74	-	0.72	-
Number of fruits on the fourth hand	0.51	-	0.50	-
Mass of the fruit of the external row on the fourth hand	0.73	-	0.75	-
External length of the fruit of the external row on the fourth hand	0.64	-	0.66	-
Diameter of the fruit of the external row on the fourth hand	0.41	-	0.42	-
Number of hands	-	0.59	-	0.59
Number of fruits	-	0.73	-	0.73
Number of fruits on the fifth hand	-	0.65	-	0.65
Mass of the fruit of the external row on the fifth hand	-	0.66	-	0.66
External length of the fruit of the external row on the fifth hand	-	0.57	-	0.57
Diameter of the fruit of the external row on the fifth hand	-	0.34	-	0.34

** $P<0.01$.

procedure called backward elimination was used for choosing the best fit of the predicting equations.

Correlations between masses of bunch and hands with the number of bunches and number of fruits in the bunch, and the characters of the fourth and fifth bunch, respectively, (number of fruits; mass, external length and diameter of the fruit in the external row), for 'Dwarf Prata' and 'BRS Platina', were significant and positive for predicting the yield of bunch and hands (Table 3). The number of hands and fruits per bunch are the components that most influence on the mass of the bunch. However, the amount of fruits and fruit diameter of the extrenal row of the fourth bunch had the lowest correlation with the masses of the bunch and hands. Soto Ballestero (2008) reports on the relationship between the diameter of the central fruit of the external row of the second hand and the age of bunches for harvesting, and therefore, this fruit is used as a reference for indicating the harvest time of the bunch.

The associations between the masses of bunch and hands with the mass of the fruit of the external row of the fourth hand were statistically significant, positive and with highest value in relation to other variables for 'Dwarf Prata' bananas, being respectively, 0.73 and 0.75 (Table 3), indicating that the fourth hand of the bunch is strongly related to production (Meyer, 1975). However, still with the analysis of Table 3, it appears that the yield descriptors, fruit number and mass of the fruit of the external row of the fifth hand, showed the highest correlation coefficients in relation to other variables for hybrid 'BRS Platina'. This fact accredits these variables to compose the harvest prediction model.

In summary, the variable with the highest correlation for the bunch mass and mass of the hands was the number of fruits, for both genotypes. Despite the mass of the fruit of the external row have presented interesting correlation coefficient values for the Prata type, there was divergence for the genotypes according to the position of the hand. The fifth hand showed the second highest expression for the hybrid 'BRS Platina', while for the 'Dwarf Prata', the highest correlation was obtained for the mass of the fruit of the external row of the fourth hand (Table 3). The second hand has been considered a reference for studies and the classical procedure for determination of the harvest time for some banana genotypes (Jaramillo, 1982). Based on data from Meyer (1975) and in the present work, it can be inferred that the fourth and fifth hands, respectively, can be just as suitable for the development of postharvest studies and programming the harvest time as the second hand, which suggests the need of specific research to prove this hypothesis.

According to the correlation study, it can be suggested that the yield descriptors have the potential to compose the prediction model. In this context, the highest determination coefficients obtained for the harvest estimates for both mass of bunch and hands were observed with yield variables fruit number and mass of the fruit of the external row of the fourth hand for the progenitor 'Dwarf Prata'. For the 'BRS Platina', the number of fruits and fruit mass of the external row of the fifth hand showed the highest determination coefficients. By the methodology used, the harvest prediction equation which obtained the best fit for 'Dwarf Prata', both for the mass of the bunch, and for mass of hands, showed a determination coefficient (r^2) of 0.87 (Table 4). For the yield characteristics measured, considering the 'Dwarf Prata', the best fitted equations to determine the mass of

Table 4. Components of the equation for prediction of the masses of bunch and hands of 'Dwarf Prata' bananas, according to yield characteristics. Guanambi, BA, 2009.

Yield characteristic	Constant**	Coefficients							
		[a]NH	NFR**	NFR4	MFER4**	ELFE4	DFER4	r^2	CV(%)
Mass of bunch	-18.77	-	0.14	-	0.14	-	-	0.87	22.87
Mass of hands	-16.63	-	0.12	-	0.13	-	-	0.87	23.15

[a]NH, number of hands; NFR, number of fruits; NFR4, number of fruits on the fourth hand; MFER4, mass of the fruit on the external row of the fourth hand; ELFE4, external length of the fruit on the external row of the fourth hand; DFER4, diameter of the fruit on the external row of the fourth hand; -, non-significant variables by the *backward elimination* procedure. ** P<0.01.

Table 5. Components of the equation for prediction of the masses of bunch and hands of 'BRS Platina' bananas, as a function of yield characteristics. Guanambi, BA, 2009.

Yield characteristic	Constant**	Coefficients							
		[a]NH	NFR**	NFR4	MFER4**	ELFE4	DFER4	r^2	CV(%)
Mass of bunch	-14.25	-	0.16	-	0.12	-	-	0.80	22.87
Mass of hands	-13.87	-	0.14	-	0.11	-	-	0.79	23.15

[a]NH, Number of hands; NFR, number of fruits; NFR4, number of fruits on the fourth hand; MFER4,mass of the fruit on the external row of the fourth hand; ELFE4, external length of the fruit on the external row of the fourth hand; DFER4, diameter of the fruit on the external row of the fourth hand; -, non-significant variables by the backward elimination procedure. ** P<0.01.

bunch and hands at the harvest time were, respectively, $MB = -18.78 + 0.14NFR + 0.14MFER4$; and $MH = -16.63 + 0.12NFR + 0.13MFER4$. In this model, MB and MH indicate, respectively, the bunch mass and the mass of hands; NFR = number of fruits and $MFER4$ = mass of the fruit of the external row of the fourth hand. The other variables measured at the harvest time were not significant by the adopted procedure (Table 4). The fact that the determination coefficient of this model is highest is an indication that the descriptors measured at the time of harvest can represent production more accurately. This event takes on greater significance when associated with the occurrence of appropriate coefficient of variation (Table 4).

Meyer (1975) estimated a harvest prediction equation for Cavendish type bananas. The author investigated the relations between liquid bunch mass (mass of the bunches, without the stalk) and some parameters easily measured on the harvest day. The best fit found by the author was for the mass of the middle finger of the fourth hand, which is similar to this work. The equations were: $Y = 15.30X1 + 9.84X2 + 13.55$ and $Y = 8.41Z + 4.31$, in which Y = net mass of the bunch, expressed in kg; $X1$ = number of fingers per hand x 10-2; $X2$ = mass of the medium finger on the fourth hand, in hectograms; $Z = X1$ x $X2$.

Soares et al. (2012), studying the 'Tropical' cultivar banana, for harvest prediction, worked with the stepwise statistical procedure for multiple linear regression, using yield and vegetative characteristics, and determined the best fit for the prediction equation of bunch mass on harvest time, adding the variables that composed the prediction model. The equation was $MB = -5.25 + 0.11NLH + 0.07NFB + 0.05FW + 0.18LF + 2.04RW - 0.01LS$, in which: MB = Bunch weight, expressed in kg; NLH = Number of leaves at harvest; NFB = Number of fruits per bunch; AFW = Average fruit weight; LF: Length fruit; RW = Rachis weight. In this case, the determination coefficient obtained was 0.71. However, the variables are hard to be measured or are destructive to the samples, for the average mass of the fruit and mass of the rachis, respectively, a fact that complicates or little contributes to the efficiency and practicality of the prediction process, as reason Walpole et al. (2009). When analyzing the determination coefficient, for the same harvest estimates, the hybrid BRS Platina expressed inferior values to the presented by the genitor (Table 4), however, considered adequate, being r^2= 0.80 and 0.79, respectively, for the masses of bunch and hands (Table 5). That is justifiable, because the associations between bunch massand the other characters in banana can vary between genotypes and cycles and also between hybrids and their respective genitors (Arantes et al., 2010; Donato et al., 2006).

The coefficients of variation showed similarity between genotypes. The best equations to determine the masses of bunch and hands on the harvest time, for the 'BRS Platina', were, respectively, $MB = -14.25 + 0.16NFR + 0.12MFER5$, and $MH = - 13.87 + 0.14NFR + 0.11MFER5$, in which: BM and BH indicate, respectively, the bunch mass and the mass of the hands; NFR = number of fruits and $MFER5$ = mass of the fruit of the external row of the fifth hand. The other variables measured on harvest time were not significant by the procedure used to choose the models (Table 5). Jaramillo (1982) states that the bunch

Table 6. Prediction models of the masses of bunch and hands on type Prata bananas, 'Dwarf Prata' and 'BRS Platina', as a function of the number of hands. Guanambi, BA, 2009.

Genotype	[1]Estimate ($\hat{Y}$)	Simple linear regression equation**	(r²)
Dwarf Prata	MB	$\hat{Y} = -12.3804 + 3.55NH$	0.99
Dwarf Prata	MH	$\hat{Y} = -10.0485 + 3.08NH$	0.99
BRS Platina	MB	$\hat{Y} = -1.36638 + 3.02NH$	0.97
BRS Platina	MH	$\hat{Y} = -2.73380 + 2.88NH$	0.97

[1]Yield estimate ($\hat{Y}$): *MB* = mass of bunch; *MH* = mass of hands; [2]Independent variable (X): *NH* = number of hands. ** P<0.01.

mass of a genotype is closely related to the number of fruits, which, in turn, is directly proportional to the number of hands. Still, this author estimated regression equations and found that the number of hands per bunch is strongly related to the bunch mass. Other authors (Fernandez-Caldas et al., 1977; Holder and Cumbs, 1982; Donato et al., 2006; Arantes et al., 2010) confirm the relationship of the components mass of hands and bunch with the number of hands.

Thus, it can be inferred that the characters of production and number of hands are directly correlated. Donato et al. (2006) found correlation coefficients of 0.94 and 0.92, respectively, for 'Dwarf Prata' and 'BRS Platina', between bunch mass and number of hands, in the first production cycle. In addition, the mass of hands correlated significantly with the number of hands, as might be expected, since the hands mass is the main component of the bunch, only without the rachis. Arantes et al. (2010) found a correlation of 0.97 and 0.98 between bunch and hands mass, respectively, with number of hands in plantains. Therefore, given the high correlation usually found between the production and the number of hands, combined with ease of measurement and its non-destructive character, this variable was tested as a component of the equation fit to predict the masses of bunch and hands for the 'Dwarf Prata' and 'BRS Platina'. In this sense, simple linear regression models were adjusted between these variables from the average of the replicates, taking as independent variable the number of hands (Table 6).

The number of hands is an easy-assessment characteristic, by simple counting, and can be obtained at the stage of flowering, well before the harvest of the bunch of 'Dwarf Prata' and 'BRS Platina', about 120 to 150 days (Donato et al., 2009). When associated with the technique of marking of the bunch by age as a criterion for harvest time (Lichtemberg et al., 2008; Soto Ballestero, 2008), it allows to predict the timing and amount of harvest with more accuracy. The simple linear regression models that estimate the masses of the bunch and hands for the 'Dwarf Prata' and 'BRS Platina' showed significant values for the yield variable number of hands. Thus, the number of hands showed linear dependence on the harvest prediction model for the Prata type genotypes. Thus, for the progenitor 'Dwarf Prata', both for the mass of hands as for the mass of the bunch, with the variable number of hands, a r² of 0.99 was fitted (Table 6). For the hybrid 'BRS Platina', the determination coefficient was 0.97 for the masses of bunch and hands too, with the variable number of hands (Table 6). In this sense, Jaramillo (1982) observed, by means of linear regression, a fit of high linear dependence (r² = 0.99) between the number of hands and the bunch mass for Cavendish type bananas. Therefore, by the simple linear regression equations it was possible to estimate with high accuracy the yield in masses of bunch and hands, for 'Dwarf Prata'. In this order, the increase rate of mass of hands was estimated at 3.55 and 3.08 kg, and the yield in mass of bunch and mass of hands, for hybrid and 'BRS Platina', presented by hand, respectively, an increase estimated at 3.2 and 2.88 kg (Table 6).

To compose the equation, it was sought the ease in obtaining the values of yield variable coupled with a high correlation coefficient with production. In this set, the fit of the data provided r² 0.99 and 0.97, respectively, for 'Dwarf Prata' and 'BRS Platina'. This result suggests that the prediction models found in Table 6 are significant and of great scientific and practical application. Moreover, as these models use the variable number of hands of the bunch, which is defined in flowering, that is, three to four months before harvest, they allow the producer a more efficient planning.

Conclusions

The simple regression linear models estimate with a relative accuracy the masses of bunch and hands based on the pseudostem perimeter measured at ground level, for 'Dwarf Prata' and 'BRS Platina'.

The multiple linear regression models estimated with adequate accuracy the masses of the bunch and hands, as a function of the characteristics number of fruits and mass of fruit of the fourth hand for 'Dwarf Prata', and fruit number and mass of fruit of the fifth hand for 'BRS Platina'.

The simple linear regression models allow prediction of the masses of the bunch and hands, according to the number of hands in advance of at least 120 days to harvest and high precision, for 'Dwarf Prata' ($r^2 = 0.99$) and 'BRS Platina' ($r^2 = 0.97$), which ensures the consistency of the regressions to estimate the production Prata type bananas.

ACKNOWLEDGMENT

The authors express thanks to EPAMIG - Agricultural Complex of Minas Gerais and Brazil, for the support and technological solutions provided during this research.

Abbreviations: AAB, Dwarf Prata, triploid; **AAAB,** BRS Platina, tetraploid; **AA,** AAB and M53 diploid; **MB,** bunch weight; **MH,** mass of hands; **NFR,** number of fruits; **MFER4,** mass of the fruit of the external row of the fourth hand; **NLH,** number of leaves at harvest; **NFB,** number of fruits per bunch; **AFW,** average fruit weight; **LF,** Length fruit; **RW,** Rachis weight.

REFERENCES

Arantes AM, Donato SLR, Silva SO (2010). Relação entre características morfológicas e componentes de produção em plátanos. Pesq. Agropec. Bras. 45(2):224-227.

Bíscaro GA (2007). Meteorologia agrícola básica. Cassilândia: UNIGRAF. P. 86.

BRASIL. Ministério da Agricultura, Pecuária e Abastecimento. Secretaria de Política Agrícola, Instituto Interamericano de Cooperação para a Agricultura (2007). Cadeia produtiva de frutas. Série agronegócios, v. 7. Antônio Márcio Buainain e Mário Otávio Batalha (Coord.). Brasília: IICA: MAPA/SPA, 2007.

Cordeiro ZJM, Moreira RS (2006). A bananicultura brasileira. Bananicultura: um negócio sustentável. XVII REUNIÃO INTERNACIONAL ACORBAT 2006. 15 a 20 de outubro de 2006, Joinville – Santa Catarina – Brasil. Anais... XVII Reunião Internacional da Associação para a Cooperação nas Pesquisas sobre Banana no Caribe e da América Tropical, pp. 36-47.

Dantas JLL, Soares Filho WS (2006). Classificação, origem e evolução. Frutas do Brasil, março 2006. Available in: <www.ceinfo.cnpat.embrapa.br/arquivos/artigo_2317.pdf>. Acess in: may 15 2011.

Donato SLR, Silva SO, Lucca Filho AO, Lima MB, Domingues H, Alves JS (2006). Correlações entre caracteres da planta e do cacho em bananeira (Musa spp). Ciênc. agrotec. 30:21-30.

Donato SLR, Arantes AM, Silva SO, Cordeiro ZJM (2009). Comportamento fitotécnico da bananeira 'Prata-Anã' e de seus híbridos. Pesq. Agropec. Bras. 44(12):1608-1615.

FAO-Food and Agriculture Organization (2011a). Banana. Available in: <http://faostat.fao.org/site/567/DesktopDefault.aspx?PageID=567#ancor>. Acess in: Apr. 08, 2013a.

FAO –Food and Agriculture Organization (2011b). Consumo. Available in: <http://faostat.fao.org/site/609/DesktopDefault.aspx?PageID=609#ancor>. Access in: Apr. 08, 2013b.

Fernandez-Caldas E, Garcia V, Perez-Garcia V, Diaz A (1977). Análisis foliar del plátano en dos fases de sudesarrollo: floración y corte. Fruits 32(11):665-671.

Holder GD, Cumbs FA (1982). Effects of water supply during floral initiation and ifferentation on female flower production by robusta banana. Exp. Agric. 18(2):183-193.

Hernández MS, Martínez O, Fernández-Trujillo JP (2007). Behavior of arazá (Eugenia stipitata Mc Vaugh) fruit quality traits during growth,

development and ripening. Sci. Hortic. 111:220-227.

IBRAF - Instituto Brasileiro de Frutas (2005). Estudo da cadeia produtiva de fruticultura do estado da Bahia: Análises. São Paulo.

IPGRI - International Plant Genetic Resources Institute (1996). Descriptors for banana (Musa spp.). Roma: IPGRI, P. 55.

Jaramillo RC (1982). Lasprincipales características morfológicas del fruto de banano, variedade Cavendish Gigante (Musa AAA) em Costa Rica. Upeb-Impretex, P. 42.

Lichtemberg LA, Vilas Boas EVB, Dias MSC (2008). Colheita e pós-colheita da banana. Inf. Agropec. 29(245):92-110.

Lima Neto FP, Silva SO, Flores JCO, Jesus ON, Paiva LE (2003). Relações entre caracteres de rendimento e de desenvolvimento em genótipos de bananeira. Magistra 15(2):275-281.

Maia E, Siqueira DL, Silva FF, Peternelli LA, Salomão LCC (2009). Método de comparação de modelos de regressão não-lineares em bananeiras. Cienc. Rural. 39(5):1380-1386.

Meyer JP (1975). Estimation de productivité: calculdupoidsdes régimes de bananier em function Du nombre de doigts et dupoidsd"undoigt. Fruits 30(12):739-744.

Ortiz R (1997). Morphological variation in Musa germplasm. Genet. Resour. Cropevol. 44:393-404.

Pimentel-Gomes F (2000). Curso de estatística experimental. 14.ed. Piracicaba: Nobel. P. 477.

Ribeiro Júnior JI (2001). Análises estatísticas no SAEG. Viçosa: UFV. P. 301.

Robinson JC, Galán SV (2010). Bananas and plantains. 2nd ed. Oxford: CAB International (Crop production science in horticulturae series, 19:311.

Roberto SR, Sato AJ, Brenner EA, Jubilei BS, Santos CE, Genta W (2005). Caracterização da fenologia e exigência térmica (graus-dia) para a uva 'Cabernet Sauvignon' em zona subtropical. Acta. Sci. Agron. 27(1):183-187.

SAEG (2007). Sistema para análises estatísticas. Versão 9.1. CD-ROM. Viçosa: FUNARBE, UFV, 2007. [CD-ROM].

Rocha J (2010). Avaliação do coeficiente de variação e relações entre caracteres de rendimento e desenvolvimento na cultura da bananeira (Magister Scientiae Dissertation). Cruz das Almas, BA: Universidade Federal do Recôncavo da Bahia. P. 46.

Savin IY, Stathakis D, Negre T, Isaev VA (2007). Prediction of crop yields with the use of neural networks. Rus. Agric. Sci. 33(9):361-363.

Silva SO, Rocha AS, Alves EJ, Credico MDI, Passos AR (2000). Caracterização morfológica e avaliação de cultivares e híbridos de bananeira. Rev. Bras. Frutic. 22(2):161-169.

Stenzel NMC, Neves CSVJ, Marur CJ, Scholz MBS, Gomes JC (2006).Maturation curves and degree-days accumulation for fruits of 'Folha Murcha' orange trees. Sci. Agric. 63(3):219-225.

Scarpari MS, Beauclair EGF (2009). Physiological model to estimate the maturity of sugarcane. Sci. Agric. 66(5): 622-628.

Siqueira DL (1984). Variabilidade e correlações de caracteres em clones da bananeira 'Prata'. Lavras, P. 68. Magister Scientiae Dissertation – Escola Superior de Agricultura de Lavras, 1984.

Soares JDR, Pasqual M, Lacerda WS, Silva SO, Donato SLR (2013). Utilization of artificial neural networks in the prediction of the bunches weight in banana plants. Sci. Hort. 155:24-29.

Soares JDR, Pasqual M, Rodrigues FA, Lacerda WS, Donato SLR, Silva SO, Paixão CA (2012). Correlation between morphological characters and estimated bunch weight of the Tropical banana cultivar. Afr. J. Biotechnol. 11(47):10682-10687.

Soto Ballestero MS (2008). Bananos: Técnicas de Producción, Poscosecha y Comercialización. 3a.ed. San José, Costa Rica: Lil,. 1 CD – ROM.

Streck NA, Michelon S, Bosco LC, Lago I, Walter LC, Tellesrosa H, Paula G (2007). Soma térmica de algumas fases do ciclo de desenvolvimento da escala de counce para cultivares sul-brasilerias de arroz irrigado. Bragantia 66(2):357-364.

Walpole RE, Myers RH, Myers SL, Ye K (2009).Probabilidade e estatística para engenharia e ciências. 8. Ed. Americana. ISBN 978-85–7605–199-2. São Paulo: Pearson Prentice Hall.

Wyzykowski J (2009). Modelos de regressão para a descrição do crescimento do cafeeiro irrigado e não irrigado após recepa. 2009. P. 80. Dissertação (Master of Science in Statisticsand Agricultural

Experimentation), Universidade Federal de Lavras, Lavras.

Zhang W, Bai XC, Liu G (2007). Neural network modeling of ecosystems: a case study on cabbage growth system. Ecol. Model. 201(3):317-325.

26

Production preference and importance of fruit species in home garden among rural households in Igbo-Eze North Agricultural Zone of Enugu State, Nigeria

Dimelu, M. U. and Odo, R. N.

Department of Agricultural Extension, Faculty of Agriculture, University of Nigeria, Nsukka, Enugu State, Nigeria.

The study examined production preference and importance of fruits in home garden using one hundred randomly selected household heads. Data were collected by use of structured interview schedule and analysed using descriptive statistics. Household produced *Treculia Africana* (100%), *Anacardium occidentalis* (100%), *Psidium guajava* (100%), *Citrus* spp. (100%), *Carica papaya* (93%) and *Manifera indica* (90%) fruits for nutritional purpose, while fruits of major economic importance were *Irvingia gabonensis* (98.0%), *Kola acuminate* (97.0%), *Persea Americana* (88.0%), *Spitium sativum* (84.0%), *Citrus* spp (80%), *Pentaclethra macrophyllum* (78%), *Musa-sapientum* (71.0%) and others. Fruits of social importance were *Kola macrophyllum* (100%), *Garcina kola* (100%), *A. occidentalis* (100%) and *Cocos nucifera* (58%); and only *G. kola* (100%) was of medicinal benefit to households. The most preferred fruits for production in home garden were *Irvingia gabonensis* (1st), *K. acuminate* (2nd), *Citrus* spp. (3rd), *Persea americana* (4th), *Dennettia tripatale* (5th) and the least preferred was *Manifera indica* (15th) fruit species. Preference was based on input requirement, resistant to pest and diseases, frequency of fruiting, availability of market and others. Extension and research should promote, intensify research/training to increase awareness on nutritional and medicinal importance of most fruit species particularly the less preferred to guide against extinction.

Key words: Fruit, production, home garden, economic, nutritional, medicinal, social.

INTRODUCTION

Fruits are widely accepted as important component of a healthy diet and adequate consumption could help to reduce a wide range of diseases. They play a significant role in human nutrition, especially as sources of vitamins C (ascorbic acid), A, thiamine (B_1), niacin (B_3), pyridoxine (B_6), Folacin (also known as folic acid or folate), (B9), E, minerals, and dietary fiber (Craig and Beck, 1999; Quebedeaux and Eisa, 1990). According to Food and Agriculture Organization/ World Health Organization (FAO/WHO) (2004), approximately 16.0 million (1.0%) disability adjusted life years (DALYs; a measure of the potential life lost due to premature mortality and the years of productive life lost due to disability) and 1.7 million (2.8%) of deaths worldwide are attributable to low fruit and vegetable consumption. The report showed that insufficient intake of fruit and vegetables is estimated to cause around 14% of gastrointestinal cancer deaths, about 11% of ischaemic heart disease deaths and about 9% of stroke deaths globally. Thus, promoting increased production (for availability, affordability, and access) and consumption for maximum health benefits is a global concern.

The FAO (2007) reported that the production of high value agricultural commodities such as vegetable, fruits, and milk is growing at a fast rate. According to the report, annual growth of high value agricultural production between 2004 and 2006 is 2.9% (vegetable), 3.0% (fruits), 4.0% (meat), and 4.0% (milk). Developing countries account for about 98% of total production, while developed countries account for 80% of world import trade (FAO, 2004). Also, the major tropical fruits account for approximately 75% of global tropical fresh fruit production.

In the same vein, worldwide food demand is shifting from such basic commodities as cereal and rice to products with higher value added such as vegetables, fruits, fats, meats and oil (von Braun, 2007). Hence the composition of food budget is shifting from the consumption of grains and other staple crops to vegetables, fruits, meat, dairy and fish. However in many developing countries like Nigeria and Brazil the shift to more value added products is less pronounced and statistics showed a decline. Von Braun (2007) reported that in Brazil, Kenya and Nigeria, the consumption of some high value products declined probably due to growing inequality in some of these countries. More than three-quarters of adults in less developed countries consume less than the minimum recommended five daily servings of fruits and vegetables. In Sub-Saharan Africa the level of fruit and vegetable consumption ranges from 27 to 111 kg per capita per year, far below the WHO/FAO minimum recommendation of 146 kg per capita per year. Moreover, while vegetable consumption is almost universal in most Sub-Saharan Africa; the consumption of fruits is much less common and varies across countries. Also the average consumption (in kg per capita per year) is lower for fruits than vegetables in most countries (Ruel et al., 2004). The authors further reported that consumption of both products is generally higher in urban areas compared to rural areas.

Generally, situations suggest a significant gap in mean consumption of fruits and vegetable across countries, sectors/locations and economic groups. Thus, in spite of the growing body of evidence on the protective effect of fruits and vegetable, their consumption/intake is still grossly inadequate. The global production as well as consumption is expected to grow to meet WHO/FAO minimum recommendation for fruit and vegetable intake. Nigeria is credited with production of variety of fruits such as mangoes, watermelon, guava, pineapples, pawpaw, oranges, tomatoes, tangerines, and many other indigenous fruits (Adenegan and Adeoye, 2011). Specifically, Enugu State with its characteristics temperate ecology is known to favor the production and growth of several tropical fruits. Many rural farmers in the state explore this advantage particularly in the face of increasing crop failure perceived to be associated with climate change. This is because climate change alters planting pattern, reduces yield of crops and animals, affect flowering periods of crops, gestation and

reproduction in wild life (Akpan et al., 2010). Consequently, there is increased rural household interest in home garden and also establishment of urban - intensified home production of fruits. Different type of fruits that are unevenly distributed, both exotic and indigenous fruits are found either protected or cultivated in home gardens. It is therefore a paradox that though most of the fruits are produced in the rural communities, the consumption levels has remained below the FAO recommendation and lower than the consumption level in urban cities. Moreover, distribution, access and availability of the fruit species largely varied in the home gardens, hence the volume of production and supply differ. Therefore the study aimed to ascertain the perceived importance of fruits species in home gardens, production preference of fruits species and reasons for preference of fruit species in home garden among households.

METHODOLOGY

The study was carried out in rural communities of Enugu-Ezike North Agricultural Zone of Enugu State. Household heads constituted the population for the study. Out of three local government areas in the zone, one local government area (Igbo-Eze North Local Government Area) was purposively selected because of the volume and intensity of fruit production in the area (Figure 1). Igbo-Eze North Local Government Area comprised 4 autonomous town communities namely: Umuozzi, Umuitodo, Essodo, and Ezzodo. A proportionate random sampling technique was used to select 5, 2, 2, and 1 villages from Umuozzi, Umuitodo, Essodo, and Ezzodo town communities, respectively. Thus, a total of ten (10) village communities was used. From each of these villages, ten (10) households' heads were randomly selected from list of households provided by informants using simple random sampling techniques. A total of 100 respondents were used for the study. Structured interview schedule was used to obtain relevant information based on the objectives. Respondents were asked to indicate the perceived importance of fruits in their home garden under four major categories-economic, medicinal, nutritional, and social. Data on the households' production preference were collected by asking the respondents to arrange fruits in their home garden in order of preference using-1, 2, 3, 4, 5 etc. The mean preference by fruits was used to determine the most preferred fruits. Also the respondents indicated against eight listed variables (frequency of fruiting, ease of processing, early maturity, etc) the perceived reasons for preference. Data collected were analysed using percentage and mean.

RESULTS AND DISCUSSION

Perceived importance of fruits in home garden

Economic importance

The major fruits cultivated in home garden for economic purposes were ogbono (*Irvingia gabonensis*) (98.0%), kola (*Kola acuminate*) (97.0%), avocado (*Persea americana*) (88.0%), pear (84.0%), orange *Citrus* spp.) (80%), ukpaka (*Pentaclethra macrophyllum*) (78%),

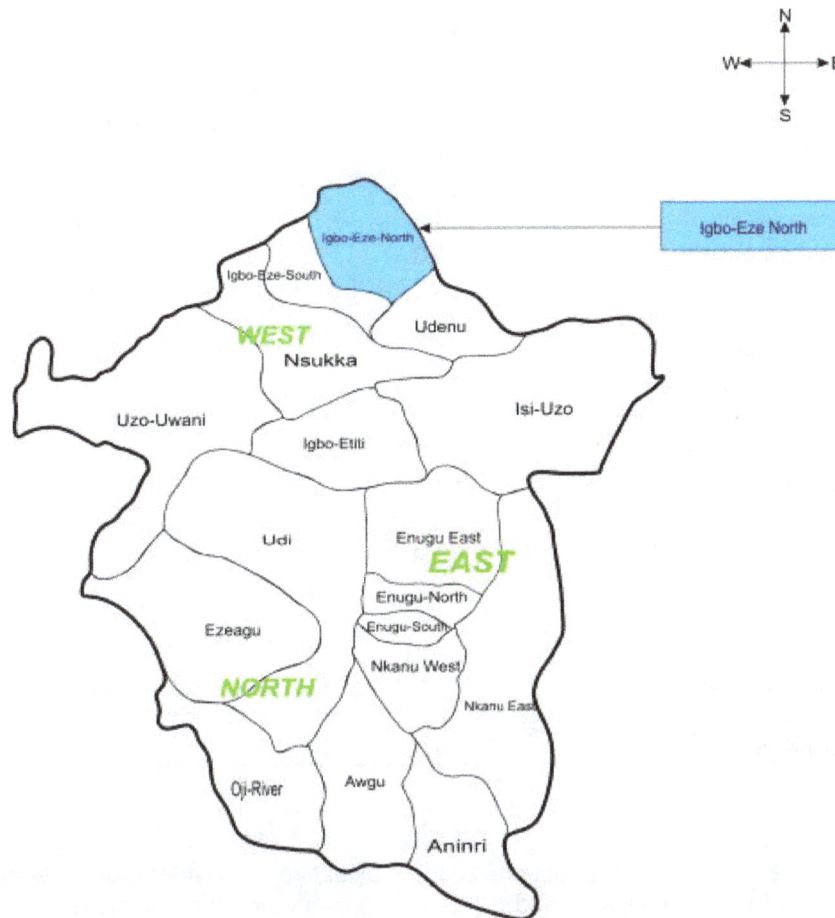

Figure 1. Map of Enugu State showing Igbo-Eze North Local Government Area.

banana (*Musa sapientum*) (71.0%) and plantain (*M. paradisiacal*) (59.0.0%) (Table1). A lesser proportion (39.0, 30.0, 27.0, 24.0, 23.0 and 16.0%) of the respondents indicated that udala (*Chrysophyllum albidum*), pawpaw (*Carica papaya*), black pear (*Dacryodes edulis*), mmimi (*Dennettia tripatale)*, pineapple (*Ananas comosus*), and mango (*Manifera indica*) were domesticated for economic purposes, respectively. Only one percent had coconut in their home garden for the same reason. This collaborate with the general view that fruits and fruits vegetables play a key role in income generation (Adebooye, 2003), contribute to livelihood enhancement (Poulton and Poole, 2001) and apart from consumption, they are increasingly being sold in the market. For instance, Adebisi (2004) reported that income from *Garcinia cola* (bitter cola) in south eastern parts is used to cover schooling cost and other social obligations. In 1999 export of *D. edulis* (black pear) from central Africa and Nigeria to France, Belgium and U.K were worth more than US$ 2 million per annum (Awono, 2002). Moreover, in Cameroon, trade in banana, kola spp, *D. edulis* and others within Cameroon and neighboring countries was worth US$ 1.75 million in the

first half of 1990 (Ndoye, 1997; Temple, 2001).

In addition to financial gain accrued from fruits, other attractions of these fruits may have root in the fact that they are less labour-intensive than conventional crops and so are planted where labour is a limiting factor. Labour shortages are common especially in areas experiencing rural-urban youth migration in search of white cola job. Besides, fruit production is done with little capital investment. Furthermore, depending on areas, some fruit species may be more economically important than others to local livelihoods. For example, in a village in southern Nigeria, the sale of ogbono is considered to be the primary cash income source for 30% of households and secondary cash income source for another 55% of households (Adebisi, 2004).

On the other hand, Ukwa (*Treculia africana)* and bitter kola (*G. kola*) were not perceived by respondents as economic fruits contrary to assertion made by Adebisi (2004) that the sale of *G. cola* (bitter cola) nuts bring about 8% of household's annual total income in south western parts of the country and also in foreign exchange. Similarly, Ukwa according to Ebuehi (2006) constitutes a strategic food reserve of essential food

Table 1. Percentage distribution of respondents by perceived importance of fruits in home garden (n=100).

Fruits	Percent importance			
	Economic	Medicinal	Nutritional	Social
Mango (*Manifera indica*)	16.0	-	90.0	53.0
Orange (*Citrus spp*)	80.0	-	93.0	64.0
Banana (*Musa-sapientum*)	71.0	-	5.0	-
Plantain (*M. paradisiacal*)	59.0	-	17.0	-
Pineapple (*Ananas comosus*)	23.0	1.0	49.0	3.0
Ukpaka (*Pentaclethra macrophyllum*)	78.0	-	64.0	-
Udala (*Chrysophyllum albidum*)	39.0	-	25.0	2.0
Ogbono (*Irvingia gabonensis*)	98.0	-	1.0	-
Ukwa (*Treculia africana*)	-	-	100.0	-
Mmimi (*Dennettia tripatale*)	24.0	-	-	17.0
Kola (*Kola acuminate*)	97.0	-	-	100.0
Avocado (*Persea Americana*)	88.0	-	87.0	-
Black pear (*Dacryodes edulis*)	27.0	-	-	-
Pear (*Spitium sativum*)	84.0	-	63.0	-
Cashew (*Anacardium occidentalis*)	-	-	100.0	100.0
Guav (*Psidium guajava*)	-	-	100.0	-
Coconut (*Cocos nucifera*)	1.0	-	56.0	58.0
Bitterkola (*Garcina kola*)	-	100	--	100.0
Pawpaw (*Carica papaya*)	30.0	-	93.0	55.0

nutrient that have become delicacies and specialized meals which are available at certain critical period of the year when these nutrients are very scarce. Ideally, the above scenario is expected to have stimulated sustained interest for its commercial production and distribution in home garden. The finding however, suggests that the fruit species mainly exist as protected fruits in home garden; and production is limited to home consumption.

Medicinal importance

Table 1 shows that all (100%) the respondents grew bitter kola for medicinal reason and only one percent had pineapple in their home garden for similar purpose. Other fruits were of no medicinal importance to the respondents. Largely, this confirms Adebooye (2003) who asserted that in addition to contributing to food security, fruits are also sources of folk medicine. For instance the seed of *G. kola* (bitter kola) are used in folk medicine and in many herbal preparations for the treatment of ailments because according to Terashima (1999), it contains a complex mixture of biflavonoids, xanthones, and prenylated benzophenones, and has antioxidant activities (Terashima, 2002). The major component of *G. kola* is kolaviron and this has been reported to significantly prevent hepatotoxicity and has a chemo-preventive effect against carbon-tetrachloride and potassium bromate (Farombi, 2002). The kolaviron may also protect against carcinogen and drug induced

oxidative and membrane damage and such may be relevant in the chemotherapy of liver and kidney diseases. Adodo (2000) reported that when chewed raw, it helps to soften cough and remove tronsilltis.

Similarly, Oguntola (2008) reported that pineapple has high amount of mineral, vitamins, and small amount of fats and protein, an anti-oxidative vitamin C and enzymes bromelin which can protect against breast cancer. Contrary to the findings, other fruits perceived of no medicinal importance by the respondents such as *D. eduli (*black pear*), Xylopia aethiepica (uda), Spondic mombin (*ijikara) and others were reported to have demonstrated their contribution to the treatment of diseases such as sickle cell anemia, snake bite etc. Also Mango is rich in beta-carotene, an antioxidant potent in helping to cure various diseases, including dermatitis, flu, asthma, vision problems, bleeding gums, sore throat, inflammation of the airways, shortness of breath and ulcers. It can also handle boils, scabies, eczema, abdominal colic, diarrhea, motion sickness, worms, loss of appetite, vaginal discharge, menstrual disorders, hernias and rheumatism. In the same vein, banana fruit contains health promoting flavonoid poly-phenolic antioxidants such as *lutein, zea-xanthin, ß and α-carotenes* in small amounts which help act as protective scavengers against oxygen-derived free radicals and reactive oxygen species (ROS) that play a role in aging and various disease processes (www.localharvest.org/blog/20618/entry/medicinal.values. of.the.tropical).

Generally the results underpin the report by WHO (2001) that 80% of the world population use medicinal plants in the treatment of diseases and in Africa, the rate is much higher. According to Iwu (1999), they are relatively safe than their synthetic alternatives, thus offering more affordable treatment.

Nutritional importance

Table 1 shows that 100, 100 100, 93, 93 and 90% of the respondents domesticated ukwa (*T. africana)*, cashew (*Anacardium occidentalis*), guava (*Psidium guajava*), orange (*Citrus* spp), pawpaw (*C. papaya*) and mango (*M. indica*) fruits for nutritional purpose, respectively. Also ukpaka, (*P. macrophyllum* (64.0%), cashew (63.0%), coconut (*Cocos nucifera*) (56.0%), and pineapple (*A. comosus*) (49.0%) were cultivated/protected in home garden for nutritional reason. However, only 1, 5, 17.0 and 25.0% of the respondents had ogbono (*I. gabonensis*), banana (*Musa sapientum*), plantain (*M. paradisiacal*), and udala (*C. albidum*) for nutritional purposes, respectively. Other fruits like mmimi (*D. tripatale)*, kola, black pear, and bitter cola were of no nutritional importance to the households. The results confirm the popular assertion that fruits play key role in the nutritional livelihood of the Nigeria population especially in the rural area where people could scarcely pay for meat, milk, and egg (Adebooye, 2003). The author further reported that studies on chemical composition of indigenous fruits have shown that they contain an appreciable amount of crude protein, fat, and oil, energy, vitamin and minerals. Above all, fruits have no cholesterol unlike meat and egg thus providing the body with essential micro nutrients that increase body's anti-oxidative potential as well as dietary fibre which substantially reduce the risk of excessive body weight gain and obesity. Thus, sufficient intake of them provides the body with essential micro-nutrient that increases the body's antioxidative potential as well as dietry fibre. Hence most fruits such as ukwa and ukpaka have today become popular delicacies and specialized meals not only for the rich and urban dwellers but also for rural people and export food.

Social importance

All (100%) of the respondents domesticated kola, bitter kola and cashew for social reasons, while 58% domesticated coconut for similar reasons (Table 1). This corroborated with Terashima (1999) that seeds of bitter kola are major kola substitutes offered to guests in homes and social gathering. Traditionally, kola, bitter kola and coconut play important role in social life of the people. It is widely used across cultures during different ceremonies such as burial, marriage ceremonies, child naming, title taking, and other related traditional gatherings. They are also popular in pure African traditional religion for it is believed that one cannot talk to gods without a kola nut. When used in ceremonies it shows good reception and acceptance. Thus, it is valued more than anything by guest and virtually no ceremony is performed in most cultures without kola, or bitter kola.

Similarly, mango (53%), orange (64%), pawpaw (55%) were also of social importance to the respondents but are commonly used for entertainment of visitors and friends. They are one of the hungry fruits consumed during relaxation from long journey or drudgery task particularly among farming communities.

Production preference of fruits in home garden

The most preferred fruits by the respondents were ogbono (1st), kola (2nd), orange (3rd), avocado (4th), mmimi (5th), banana (6th) and others (Table 2). Also the least preferred fruits were pineapple (14th) and mango (15th). The reasons for the observed preference are discussed as follows:

Reasons for preference

Frequency of fruiting

This means how often a particular fruits produces in a year. The respondents preferred ogbono (97%), ukpaka (80%), orange (70%), avocado (68%), coconut (65%), plantain (60%), banana (58%), and kola (50%) for the frequency of fruiting (Table 3). For instance, ogbonno (Bush mango tree) produces two times every year. The first time of fruiting is between April and June while the second time is between August and October. Also, ukpaka fruits produce continuously throughout the year. Similarly, orange tree fruits produce at least two times every year depending on geographical location. However, a lesser proportion (25, 25, 20, 20, 15, 13 and 10%) of the respondents preferred mmimi, pineapple, black pear, udala, ukwa, pear, and ukwa in the home garden for the same reason. High frequency of fruiting is a crucial factor in production decision because it offers households opportunities for regular income, and resilience especially at off-farm season or crop failure.

High productivity

This has to do with the quantity of fruits per tree. Involvement of the respondents in production of ogbono (100%), ukpaka (96%), black pear (70%), plantain (67%), pear (63%), pawpaw (60%), banana (59%), coconut (58%), avocado (52% and orange (50%) in home garden was on the basis of productivity. These fruits turn out heavy produce each year, ensuring greater income. Only 32 and 20% preferred udala and mango because of high

Table 2. Fruit preference among respondents (n=100).

Fruits	Order of preference
Ogbono	1st
Kola	2nd
Orange	3rd
Avocado	4th
Mmimi	5th
Banana	6th
Pear	7th
Udala	8th
Plantain	9th
Ukpaka	10th
Black pear	11th
Pawpaw	12th
Coconut	13th
Pineapple	14th
Mango	15th

productivity. In reality, mango and udala often produce large quantity of fruits, though not all round. Productivity is an important determinant of choice of crops/fruits and farming system. However it is a function of factors such as varieties, fertility of soil, disease and pest, edaphic factors, agronomic practices etc.

High pricing of produce

This is concerned with how much is realized from the sale of each fruit produced. Ogbono (100%), kola (100%) and ukpaka (96%) were highly preferred in home gardens because of high pricing (Table 3). The respondents indicated preference for banana (61%), plantain (61%), black pear (48%), pear (43%) and avocado (40%) for the pricing. Other fruits including mmimi (20%), udala (28%), mango (18%) were less priced. Currently, the prices and demand for ogbono, ukpaka, kola, banana and plantain are very impressive and attractive. Largely the profit accrue from them are the driver of household interest in their production in home garden. Strong market structure is one of the major institutional factors that drive supply-demand chain. It is even more serious for fruits because of the relatively short ripening period and reduced post-harvest life particularly where facilities and capability for processing and preservation are limited.

Early maturity

This refers to the time between planting and first harvest. Table 3 shows that production of ogbono (96%), banana (94%), plantain (94%), ukpaka (93%), pawpaw (89%), pineapple (86%), and pear (67%) were produced because of early maturity features. Also, production of

guava (46%), mango (23%, udala (18%), coconut (14%), and orange (10%) were less preferred as early maturing fruits. Certainly, some fruits mature early and others take a longer time to mature. Most fruit plants mature within the period of six months to five years before fruiting. Usually farmers are attracted to fruit with shorter period of maturity for quick economic return. Besides, fruits with short fruiting period lend easily to home garden.

Low input requirements

The respondent indicated that their involvement in production of all the fruits in home garden was because of the low input requirements (Table 3). Unlike conventional crops such as yam, cassava and maize, most fruits like ogbono, avocado, pear and ukpaka rarely require inorganic input (example fertilizer and herbicide). It is not surprising because according to Odebode (2006), home gardening is the cultivation of small portion of land which may be at the back of the home or within a walking distance from home. Hence soil fertility largely depends on the use of kitchen waste, animal dung and compost manure. It is both environmentally friendly and sustainable. Nevertheless, insecticide and pesticide are sometimes used for treatment of pest and diseases, particularly in cases of persistent fruit abortion and disease outbreak.

Ease of harvest

The respondents preferred pawpaw (89%), guava (76%), pineapple (67%), cashew (63%), mmimi (56%), banana (50%), plantain (50%) because of easy of harvest (Table 3). Lesser proportion (48, 43, 20%, 10, 2.0%) of the respondents cultivated/protected pear, kola, avocado, ogbono and mango in home due to ease of harvest, respectively. Naturally, some of these fruits such as black pear, ogbono, ukwa and udala, avocado and other standard-sized fruit trees grow too tall that harvesting them becomes very difficult, particularly where they have grown for many years. In such situation, the cost of labour becomes very high and often results to waste of ripe fruits, low quality products, environmental pollution from over ripped fruits and subsequently poor return for the farmers This discourages households from venturing into cultivating/protecting such fruits.

Resistance to pest and disease

About 79% of the fruits identified in home garden were preferred by the respondents because of the high resistance to pest and diseases. They include pineapple (100%), ukpaka (100%), udala (100%), ogbono (100%), ukwa (100%), mmimi (100%) avocado (100%), black pear (100%), cashew (100%), guava (100%) and others.

Table 3. Percentage distribution of respondents based on reasons for preference (n=100).

Fruit	Reasons for preference									
	Hff (%)	GP (%)	Eh (%)	Ep (%)	Am (%)	Hpp (%)	Em (%)	Lir (%)	Rpd (%)	Sv (%)
Mango	-	20.0	2.0	-	10.0	18.0	23.0	100.0	53.0	53.0
Orange	70.0	50.0	-	-	28.0	4.0	10.0	100.0	13.0	64.0
Banana	58.0	59.0	50.	72.0	98.0	61.0	94.0	100.0	65.0	-
Plantain	60.0	67.0	50.	72.0	98.0	14.0	94.0	100.0	65.0	-
Pineapple	25.0	-	67.	-	52.0	96.0	86.0	100.0	100.0	3.0
Ukpaka	80.0	96.0	-	100.	98.0	28.0	93.0	100.0	100.0	-
Udala	15.0	32.0		-	29.0	100.	18.0	100.0	100.0	2.0
Ogbono	97.0	100.0	10.	-	29.0	-	96.0	100.0	100.0	-
Ukwa	10.0	-	-	6.0	100.	23.0	-	100.0	100.0	-
Mmini	25.0	-	56.	-	-	100.	-	100.0	12.0	17.0
Kola	50.0	-	43.	2.0	60.0	40.0	-	100.0	100.0	100.0
Avocado	68.0	532.0	20.	-	100.	48.0	-	100.0	100.0	-
Black pear	20.0	70.0	-	-	62.0	46.0	-	100.0	100.0	-
Pear	13.0	63.0	48.	-	75.0	-	67.0	100.0	100.0	-
Cashew	-	-	63.	-	67.0	-	-	100.0	100.0	100.0
Guava	-	-	76.	-	-	-	46.0	100.0	100.0	-
Coconut	65.0	58.0	-	-	-	-	14.0	100.0	100.0	58.0
Bitter kola	-	-	-	-	-	-	-	100.0	100.0	100.0
Pawpaw	-	60.0	89.	-	-	-	89.0	100.0	41.0	55.0

* Multiple response; HFF = High frequency of fruiting; Am = availability of market; Lir = low input requirements; GP = greater productivity; PP= high price of produce; Rpd = resistance to pests and diseases; EP = ease of processing; EM = early maturity; SV= social value.

Only 13 and 12%, respectively of the respondents indicated preference for orange and kola because they are less resistant to pest and diseases. It is possible that most of the fruits are indigenous fruit trees which characteristically are more adaptable to the environment. Pest and diseases stands as one of the major threats to fruit production. It results to delayed maturity, low yield, poor quality products and low income for households. In this case the use of improved resistance varieties, application of routine agronomic practices becomes imperative.

Ease of processing

Majority (100, 72 and 72%) of the respondents preferred ukpaka, banana, and plantain because they are easy to process, respectively, while only 6 and 2%, respectively expressed preference for ukwa and kola for the same reason. These are actually fruits that undergo some processes, sometimes rigorous before consumption. Preferences of other fruits (73.7%) were not influenced by ease of processing. Most of these fruits (e.g pawpaw, avocado, orange, etc) are edible in their natural form and as a result no processing is required. However, due to high perishability of some like banana, pawpaw, pineapple, mango, orange, and many others, some are often, processed into different products such as juice,

wine, vitamin supplements, flavours, laxative and others (Rallof, 2000). On the other hand, some like oil bean, ukwa, African black pear etc cannot be consumed unless they are subjected to either traditional or industrial processing.

Conclusion

The results show that fruits in home garden were of economic, medicinal, nutritional, and social importance to households, but the major attraction to their cultivation in home garden of household is economical in terms of income generation, labour, market e.t.c. Households preferred some fruits to others. The most preferred fruits were ogbono (*I. gabonensis*) followed by kola (*Kola acuminata*), orange (*Citrus* spp), avocado (*P. americana*) and others, while the least preferred was mango (*M. indica*). Preferences were based on input requirement, resistant to pest and diseases, frequency of fruiting, availability of market, maturity and others. Extension should intensify awareness on the nutritional and medicinal benefits of less preferred fruits species to increase interest, production and consumption, particularly among rural households. Above all, building capability, skill, knowledge on processing and preservation of the fruits, particularly the indigenous fruits to increase interest, production and guard against extinction is pertinent.

REFERENCES

Adebisi AA (2004). A Case study of Garicinia cola production in area of Omo forest reserve, South-west Nigeria. pp. 115-132.

Adebooye OC (2003). Ethnobotany Of indigenous leafy and fruit vegetables of southwest Nigeira. Italy, Delpinoa, University of Naples 45:295-299.

Adenegan KO, Adeoye IB (2011). Fruit consumption among University of Ibadan students, Nigeria. ARPN J. Agric. Biol. Sci. 6(6):18-21.

Adodo A (2000). Herbs for healing:Receiving Gods healing through nature. Decency Publisher, Ilorin, Nigeria.

Akpan UE, Eric EE, Udoh CE, Afolarin TA, Edeh SG (2010). Forest conservation and food security in an unstable climatic condition. Proceedings of the 44th Annual Conference of Agricultural Society of Nigeria, Nigeria.(Date Unknown)"LAUTECH 2010". pp. 633-635.

Awono A (2002). Production and Marketing of Safou (*Dacryodes edulis*) in Cameroon and Internationally: Market Development Issues. Fst Trees and livelihds 12:125-147.

Craig W, Beck L (1999). Phytochemicals: health protective effects. Can. J. Diet. Pract. Res. 60:78-84.

Ebuehi OA (2006). Physico-chemical and fatty acid content of water melon seed oil, Nigeria. Food J. 24(1):17-24.

FAO (2004). Food and Agriculture Organization of the United Nations. 2004. FAOSTAT.

FAO (2007) FAOSTAT database. www.faostat.fao.org/default.aspx.

FAO/WHO (2004). The state of food insecurity in the world: Monitoring Progress towards the World Food Submit and MDGs. Rome. FAO.

Farombi EO (2002). Kolaviron modulates Cellular redox status and impairment of membrane protein activities by Potassium bromoate. Pharm. Res. 45:613-668.

Iwu MM (1999). Perspective in new crops and new uses. ASHS Press Alexandria V.A., pp. 457-462.

Ndoye O (1997) Marketing of non-timber forest products in the humid zones of Cameroon. Rural Development Network ODI, London. P. 22C,

Odebode OS (2006). Assessment of home gardening as a potential source of household income in Akinyele Local Government Area of Oyo State, Nigerian. J. Hort. Sci. 2:47-55.

Oguntola S (2008). Paw paw, Nature's natural Antibiotics for Typhoid. Htt.www.tribute.com.ng/2802008/thr/hlt2.html. Accessed 3/3/2010.

Poulton C, Poole N (2001). Poverty and fruit tree research. FRP Issues and Options papers No. 6. Forestry Research Programme, U.K.

Quebedeaux B, Eisa HM (1990). Horticulture and human health: Contributions of fruits and vegetables. Proceedings of 2nd Intl. Symp. Hort. and Human Health, Hort. Sci. 25:1473-1532.

Rallof J (2000). Detoxifying Desert's Manna; Farmers need no longer fear the sweet pea's dry land cousin. In Science News, The weekly news magazine for Science: 158(5). www.science.news.org/20000729/nobl.asp.

Ruel MT, Minot N, Smith L (2004). Pattern and determinants of fruit and vegetable consumption in Sub Saharan Africa: A multi country comparism. Background Paper for the Joint FAO/WHO Workshop on Fruit and Vegetable for health, 1-3September, Kobe, Japan, pp. 1-5.

Temple L (2001). Quantification des production des eschangos der fruits et legumes au Cameroon.Cashiers d'e tude et de researches Francophones. Agric. 10:87-94.

Terashima K (2002). Powerful antioxidative agents based on Garcinoic acid from G. Kola, Biol. Org. Chem. 10:1619-1625.

Terashima K (1999). A study of bioflavoid from the stem of G. kola. Heterocycles 50:238-340.

von Braun J (2007). The world food situation: New driving force and required actions. Food Policy Report, Washington D. C

World Health Organization (WHO) (2001). Legal status of traditional medicine and alternative medicine: A world wide view. WHO Publishing.www.localharvest.org/blog/20618/entry/medicinal.values. of.the.tropical accessed 17/10/2013.

27

Postharvest and sensory quality of pineapples grown under micronutrients doses and two types of mulching

Aiala Vieira Amorim[1], Deborah dos Santos Garruti[2], Claudivan Feitosa de Lacerda[3], Carlos Farley Herbster Moura[2] and Enéas Gomes-Filho[4]

[1]Universidade da Integração Internacional da Lusofonia Afro-brasileira, Avenida da Abolição, número 3, Centro, 62790-000, Redenção, CE - Brasil.
[2]Embrapa Agroindústria Tropical, Rua Dra. Sara Mesquita, 2270, 60511-110, Fortaleza, CE - Brasil.
[3]Departamento de Engenharia Agrícola/UFC and INCTSal, Campus do Pici, bloco 804, 60455-760, Fortaleza, CE - Brasil.
[4]Departamento de Bioquímica e Biologia Molecular/UFC and INCTSal, Campus do Pici, bloco 907, 60440-554, Fortaleza, CE - Brasil.

Mineral nutrition has a major influence on growth and, consequently, on the production and quality of fruits. However, little is known about the effect of fertilization with micronutrients on pineapple, especially under tropical conditions. So, the goal of the present study was to evaluate the effects of soil and leaf fertilization with micronutrients on the postharvest and sensory quality of pineapple cv. Vitória produced under two different soil covers. The physical (weight and length) and chemical (total sugars and ratio soluble solid/titratable acidity-SS/TA) characteristics of the fruits were determined. A sensory analysis was performed to evaluate the acceptability of the external appearance of the whole fruit and slices (color, flavor and overall acceptability). The purchase intent and intensity of acid taste were also assessed. Application of micronutrients increased sugar content and SS/TA ratio, especially when applied on leaves, and caused higher weight and length of the fruits. Therefore, these positive effects of micronutrients can be contributed to the good market acceptability of pineapple cv. Vitória, evidenced by sensory analysis. Fruits of Vitória pineapple mulched with black plastic showed higher total sugars concentrations, which may have contributed to a sweeter taste and, consequently, better consumer acceptance.

Key words: *Ananas comosus*, fertilization, soil cover, fruit quality, sensory analysis.

INTRODUCTION

Fresh-cut pineapple is greatly appreciated worldwide because of its sensory qualities such as tastes, flavor and juiciness (Calderón et al., 2008). Pineapple growing is highly profitable and assumes great importance by generating employment and income, as it is widely exploited throughout the tropical regions and requires intensive hand-farming by a rural labor force. Many of the crops worldwide are affected by *Fusarium* wilt, a disease that attacks the pineapple crop and causes losses of plants, fruits and seedlings estimated at more than 40% of total production. Therefore, the use of such resistant cultivars as Vitória seems to be one of the ways to control this disease.

Consumer requirements and demands for good-quality

fruits that are free of diseases are increasing every year. The quality attributes of pineapple include size, shape, skin color, the pulp characteristics and sensory properties, which are responsible for the fruit's market acceptance (Saradhuldhat and Paull, 2007). Sensory evaluation can serve as a quality assurance tool for final products on the market, detecting peculiarities that cannot be perceived using other instruments (Liu et al., 2011).

Mineral nutrition has a major influence on growth and, consequently, on the production and quality of fruits (Malézieux and Bartholomew, 2003; Soares et al., 2005; Agbangba et al., 2011). However, little is known about the effect of fertilization with micronutrients on tropical crops. Some studies report the importance of boron and zinc in the pineapple crop (Maeda et al., 2011; Siebeneichler et al., 2008). Currently, there seems to be some awareness among farmers about the use of micronutrients in the growth of pineapple; but, they are usually used empirically (Siebeneichler et al., 2008).

Mulching is taken into account by producers before planting. The objective in covering the soil is to minimize the effects of erosion, conserve the soil moisture by reducing evapotranspiration, control the use of irrigation water and reduce weed growth. Several materials can be used for this purpose: *bagana* of carnauba (*Copernicia prunifera*), portions of grassy and leguminous plants, and polyethylene plastics.

Given the importance of mineral fertilization to the harvest of excellent quality fruits, the goal of the present study was to evaluate the effect of applying micronutrients via soil and leaves on postharvest and sensory qualities of Vitória pineapples produced under two mulching conditions: *bagana* and black polyethylene.

MATERIALS AND METHODS

Experimental conditions

The experiment was conducted from December 2008 to October 2010 in a 0.348 ha irrigated perimeter area located in Marco county. This county is in the northern part of the state of Ceará, Brazil. According to the Köppen classification, the area is under the influence of an Aw' (tropical rain) climate, at a latitude of 3°07'13"S and a longitude of 40°05'13"W. The soil of the experimental area is classified as "Typic Quartzipsamment", with a sandy texture and a density of 1.590 kg m^{-3}. The chemical characteristics of the soil, in layers from 0 to 20 cm, are pH, 5.8; EC, 0.15 dS m^{-1}; and 0.77, 0.30, 0.08, 0.02 and 0.75 cmol$_c$ kg^{-1} of Ca, Mg, K, Na and Al, respectively.

Within 90 days after micropropagation, pineapple (*Ananas comosus* L. Merril) cv. Vitória seedlings were transferred from trays to black polyethylene plastic bags containing sand as substrate, with 800 gm^{-3} of simple superphosphate. They were acclimated for six months and, during this period, irrigated twice a week with water with an electrical conductivity of 0.44 dS m^{-1}. Transplantation was performed in April 2009; the plants were arranged in double rows, spaced 0.90 × 0.40 × 0.30 m, with an area of 38.4 m wide and 88.0 m long, totaling 14,080 plants in 0.348 ha. Each plot consisted of four subplots, with four double rows with 11 plants per row, and the two central lines indicated the usable area of the subplot.

In the experimental area, two mulching experiments were conducted: one with *bagana* (the straw resulting from the extraction of the carnauba wax sheet) of carnauba (*C. prunifera*) and other with black polyethylene plastic. In both, the plants were exposed to different doses of micronutrients applied with two different types of fertilization: soil fertilization and leaf fertilization (15 applications every 28 days).

The experimental design was a randomized block split plot with four levels of soil fertilization and four levels of leaf fertilization, with five repetitions. For the soil fertilization, the commercial micronutrient formulation FTE-12 (9.0; 1.8; 0.8; 3.0 and 3.0% of Zn, B, Cu, Fe and Mn, respectively), was used. It was applied in the pits of each plot before planting at doses of 0, 60, 120 and 180 kg ha^{-1}. The four levels of leaf fertilization were: LF0 (no fertilizer), LF 1 (15 leaf fertilization applications, using 1158.8, 844.7, 391.5, 322.7 and 216,0 gha^{-1} of Fe, Mn, Zn, Cu, and B, respectively), LF2 (15 leaf fertilization applications, using twice the quantities used in LF1), and LF3 (15 leaf fertilization applications, using three times the amount used in LF1). The leaf fertilization was performed monthly, and the concentrations were defined using a modified Murashige and Skoog (1962) nutrient solution as a base. The total volume of the solution in each application was 463 L ha^{-1}.

Macronutrients were applied to all plants via fertirrigation beginning two months after transplanting, following similar recommendations used by producers. The applied amount of each compound was 20 kg ha^{-1} Ca (NO$_3$)$_2$, 24 kg ha^{-1} MgSO$_4$, 80 kg ha^{-1} NH$_4$H$_2$PO$_4$, 98 kg ha^{-1} H$_3$PO$_4$, 688 kg ha^{-1} urea and 797 kg ha^{-1} K$_2$SO$_4$. The irrigation management was based on soil moisture sensors installed at soil depths of 15 and 25 cm, the first representing the soil depth from 5 to 15 cm, while the second band of 15 to 25 cm aimed to determine the water storage in the soil, thereby indicating how much irrigation was necessary.

The first month after transplantation, preventive control of pineapple tree rot eye was performed with the fungicide Aliette. On the sixth month, the plants suffered white mealy bug attack, which was controlled with methyl parathion 600 CE, and on the 12th month, weed control was performed using Metrimex 500. Floral induction was performed 13 months after transplanting by spraying with ethephon at a concentration of 1,500 mgL^{-1} when sampling revealed that the green mass of the largest treatment plants weighed approximately 2 kg.

Quality evaluation

Harvest took place in October 2010. The pineapple fruits were physiologically mature, presenting yellow fruitlets along the entire length, and they were classified as in the yellow maturity stage, according to the standards of pineapple classification. Based on the color of the fruitlet, pineapple may be ranked as green or greenish, painted, colored or yellow. The pineapples from the usable area of each subplot were weighed and measured. One fruit of each repetition (subplot), with an average representative weight, was taken for physicochemical evaluations. For the sensory analysis, 30 pineapples were taken from the group in each of the mulching treatments (*bagana* and black plastic) that received the greatest amount of nutrients in the soil and leaf fertilization applications. Fruits of the Pérola pineapple cultivar were also used in the sensory analysis, as Pérola is the most widely consumed and appreciated variety in Brazil. The Pérola fruits came from a producer in the Baixo Acaraú irrigated perimeter who cultivates pineapple using black plastic to cover the soil and the same mineral nutrition package used by other producers in the perimeter. After their weights and lengths were measured, the pineapples were identified, packed in plastic boxes (cap. 20 kg), and cold transported (15°C) to laboratory, where they remained in a cold chamber at the same temperature for 24 h before undergoing physicochemical and sensory assessments.

For the physicochemical property evaluation, the fruits were peeled and cut longitudinally without the central axle, using sharp knives. The slices were processed in a domestic centrifuge machine with a 650 watts engine and a microhole sieve. The resulting liquid was used for the immediate measurement of total sugars, titratable acidity (TA) and soluble solids. The total sugars were determined using the anthrone method (Yemn and Willis, 1954).

Sensory analysis

For the sensory analysis, consumers were recruited among students (undergraduate and graduate) and workers. The group was composed of 60 individuals of both sexes, aged between 18 and 55 years old. Hedonic tests were performed at the Laboratory of Sensory Analysis of Food. Two acceptance tests were simultaneously conducted regarding the external appearance of the whole pineapple and the fruit's palatability (color, flavor and overall acceptance). The acceptance test for the external appearance of the whole fruit was set up on a bench. Two fruits were placed in white trays coded with random three-digit numbers. The acceptance of the overall appearance was evaluated using a hedonic scale with nine categories varying from disliked extremely to liked extremely. The evaluation form also included a purchase intent test that used a five-point scale. Both the hedonic scale and the purchase intent scale are described in Meilgaard et al. (1999). To perform statistical analyses, each category of the hedonic scale was assigned a numerical value from one to nine (1 = disliked extremely; 5 = neither liked nor disliked; 9 = liked extremely).

Palatability tests were held in air-conditioned booths (24°C) under fluorescent daylight-type white light. The pineapples were peeled and cut into slices approximately 1 cm thick. Each slice was cut into four pieces with the central axis removed. The samples were presented to panelists in white disposable plastic containers coded with three-digit numbers, along with evaluation forms. Overall acceptance and color and flavor attributes were evaluated using the hedonic scale described above. On the same ballot, the intensity of the sample's acidity was assessed with a just-right scale according to the method described by Meilgaard et al. (1999). In this test, each subject compared the sample with his/her mental criterion of what is an ideal acidity for fresh pineapple. To eliminate the after taste influence, a piece of bread and mineral water were offered between samples. The order of sample presentation was balanced among the panelists according to a design proposed by MacFie et al. (1989) to minimize first-order and carryover effects.

Statistical analysis

The data were submitted to analysis of variance (ANOVA) and Tukey tests (p < 0.05) for the comparison of means. Regression analysis was performed for data in which significant effects occurred. The analyses were performed using the statistical software SAS® (2002). The results from the purchase intent and just-right tests were presented as frequency distribution histograms.

RESULTS AND DISCUSSION

Quality evaluation

Both fertilization procedures (soil and leaf) influenced the pineapples' weight and length (p <0.01) in both experimental conditions (*bagana* and black plastic). Leaf fertilization also influenced the total sugars (p < 0.05) and SS/TA ratios (p < 0.01) under the black plastic mulching condition and the SS/TA ratio (p < 0.05) under the *bagana* mulching condition. The interaction between these effects was significant for the weight and length of the fruits (p < 0.01) grown in *bagana* and for the weight of the fruits (p < 0.01) grown in black plastic.

As shown in Figure 1, there was an increase of fruit weight in relation to the application of FTE-12, even in the absence of leaf fertilization, in both experiments. There were positive responses in weight with the leaf applications, even in the absence of FTE-12. However, the increments provided by the micronutrient application, especially in the form of FTE-12, were higher when it was applied at the highest level of leaf fertilization, indicating an interaction between factors (p < 0.01). At the highest level of leaf fertilization in the *bagana* condition, each kg ha^{-1} of FTE-12 was associated with a 0.0026 kg weight increase in the fruits, while for the lower level of leaf fertilization, this increase was 0.0021 kg (Figure 1). The heaviest fruits in this study weighed 1.26 and 1.31 kg and came from plants subjected to higher doses of FTE-12 and leaf fertilization and mulched with *bagana* and black plastic, respectively. These results are lower than those observed by Ventura et al. (2009), who found an average weight of 1.5 kg for the Vitória pineapple. On the other hand, Silva et al. (2012) found a maximum weight of 1.0 kg for Vitória pineapples grown in doses of 409.2 kg ha^{-1} N. Siebeneichler et al. (2008), working with Pérola pineapples subjected to different boron treatments, found no changes in the weights of the fruits and observed values ranging from 1.653 to 1.696 kg. As shown in Figure 2A, pineapple fruits grown under *bagana* had an increased length with the application of FTE-12 even in the absence of leaf fertilization, and positive responses were observed in this variable with increasing levels of leaf fertilization, even in the absence of FTE-12. The increases provided by the application of micronutrients, especially in the form of FTE-12, were higher with higher levels of leaf fertilization, with values ranging from 10.23 to 12.84 cm for doses of 0 and 180 kg ha^{-1} FTE -12, respectively.

The two forms of fertilizer with micronutrients contributed to increases in the length of pineapples for the experiment with black plastic, but the effect was not interactive (Figure 2B and C). Considering the FTE-12 doses applied during the experiment with black plastic (Figure 2B), the fruit length varied from 8.54 to 11.18 cm for plants that received 0 and 180 kg ha^{-1} FTE 12, respectively. The increases provided by leaf fertilization ranged from 8.65 cm at the lowest dose of fertilization to 11.07 cm at the highest level (Figure 2C).

The lengths of Vitória pineapple fruits mulched with *bagana* and treated with the highest level of leaf fertilization (Figure 2A) are consistent with findings for this cultivar by other authors (Ventura et al., 2009; Silva et al., 2012). In the experiment with black plastic, only the

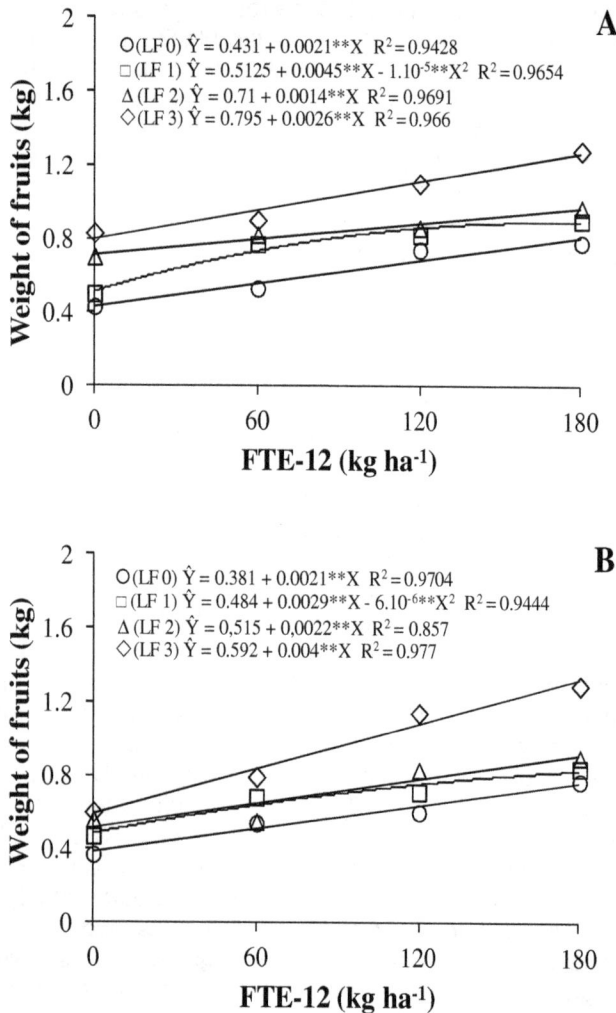

Figure 1. The average weight of Vitória pineapple fruits grown in the Baixo Acaraú irrigated perimeter with different doses of FTE-12 and levels of leaf fertilization (LF), under two types of ground cover: *bagana* (A) and black plastic (B). **p < 0.01 and *p < 0.05.

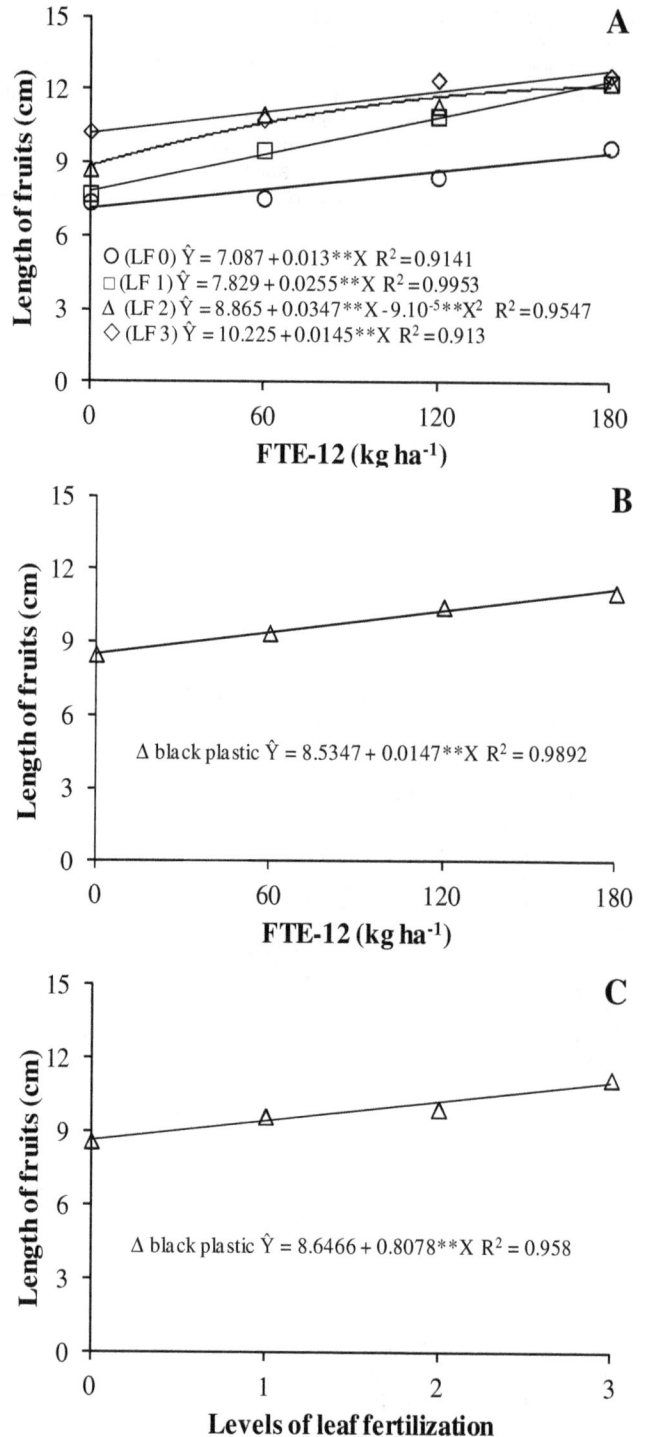

Figure 2. Length of Vitória pineapple fruits without crown grown in the Baixo Acaraú irrigated perimeter for different doses of FTE-12 and levels of leaf fertilization (LF), under two types of ground cover: *bagana* (A) and black plastic (B and C). **p < 0.01 and *p < 0.05.

fruits of plants cultivated under the doses of 120 and 180 kg ha^{-1} of FTE-12 and at leaf fertilization level 3 (Figure 2B and C) were within the range of lengths found for this cultivar of pineapple. The Vitória pineapple has a cylindrical shape and is shorter than other cultivars. For example, the average length of Smooth Cayenne pineapples subjected to different boron and zinc sources was 19.8 cm (Maeda et al., 2011), and Silva et al. (2012) observed an average length of 14.6 cm for Vitória pineapples fertilized with different ratios of N.

The soluble sugar levels in Vitória pineapple were not influenced by the FTE-12 levels in the experiments with *bagana* and black plastic (Figure 3A). Taking into account the levels of leaf fertilization in the *bagana* condition, no differences were found and the results are expressed as average (Figure 3B). However, in pineapples grown under black plastic, there was a linearly increasing correlation between soluble sugar levels and leaf

fertilization levels, with values ranging from 12.07 g (100 ml^{-1}) at leaf fertilization level 0 to 14.57 g (100 ml^{-1}) at leaf fertilization level 3 (an increase of 20.71%). In general, the total carbohydrate levels were higher in black plastic

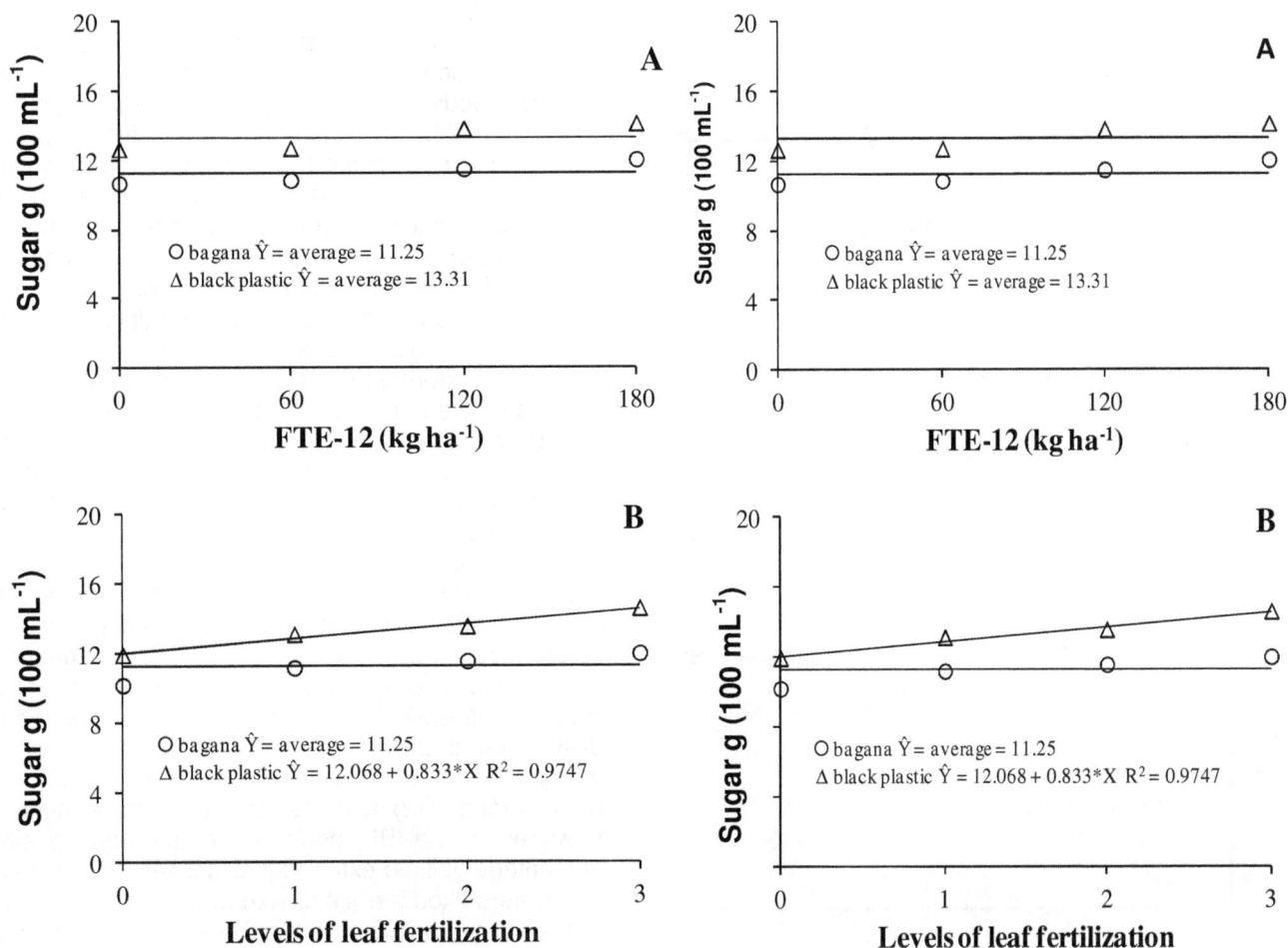

Figure 3. Sugar content in the pulp of Vitória pineapple grown in the Baixo Acaraú irrigated perimeter and treated with different doses of FTE-12 (A) and levels of leaf fertilization (B) and two types of ground cover. **p < 0.01 and *p < 0.05.

experiment; total carbohydrates averaged 18.31% higher among fruits grown with black plastic mulch than mulched with *bagana* (Figure 3).

The sugars in pineapple represent an important fraction of its edible part. The sugar values obtained in this study were within the acceptable range of total sugars in freshly picked pineapple and were suitable for consumption. These values were similar to those found by Wardy et al. (2009) in other three pineapples varieties and by Brat et al. (2004) in ripe fruits of a pineapple hybrid (FLHORAN 41 Cv.). The FTE-12 dose used did not influence the SS/TA ratio of Vitória pineapple; however, the leaf fertilization amount provided linear increases ranging from 17.50 to 21.29 for plants mulched with *bagana* and from 17.51 to 21.05 for plants mulched with black plastic (Figure 4). The SS/TA ratio found by Ventura et al. (2009) for Vitória pineapple was 19.7, an amount within the range observed in this study. Maeda et al. (2011) found SS/TA values ranging from 15.35 to 18.01 in Smooth Cayenne pineapple. This relationship is one of the most widely used indexes in flavor evaluation;

it is more representative than measuring soluble sugars and acids separately because it reflects the balance between them.

Sensory analysis

Table 1 shows the mean hedonic values consumers assigned to express their sensory acceptance of pineapple cultivars grown in the Baixo Acaraú irrigated perimeter. All attributes were well accepted, and their assigned hedonic values ranged from 6 (liked slightly) to 8 (liked very much). There was no variation among the samples for acceptance of the external appearance (p > 0.05), indicating that the consumers would accept any sample, despite differences in shape, size or other physical characteristics. The same was observed for the internal color of the fruit, as evaluated in the pineapple slices.

Table 1 shows that Vitória pineapple fruits grown under black plastic (BP) mulch presented the highest flavor

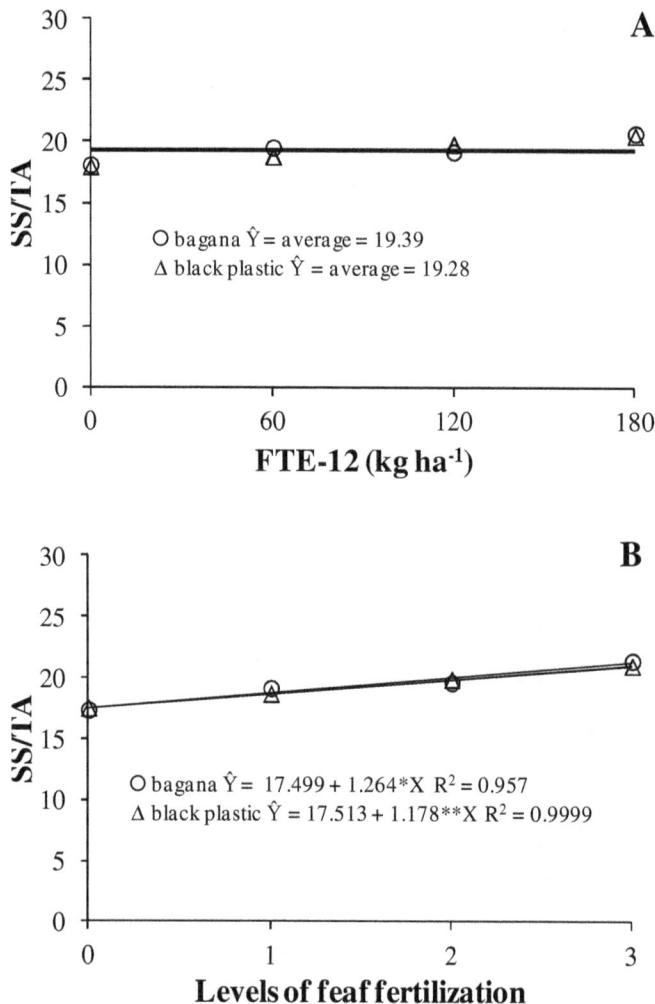

Figure 4. Titratable acidity (TA) in the pulp of Vitória pineapples grown in the Baixo Acaraú irrigated perimeter with different doses of FTE-12 (A) and levels of leaf fertilization (B) under two types of ground cover. **p < 0.01 and *p < 0.05.

considered.

The purchase intent test conducted in this study (Figure 5), we observed that approximately 31.7% of the panelists would definitely buy the Pérola pineapple, while 21.7 and 28.3% would buy the Vitória pineapple cultivated in soil covered with *bagana* and black plastic, respectively. The small difference between these percentages shows a good performance of the Vitória pineapple, meaning that consumers liked its smaller size and cylindrical shape. According to Kader (2002), 83% of the decision to purchase or reject a product is determined by the product's appearance or physical condition. Regarding external appearance, color could be the most important criteria for consumer acceptance or rejection of pineapples. This may explain why the small size of the Vitória fruits did not impair its appearance acceptability. In most cases, though fruit size is a crucial criterion at the time of purchase.

Figure 6 shows the frequencies of the hedonic values assigned to the external appearance of the pineapple, the color of slices, flavor and overall acceptance. Although no differences were detected between means for appearance (Table 1), some differences in frequency distributions were observed (Figure 6A). Vitória (B) had a distribution that was more concentrated in the higher acceptance region than the other varieties, with a mode of 8. Vitória (BP) and Pérola each had a mode of 7; however, Vitória (BP) performed better, earning a higher percentage of liked extremely responses. It was clear that consumers liked the small size of the Vitória pineapples, despite being accustomed to the elongated shape of the Pérola variety.

Frequency distributions for the color of the slices (Figure 6B) were very similar among the three pineapple samples, confirming results of Table 1. However, Vitória pineapple grown under *bagana* performed slightly lower than the others, with more than 10% of respondents indicating indifference (5 = did not like or dislike) and approximately 16% indicating rejection responses on the hedonic scale (1 to 4).

Histograms for flavor acceptance (Figure 6C) and overall acceptance confirmed the mean results (Table 1), with Vitória (B) being less accepted than the other two samples. The responses observed for this sample showed a distribution clustered around Category 7 on the hedonic scale (liked slightly) and a 25% frequency in the rejection and indifference region (Categories 1 to 5). There were some differences between the frequency distributions of the Pérola and Vitória (BP) samples, which did not differ in their flavor acceptance means (Table 1). Although Pérola showed a slightly higher percentage of responses in the maximum categories of the scale (8 and 9), Vitória presented a lower rejection rate (that is, responses in Categories 1 to 4). Furthermore, a summation of the frequencies in the last three categories (7 to 9) showed that Vitória (BP) scored slightly higher than Pérola (88 and 79%, respectively).

acceptability, which was reflected by an overall acceptance not different from that of Pérola. This was very important because Pérola and Smooth Cayenne pineapples are considered leaders in global market acceptance. The average hedonic value, however, did not always faithfully represent the opinions of the judges as a group. Univariate analyses of data implicitly assume that all subjects exhibit the same behavior and that a single value is representative of all subjects. However, individuals' opinions may vary or cluster into similar groups, and if the individuals show opposite opinions about the products, the mean values will be similar for all products. Although similar mean values would suggest that there was no difference in acceptability among products, that would clearly not be true in such cases. In this case, the frequency distribution of the responses along the hedonic scale should be

Table 1. Mean hedonic values consumers assigned for the sensory acceptance of Peróla pineapple, Vitória pineapple grown under *bagana* (Vitória B) and Vitória grown under black plastic (Vitória BP) in the Baixo Acaraú irrigated perimeter – CE.

Samples	Sensory attributes			
	External appearance	Color slices	Flavor	Overall acceptance
Pérola	6.8^a	7.4^a	7.4^a	7.4^a
Vitória (B)	7.0^a	6.8^a	6.5^b	6.6^b
Vitória (BP)	6.6^a	7.2^a	7.6^a	7.6^a
CV (%)	22.98	21.58	22.26	20.92

Means with same letters in the same column are not significantly different, according to Tukey's test set at 5% probability. CV: coefficient of variation percentage. Hedonic values: 1 = disliked extremely, 2 = disliked very much, 3 = disliked, 4 = disliked slightly, 5 = neither liked nor disliked, 6 = liked slightly, 7 = liked, 8 = liked very much, 9 = liked extremely.

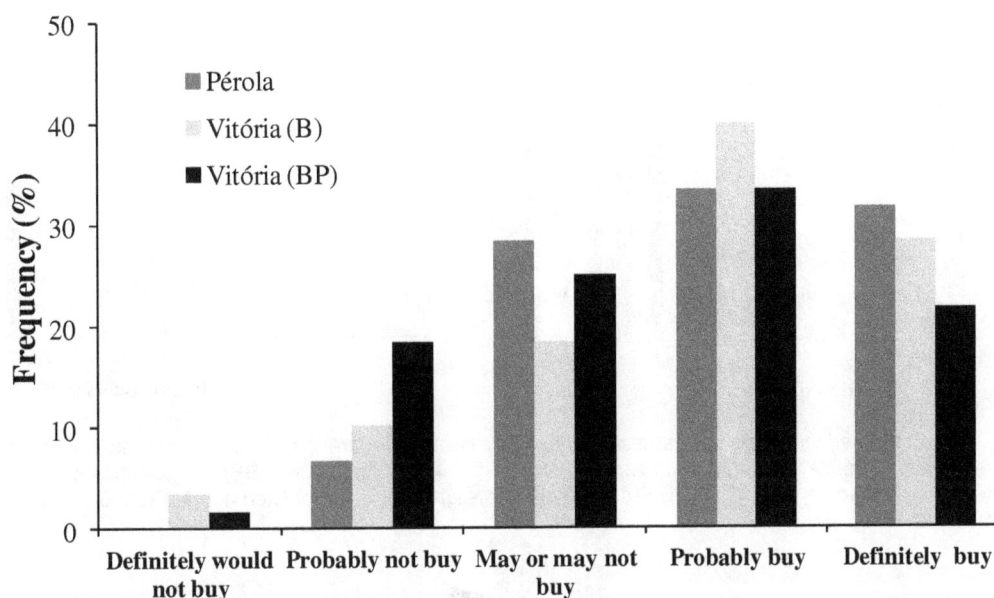

Figure 5. Consumer purchase intent based on the external appearance of "Peróla" pineapples, Vitória pineapples grown in soil covered with *bagana* (Vitória B) and Vitória pineapples covered with black plastic (Vitória BP) grown in the Baixo Acaraú irrigated perimeter - CE, 2010.

The same phenomenon was observed for overall acceptance, for which Pérola reached 76% in Categories 7 to 9 and Vitória (BP) reached 85%. The internal quality of fruit and the physicochemical constituents of the pulp (which are responsible for its characteristic flavor and aroma) are important to the fruit's overall acceptance. This was verified in this study, as Vitória (BP) pineapples showed the best chemical composition related to total sugars (Figure 3), which may have contributed to a sweeter taste and, consequently, better consumer acceptance. Regarding the intensity of acidity (Figure 7), Vitória (BP) and Pérola had a high frequency of responses in the ideal category (73 and 63%, respectively), while Vitória (B) showed a higher percentage of responses in the category stronger than ideal (40%).

Conclusion

the results of this study showed that application of the micronutrients influences some characteristic of pineapple fruit quality, especially when applied on leaves, and increases the fruit growth. Therefore, these positive effects of micronutrients can be contributed to the good market acceptability of pineapple cv. Vitória. Fruits of Vitória pineapple mulched with black plastic showed higher total sugars concentrations (Figure 3), which may have contributed to a sweeter taste and, consequently, better consumer acceptance.

ACKNOWLEDGEMENTS

We thank Banco do Nordeste do Brasil (BNB) and

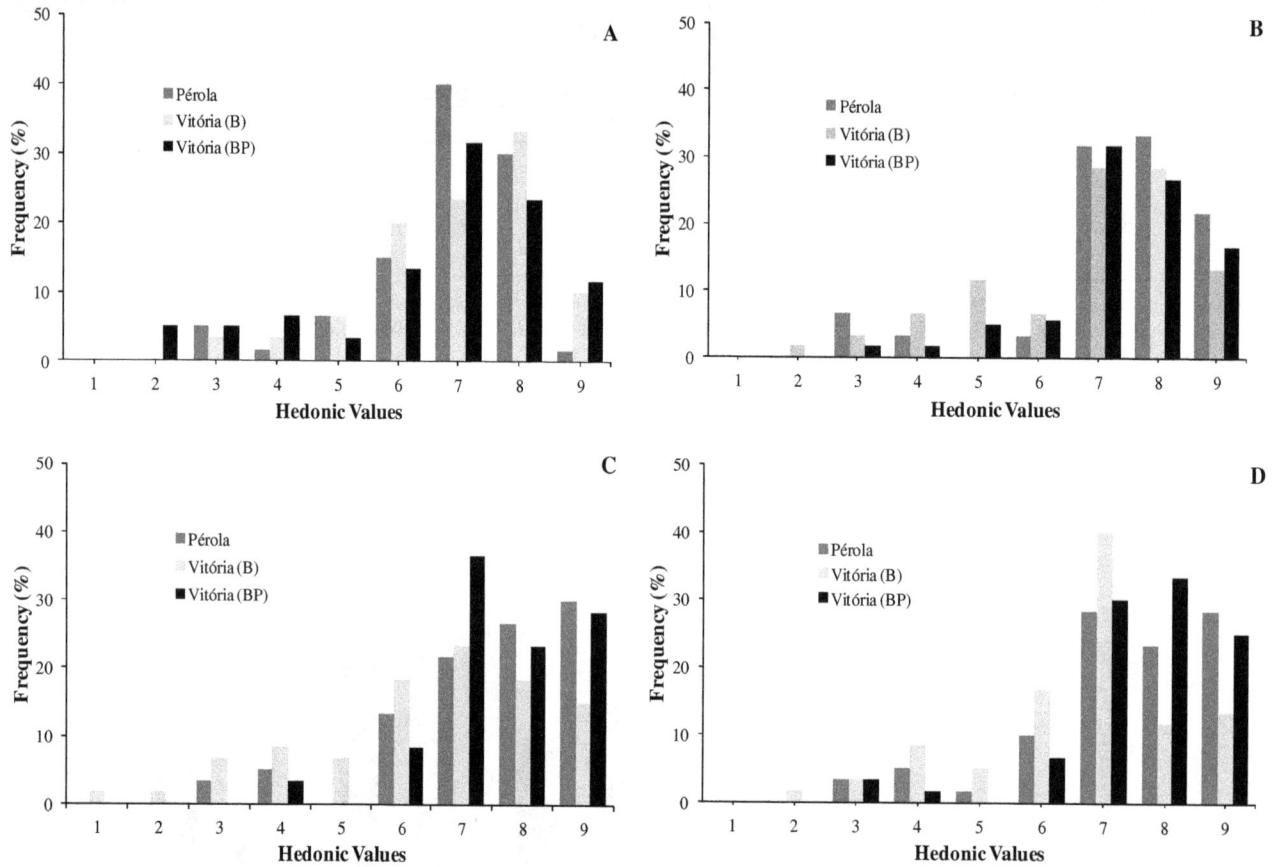

Figure 6. Frequency of the hedonic scale values assigned for (A) external appearance, (B) color of slices, (C) flavor and (D) overall acceptance in samples of Pérola and Vitória pineapples. Vitória (B) = *bagana*; Vitória (BP) = black plastic. Hedonic values: 1 = disliked extremely, 2 = disliked very much, 3 = disliked, 4 = disliked slightly, 5 = neither liked nor disliked, 6 = liked slightly, 7 = liked, 8 = liked very much, 9 = liked extremely.

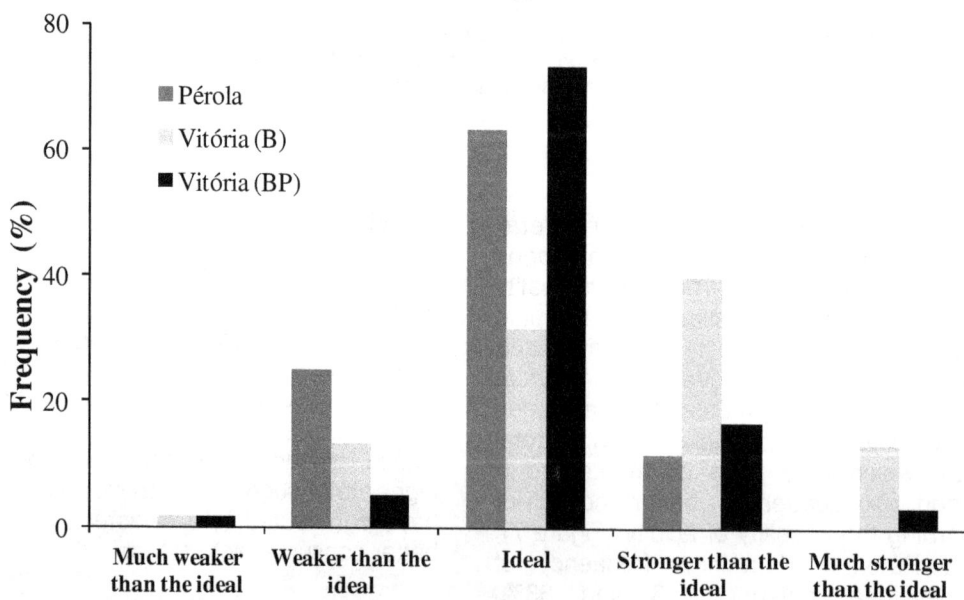

Figure 7. Consumers' assessment of the intensity of acid taste, indicated with the just-right scale, in Pérola pineapples, Vitória pineapples grown in soil covered with *bagana* (Vitória B) or black plastic (Vitória BP) in the Baixo Acaraú irrigated perimeter - CE, 2010.

Conselho Nacional de Desenvolvimento Científico e Tecnológico (CNPq) for financial support and scholarships.

ABBREVIATIONS

SS, soluble solid; **TA,** titratable acidity; **SS/TA,** ratio soluble solid/titratable acidity; **FTE,** fritted trace elements; **LF,** leaf fertilization; **B,** Bagana; **BP,** black plastic.

REFERENCES

Agbangba EC, Olodo GP, Dagbenonbakin GD, Kindomihou V, Akpo LE, Sokpon N (2011). Preliminary DRIS model parameterization to access pineapple variety 'Perola' nutrient status in Benin (West Africa). Afr. J. Agric. Res. 6(27):5841-5847.

Brat P, Hoang LNT, Soler A, Reynes M, Brillouet JM (2004). Physicochemical characterization of a new pineapple hybrid (FLHORAN41 Cv.). J. Agric. Food Chem. 52:6170–6177.

Calderón MM, Graü MAR, Belosso OM (2008). Effect of packaging conditions on quality and shelf-life of fresh-cut pineapple (*Ananas comosus*). Postharv. Biol. Technol. 50:182-189.

Kader AA (2002). Postharvest Technology of Horticultural Crops. 3ed, University of California, Davis, CA, USA.

Liu C, Liu Y, Yil G, Li W, Zhang G (2011). A comparison of aroma components of pineapple fruits ripened in different seasons. Afr. J. Agric. Res. 6(7):1771-1778.

MacFie HJ, Bratchell N, Greenhoff K, Vallis LV (1989). Designs to balance the effect of order of presentation and first-order carry-over effects in hall tests. J. Sens. Stud. 4:129–148.

Maeda AS, Buzetti S, Boliani AC, Benett CGS, Teixeira Filho MCM, Andreotti M (2011). Foliar fertilization on pineapple quality and yield. Pesqui. Agropec. Trop. 41:248–253.

Malézieux E, Bartholomew DP (2003). Plant nutrition. *In:* Bartholomew, D.P., Paul, R.E., Rohrbach, K.G., Eds. The Pineapple: Botany, Production and Uses; CAB Publishing: New York, NY, USA, pp. 143-165.

Meilgaard MC, Civille GV, Carr BT (1999). Sensory Evaluation Techniques. 3 ed, CRC Press, Boca Raton, Florida, USA.

Murashige T, Skoog F (1962). A revised medium for rapid growth and bioassays with tobacco tissue culture. Physiol. Plant. 15: 473–497.

Saradhuldhat P, Paull RE (2007). Pineapple organic acid metabolism and accumulation during fruit development. Sci. Hortic. 112:297–303.

SAS. Statistical Analysis Software for Windows, Version 9.1 (2002). SAS Institute, Cary, NC, USA.

Siebeneichler SC, Monnerat PH, Carvalho AJC, Silva JA (2008). Boro in pineapple plants 'Pérola' in the north fluminense - Contents, distribution and characteristics of the fruit. Rev. Bras. Frutic. 30:787–793.

Silva ALP, Silva AP, Souza AP, Santos D, Silva SM, Silva VB (2012). Response of 'Vitória' pineapple to nitrogen in coastal tablelands in Paraiba. Rev. Bras. Cien. Solo 36:447-456.

Soares AG, Trugo LC, Botrel N, Sousa LFS (2005). Reduction of internal browning of pineapple fruit (*Ananas comusus* L.) by preharvest soil application of potassium Postharv. Biol. Technol. 35:201–207.

Ventura JA, Costa H, Cabral JRS, Matos AP (2009). Vitória: New pineapple cultivar resistent to fusariosis. Acta Hortic. 822: 51–56.

Yemn EW, Willis AJ (1954). The estimation of carbohydrate in plant extracts by anthrone. Biochem. J. 57:508–514.

Wardy W, Saalia FK, Steiner-Asiedu M, Budu AS, Sefa-Dedeh S (2009). A comparison of some physical, chemical and sensory attributes of three pineapple (*Ananas comosus*) varieties grown in Ghana. Afr. J. Food Sci. 3(1):022-025.

28

Long-term reduction in damage by rhinoceros beetle *Oryctes rhinoceros* (L.) (Coleoptera: Scarabaeidae: Dynastinae) to coconut palms at *Oryctes* Nudivirus release sites on Viti Levu, Fiji

Geoffrey O. Bedford

Department of Biological Sciences, Macquarie University, NSW 2109, Australia.

In 1972 to 1973 the non-endemic *Oryctes* Nudivirus (OrNV) was established at three sites on Viti Levu, Fiji, where the introduced rhinoceros beetle *Oryctes rhinoceros* was causing heavy damage to coconut crowns and frequently killing the palms. The establishment of OrNV, and its dissemination by adults, was followed by a marked reduction in the beetle population and damage. When re-surveyed 35 years later, damage was still at a low level. It is postulated that *Oryctes* Nudivirus is still helping to lower damage and manage *O. rhinoceros* populations at those sites.

Key words: Oryctes rhinoceros, *Oryctes* Nudivirus, coconut, Larvae, breeding site rhinoceros beetle.

INTRODUCTION

The rhinoceros beetle *Oryctes rhinoceros* (L.) (Coleoptera: Scarabaeidae: Dynastinae) is an important invasive pest of coconut palms ranging from the Middle East and South East Asia where it is endemic, to the South Pacific where it has been accidentally introduced into island groups such as Fiji (Bedford, 1976, 1980). An adult bores a hole into the heart or crown of the palm to feed on the sap, damaging the very immature fronds so that when they later unfurl they show characteristic V or wedge-shaped cuts, reducing photosynthetic area. Repeated or heavy attacks kill the growing point, causing the death of the palm. The decaying wood at the top of such dead standing palm poles, becomes a favoured breeding site, along with decaying logs and stumps, and cow dung, compost and sawdust heaps (Bedford, 1976, 1980, 1981, 2013).

A virus that infects *O. rhinoceros* was discovered in Malaysia and its history has been summarised (Huger,

2005). It has been included in the Nudivirus group and its name was abbreviated to Oryctes Nudivirus (OrNV) (Burand, 1998, Wang and Jehle, 2009). Infection is peroral, and it attacks the midgut and fat body of larvae and the midgut of adults, curtailing life span and fecundity, but infected adults act as vectors or "flying virus factories" (Huger, 2005) disseminating it by contaminating larvae or adults in breeding sites by defaecation, or other adults during mating. Safety testing showed it harmless to beneficial insects and vertebrates. The virus was released, and established, in a number of South Pacific and Indian Ocean countries where it was not endemic. Subsequently there has been a marked reduction in the beetle population and damage (references up to that time are summarized in Bedford, 1980). In more recent years, OrNV releases and subsequent reductions in damage have been noted in the following non-endemic locations: Andaman Islands

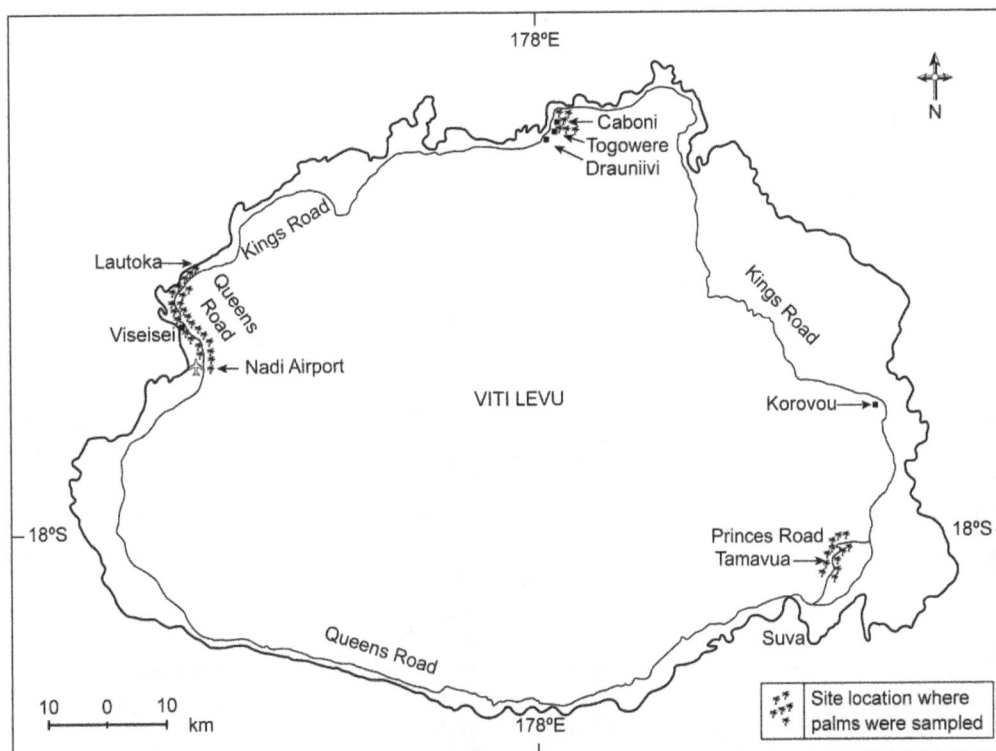

Figure 1. Map of Viti Levu, Fiji, showing locations of palm damage surveys (km = kilometres).

(India)(Jacob, 1996; Prasad et al., 2008); Maldives Islands (Zelazny et al., 1992); Minicoy Island (India) (Mohan and Pillai, 1993) and Oman (Sultanate) (Kinawy, 2004). Data on OrNV occurrence and effect in endemic regions has been provided for India (Gopal et al., 2001), Malaysia (Ramle et al., 2005) and the Philippines (Zelazny and Alfiler, 1991).

However, while it may be mentioned in various countries' Agriculture Department Annual Reports (e.g. Fiji), there have been no peer-reviewed studies of the status of *O. rhinoceros* populations or damage following OrNV establishment, in the Maldives since 1992, Mauritius since 1978, Fiji since 1976, Samoa since 1982, Tokelau Islands since 1977, and Tonga since 1981. So it may be asked: is OrNV still helping to manage rhinoceros beetle damage, or has damage resurged to levels which existed prior to its release? To explore this, in 2010 damage on the island of Viti Levu, Fiji, was re-surveyed at three locations where it had fallen markedly following OrNV release and establishment in the 1970's. This work is not a controlled trial but a follow up to those trial results published decades earlier (Bedford, 1976) and even then OrNV had begun spreading from release sites into control non-release sites. Following OrNV establishment, its incidence in adults was 68% at Tamavua, 66% between Nadi and Lautoka, and 57% at Caboni (Bedford, 1976). Occasional further releases of OrNV were made on this island from 1980 to 1982, 1984 to 1986, 1988, 1999 to

2000 (Fiji Department of Agriculture – *O. rhinoceros* - Annual Research Reports for these years). On this island coconuts produced make an important contribution to food supply and security rather than being used for copra.

MATERIALS AND METHODS

Because of the difficulty or impossibility of counting beetles directly due to their cryptic behaviour and their inaccessibility due to palms reaching a height of up to 20 – 30 m, two types of damage surveys were done (Young, 1986), which give an index or indicator of the *O. rhinoceros* population. A rapid damage survey (RDS) scans a sample of palms, taken as randomly as possible, and scores them as damaged or not in the central 3-4 fronds, giving a percentage of palms with recent damage. A detailed damage survey (DDS) counts the number of fronds above the horizontal level in the crown, and the number damaged, in a palm sample, giving the percentage of fronds damaged at that location. As a new frond is considered to unfurl on the average every 3 – 4 weeks (Zelazny, 1987), with 10 – 17 fronds available in the DDS, this gives a palm's damage history for approximately the previous 40 – 68 weeks. The three locations used were: along the Nadi (airport) - Lautoka road (188 – 265 palms sampled, OrNV had spread here by mid-1973); Caboni (157 palms sampled, OrNV released April 1972); and Tamavua (on the outskirts of Suva, 54 palms sampled, OrNV released March 1972) (Figure 1). Extensive data was available at these locations from decades previously for comparison with current readings, and the aim was to have palm sample sizes as close as possible to the size of the original samples, and representative of a substantial proportion of the palms at each locality. For the Nadi-Lautoka road

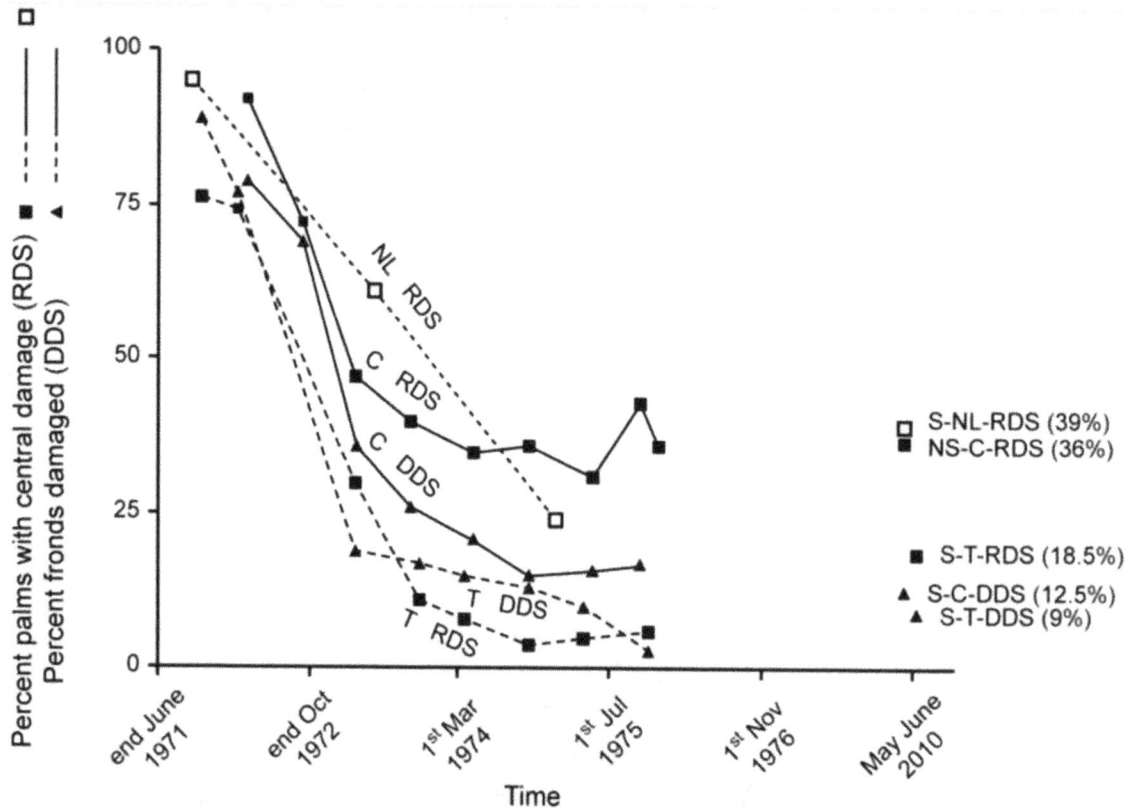

Figure 2. Change in palm damage at the 3 sites studied. C = Caboni, NL = Nadi – Lautoka Road, T = Tamavua, DDS = Detailed damage survey, RDS = Rapid damage survey, S = Significant difference p < 0.05, NS = No significant difference p > 0.05, and 2010 results are shown in brackets.

location, only a rapid damage survey was feasible. At Caboni it was found a large proportion of the original palms had been cut down so available palms up to a kilometre away had to be substituted to make up a sample representative of the location. For each location, the last old reading, and the 2010 reading were compared for significance by tests based on the normal distribution (comparing two percentages based on two large samples (Bailey, 1959: 38-40). Some rapid damage surveys were done at other locations on Viti Levu and give an impression of current damage but there are no previous readings at those locations for comparison.

RESULTS AND DISCUSSION

On the Nadi-Lautoka Road (Figure 2) damage is still well down, after 40 years, compared to the pre-OrNV level of 1971, but up marginally since February 1975. At Tamavua (Figure 2) damage in both types of survey was up (after 35 years), but still far below the level prior to OrNV release. Perhaps after the major drop-off in damage following OrNV establishment (at Nadi-Lautoka by natural spread by mid-1973) damage then fluctuates around a far lower level. At Caboni (Figure 2), interestingly there was no significant difference in the rapid damage survey from 34 years previously, and the detailed damage survey was even significantly down compared to what it had fallen to 34 years ago.

Rapid damage surveys at other locations (Figure 1) are: Nadi (from temple) along Back Road (and diverting to Nawaka) to airport - 11 palms damaged/162 sampled, 6.8% damaged; Drauniivi – 13/ 48, 27% damaged; Togowere - 13/69, 19% damaged; Vunitogoloa - 3/20, 17% damaged; Tailevu Hotel, Korovou - 1/22, 4.5% damaged; Verata – 1/33, 3% damaged.

By reducing the *O. rhinoceros* population and thus heavy damage and consequent killing of palms, OrNV indirectly reduces the number of breeding sites created and made available in the tops of the dead poles, thus reducing emergence of future generations of adults. A large rise in the number of breeding sites, from whatever cause, might lead to a rise in *O. rhinoceros* populations and damage, despite the presence of OrNV. The OrNV – *O. rhinoceros* – palms ecosystem is complex with numerous variables (Bedford, 2013). For simplicity the present study focused on damage as this is readily quantified and is the feature of concern with this pest. The OrNV genome may mutate and might be selected for less virulent forms so data on whether this might have occurred on Viti Levu, and on virus incidence, would be of interest, also data on the current situation of *O. rhinoceros* populations or damage on other parts of Fiji, preferably for comparison with data from previous years at the same locations. Sometimes OrNV effectiveness may be queried, because fresh damage can always be found everywhere, as the pest never disappears entirely, so creating a subjective impression of ongoing damage, but the key consideration which needs to be recognised, is that the overall damage had become less, often much less, compared to previous levels, (and which then might fluctuate) as shown in the present work.

In summary, it is postulated that OrNV is still helping, many years after its establishment, to hold down and manage *O. rhinoceros* populations and damage on Viti Levu at the sites studied. Similar assessments should be undertaken in other countries where OrNV has been released and established.

ACKNOWLEDGMENTS

I thank, Professor M. Zalucki, University of Queensland, Integrative Biology, for helpful comments on the manuscript and graph, Dr.S. Phillips, Faculty of Agriculture, University of Sydney, for comments on data analysis, and Mr. T. Claridge for earlier help with the graph.

REFERENCES

Bailey NTJ (1959). Statistical methods in biology. English Universities Press, London, P. 200.

Bedford GO (1976). Use of a virus against the coconut palm rhinoceros beetle in Fiji. PANS 22:11-25.

Bedford GO (1980). Biology, ecology, and control of palm rhinoceros beetles. Annu. Rev. Entomol. 25:309-339.

Bedford GO (1981). Control of the rhinoceros beetle by baculovirus, in: Burges HD (ed) Microbial control of pests and plant diseases 1970-1980, Academic Press, London, pp. 409-426.

Bedford GO (1986). Biological control of the rhinoceros beetle (Oryctes rhinoceros) in the South Pacific by baculovirus. Agric. Ecosyst. Environ. 15:141-147.

Bedford GO (2013). Biology and management of palm dynastid beetles. Annu. Rev. Entomol. 58:353-372.

Burand JP (1998). Nudiviruses, in: Miller LK, Ball LA (eds) The Insect Viruses, Plenum Publishing, New York, pp. 69-90.

Gopal M et al. (2001). Control of the coconut pest Oryctes rhinoceros L. using the Oryctes virus. Insect Sci. Appl. 21: 93 -101.

Huger AM (2005). The Oryctes virus: its detection, identification, and implementation in biological control of the coconut palm rhinoceros beetle, Oryctes rhinoceros (Coleoptera: Scarabaeidae). J. Invertebr. Pathol. 89:78-84.

Jacob TK (1996). Introduction and establishment of baculovirus for the control of Oryctes rhinoceros (Coleoptera: Scarabaeidai) in the Andaman Islands (India). Bull. Entomol. Res. 86:257-262.

Kinawy MM (2004). Biological control of the coconut palm rhinoceros beetle (Oryctes rhinoceros L. Coleoptera: Scarabaeidae) using Rhabdionvirus oryctes Huger in Sultanate of Oman. Egypt. J. Biol. Pest Control 14:113-118.

Mohan KS, Pillai GB (1993). Biological control of Oryctes rhinoceros (L.) using an Indian isolate of Oryctes baculovirus. Insect Sci. Appl. 14:551-558.

Prasad G, Jayakumar V, Ranganath HR, Bhagwat VR. (2008). Bio-suppression of coconut rhinoceros beetle, Oryctes rhinoceros L. (Coleoptera: Scarabaeidae) by Oryctes baculovirus (Kerala Isolate) in South Andaman India. Crop Prot. 27:957-964.

Ramle M, Wahid MB, Norman K, Glare TR, Jackson TA (2005). The incidence and use of Oryctes virus for control of rhinoceros beetles in oil palm plantations in Malaysia. J. Invertebr. Pathol. 89:85-90.

Wang Y, Jehle JA (2009). Nudiviruses and other large, double-stranded circular DNA viruses of invertebrates: New insights on an old topic. J. Invertebr. Pathol. 101(3):187-193.

Young EC (1986). The rhinoceros beetle project: history and review of the research programme. Agric. Ecosyst. Environ. 15:149-166.

Zelazny B (1987). Ecological methods for adult populations of Oryctes rhinoceros (Coleoptera, Scarabaeidae). Ecol. Entomol. 12:227–238.

Zelazny B, Alfiler AR (1991). Ecology of baculovirus-infected and healthy adults of Oryctes rhinoceros (Coleoptera: Scarabaeidae) on coconut palms in the Philippines. Ecol. Entomol. 16:253-259.

Zelazny B, Lolong A, Pattang B (1992). Oryctes rhinoceros (Coleoptera: Scarabaeidae) populations suppressed by a baculovirus. Journal of Invertebrate Pathology 59:61-68.

29

Limiting the rose-ringed parakeet (*Psittacula krameri*) damage on guava (*Psidium guajava*) and mango (*Mangifera indica)* with an ultrasonic sound player in a farmland of Faislabad, Pakistan

Hammad Ahmad Khan, Muhammad Javed, Ammara Tahir and Madeeha Kanwal

Department of Zoology and Fisheries, University of Agriculture, Faisalabad, Pakistan 38040.

This paper provides information regarding the damage of rose-ringed parakeet (*Psittacula krameri*: Scopoli*)* on guava (*Psidium guajava* L.) and mango (*Mangifera indica* L.) with an ultrasonic sound player in an urban garden of Faisalabad. The rose-ringed parakeet, with the status of a serious vertebrate pest throughout the region of Central Punjab, Pakistan, with the availability of suitable roosts and nests on various trees, damages and destroys both cultivated and fruit crops and incurs in substantial damage to farmers and commercial fruit growers. There was also a marked reduction in the numbers of parakeet visitations on the two fruits in the presence of the acoustic player. Seemingly, the roosts of the rose-ringed parakeet occur closely to the food sources and as such, lower levels of energy budgets are required to manifest with frequent visitations from and to their roosts throughout the day, inflicting damage and economic losses to them. Therefore, it seems plausible to use similar bird repellents on a variety of crops to reduce the bird depredatory attacks to improve on economic losses and augment the crop quality and production in the productive agro-ecosystems of Punjab, Pakistan.

Key words: Rose-ringed parakeet, guava, fruit orchards, sound player, management.

INTRODUCTION

Worldwide bird damage to various food crops has been reported to cause substantial economic losses (De Grazio, 1978; Bruggers et al., 1981; Manikowski, 1984; Tracy et al., 2007). Being a major crop pest, the rose-ringed parakeet has become naturalized in many parts of the world in its native range (Forshaw, 1989; Juniper, 1998). Populations of this parrot have significantly increased in England, with alarmingly more than 2,500 individuals, per roost (Butler, 2003). The rose-ringed

parakeet (*Psittacula krameri*), also popularly called as the green or ringed parrot, belongs to the family 'psittacidae' and order 'psittaciformes'. Due to its wide feeding niche, it is regarded as one of the most destructive vertebrate pests, particularly in the region of Punjab, Pakistan. It affects the farm crops and horticultural practices and sporadically the stored grains in good proportions (Khan et al., 2011; Ahmad et al., 2012). Of the four recognized subspecies viz. *P. k. borealis*, *P. k. manillensis*, *P. k.*

neumann and *P. k. parvirostris*, the earlier two, are in abundance throughout South East Asia, mostly in Pakistan and India, while that of the large Indian parakeet (*P. k. eupatria*), has only been rarely reported from Pakistan (Ali and Ripley, 1969; Whistler, 1986; Roberts, 1991; Forshaw, 2006).

Of the orchards, mango, guava and citrus are economically important as they not only provide nutritious food, but their export to Asia and Europe, fetch lucrative income (Shafi et al., 1986). One of the major inhibiting factors for lower fruit production in Pakistan is due to an intensive damage by the rose-ringed parakeets, particularly in the unguarded crop conditions (Khan and Beg, 1998; Iqbal et al., 2001). Farmer's frequent choice for the multiple cropping practices over a relatively small area of 12 ha, throughout the region of Punjab, for convenience, augments in severe damage by the parakeets, crows, sparrows, mynas and staling besides, the small and large mammals (Akande, 1986; Khan and Hussain, 1990; Roberts, 1991; Iqbal et al., 2001).

By far, the ringed parakeets appear to be more tenacious, as they not only spoil the food sources, but also cause substantial economic losses to local farmers and the stakeholders, establishing their roosts and nests among suitable old and tall trees, located closer to the crops (Khan and Beg, 1998; Strubbe and Mathysen, 2009). It is worth pointing out here that many of their roosts become permanent and are in their use year after years, along side the canal irrigation, road side avenues, urban gardens and the undisturbed habitats of the college and university campuses (Khan, 2002). Butler (2003) reported that roosts of the rose-ringed parakeet are considered broader and stable than that of any other global roosting bird. Potential economic impacts of *P. krameri* on agriculture, conservation concerns, and mixed public opinion regarding the species have highlighted the need to expand effective and human management options (Lambert et al., 2010). Richness for parakeet species densities have been recorded higher for both harvested sunflower and corn fields than for the small-grain and soybean fields, and the application of broad spectrum herbicides with enhanced harvesting effectiveness of crops have reduced the accessibility of weed seeds and waste grains for game and non-game wildlife (Rao and Shivanaryanan, 1981; Galle et al., 2009).

According to De Grazio (1978), worldwide avian damage to economically important crops not only brings about the damage, but also raises certain health and safety issues to man. Malhi (2000) suggested that the damage to seedlings and the mature stages of the sunflower crop is inflicted mainly by the house crow (*Corvus splendens* L.), and that of the *P. krameri*. Studies conducted by Gilsdorf et al. (2002) on sunflower crop in Ludhiana, India for a period of five months viz. November, December, January, February and March, pointed out that in the early sowing season, seemed more suitable for bird depredations. At this stage,

induction of frightening and mechanical devices, augmented in reduced bird damage and, without any serious impact on the agro-ecosystems. Farmers have mostly been relying on traditional methods of control like sling shorts, beating of metallic drums and hurling voices to disperse the attacking birds on crops, and with a least success (Whistler, 1986; Roberts, 1991; Anderson et al. 2013). Ecological friendly means coupled with intermittent avicides, have been largely recommended in view of fast deteriorating environment, particularly to do away with the predicament of pest resistance among birds (Day et al 2003; Avery et al., 2002; 2005). For the present studies, it was hypothesized that, incorporation of distress sound player would considerably reduce the rose-ringed parakeet depredations for both the fruits in terms of their production and economics of a horticultural fruit farm in the study area.

MATERIALS AND METHODS

Study sites

Observations on using ultrasonic sound player to reduce rose-ringed parakeet (*Psittacula krameri*) depredations on both guava and mango were extended in agricultural farmland of Faisalabad, with latitude 31°25 north and longitude 73°04, of Central Punjab. This region is characterized by the dry and humid hot summer (42±5°C) in May through August and exceedingly cold winters (2±5°C) in December through February. Main agricultural and horticultural crops in this region comprise wheat, maize, rice, fodders, sugarcane, sorghum and millet; citrus, dates, guava, mango and mulberry. The region is canal irrigated with three main canals viz. Jhang branch, Gogera branch and Rakh branch, with their water tributaries, irrigate bulk of the crops along side the canal rest houses with more or less modest populations on their banks.

Present observations continued for a period of twelve weeks with an ultrasonic player placed inside the both orchards during May through July, 2011 of an orchard fruit farm at the University of Agriculture, Faisalabad, of Central Punjab province of Pakistan. Observations were extended in the evening and of a total 108 fruits each observed for damage, in the unprotected conditions, for guava, it was 108±14.97SE, and for mango, 108±15.31SE, numerically assessed through manual count methods throughout the studies, for both unprotected and protected conditions. For the protected conditions using ultrasonic sound player, it were 108±7.60 SE and 108±1.64 SE. A sufficiently large area of this campus has an enrich flora of mainly experimental crops throughout the year. As such, there appears no dearth of food for pest faunistics round the year. Occurrence of a variety of invertebrate and vertebrate pests occurs in fairly good numbers. Guava (*Psidium guajava* L.) has two crops in summer and winter, while the mango (*Mangifera indica* L.), comes about during late spring (April) till early fall (September). Guava and mango were sampled using the randomized sampling design, in square nine fruit orchard of the Institute of Horticultural Sciences, University of Agriculture, Faisalabad. Observations were recorded consecutively for three hours (with a 10 minute time intervals) in the evening for the unprotected phase, and again for the similar duration, in the protected conditions (with ultrasonic sound player). The sound player is a bird repellent has au audio play back sound of alarming noises of some fearsome animals, is equipped with a chargeable battery, to disperse attack of birds. It was placed in the middle of the field. Numbers of parakeet visits per stipulated time interval were recorded during the same durations on both fruits. Data

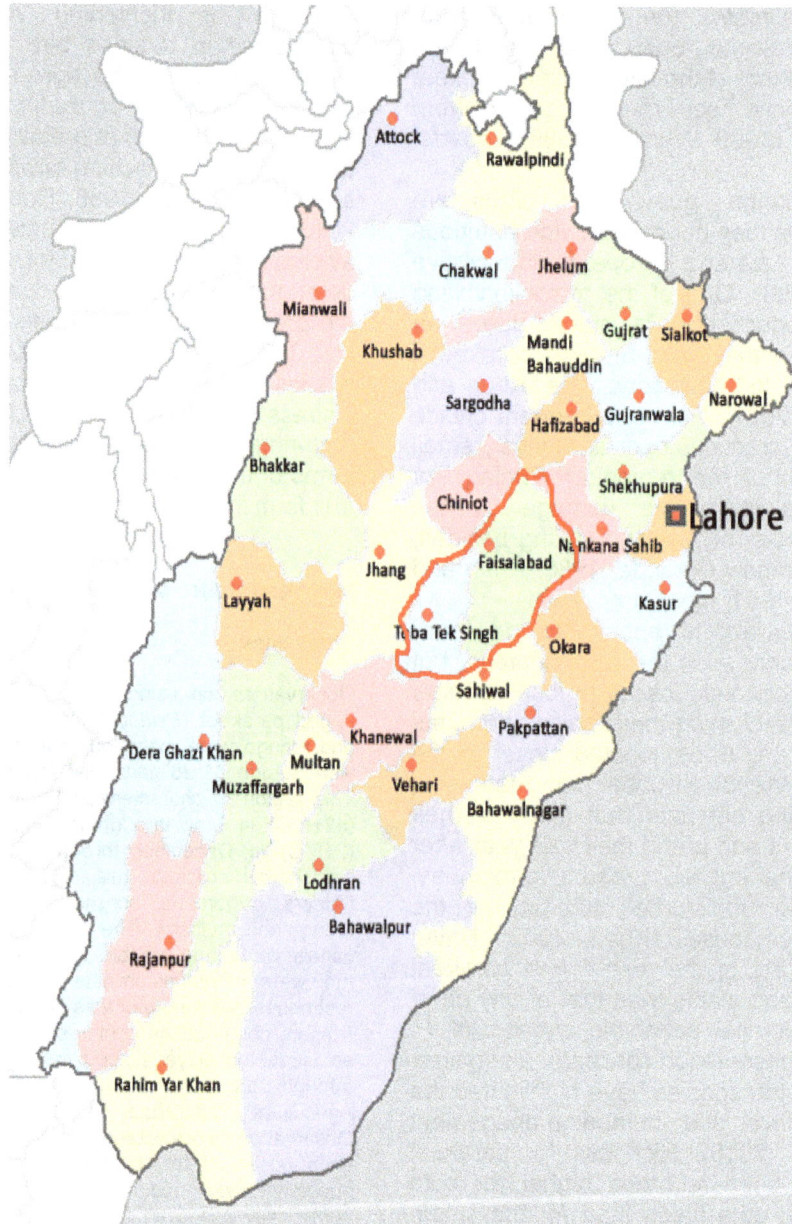

Figure 1. Map of the Punjab Province, Pakistan.

obtained was statistically analyzed with one way Analysis of Variance (ANOVA) for a comparison of means for the unprotected and protected situations with correlation drift for the fruit damage under both conditions (Figures 1 and 2).

RESULTS AND DISCUSSION

From the present data, it was evident that both the fruits were extensively damaged by the rose-ringed parakeet in an unprotected phase. However, there remained a considerable decline in the damage proportions in the presence of distress sound player (Table 1) in both of them which also impacted on the mean damage for guava and mango, and that there was a strong impact of the playback of loud and intermittent noise produced by fearsome animals (Table 2), and kept the attacking parakeets at a bay from both fruits (Figures 3 and 4). There was also a fairly strong correlation between the numbers of parakeet visitations and fruits in the orchard. Depredations were augmented without any shielding impact from that of the sound player (Table 3) and high values of coefficient of regression R^2, 0.997 and 0.999 (Figures 5 and 6). In another study conducted at the same facility (Ahmad et al., 2011, 2012), frequency of parakeet visitations for unprotected conditions for maize and sunflower also remained highly significant, and comparable findings to these, regarding the pestiferous

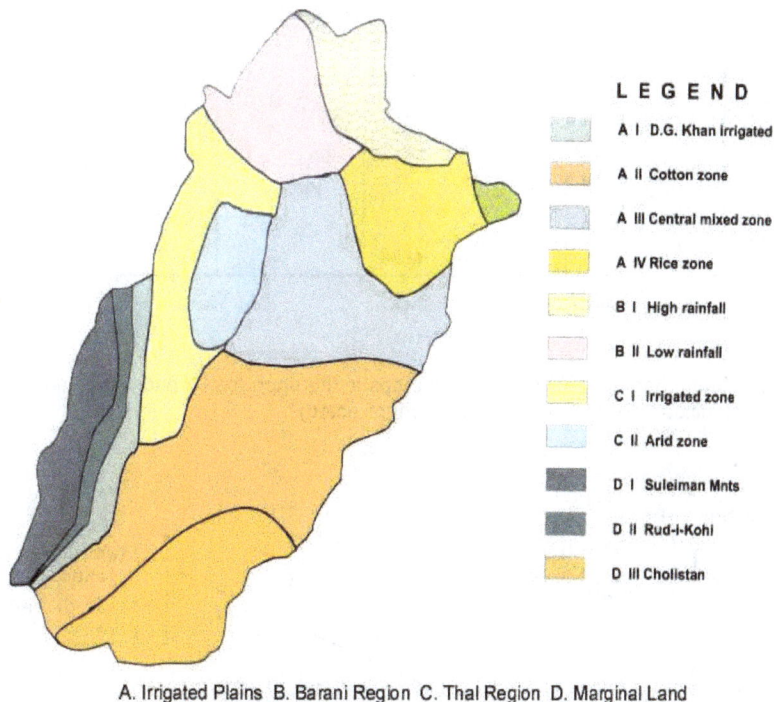

A. Irrigated Plains B. Barani Region C. Thal Region D. Marginal Land

Figure 2. Major Ecological Regions of Punjab Province, Pakistan.

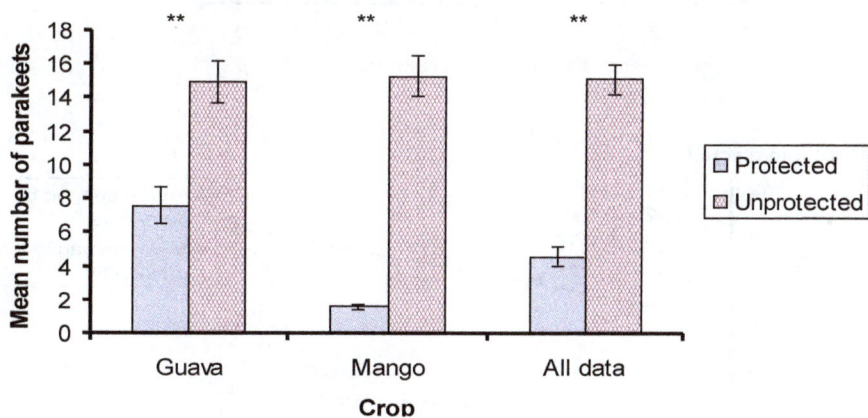

Figure 3. Mean number of rose-ringed parakeet visitations on guava and mango in an urban garden of Faisalabad. ** = Highly significant P<0.01 (Protected vs Unprotected).

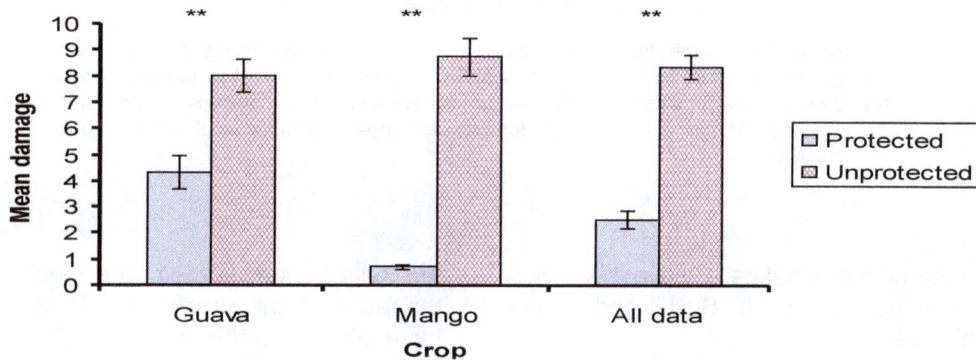

Figure 4. Damage recorded to fruits by the rose-ringed parakeet to guava and mango in a fruit farm of study area. ** = Highly significant P<0.01 (Protected vs Unprotected).

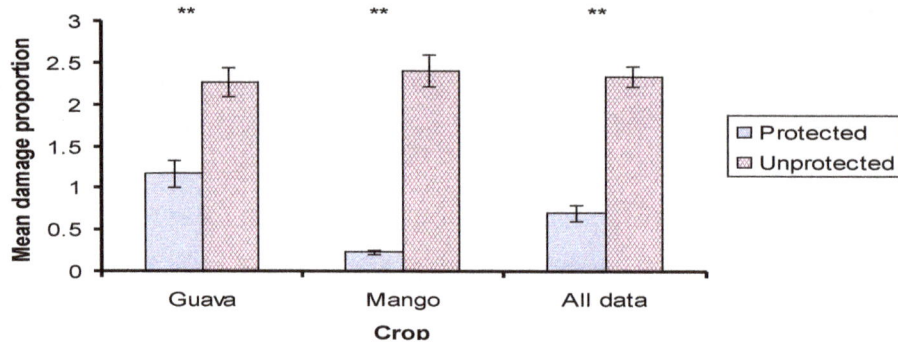

Figure 5. Mean damage recorded to fruit crops in the unprotected and protected conditions. ** = Highly significant P<0.01 (Protected vs Unprotected).

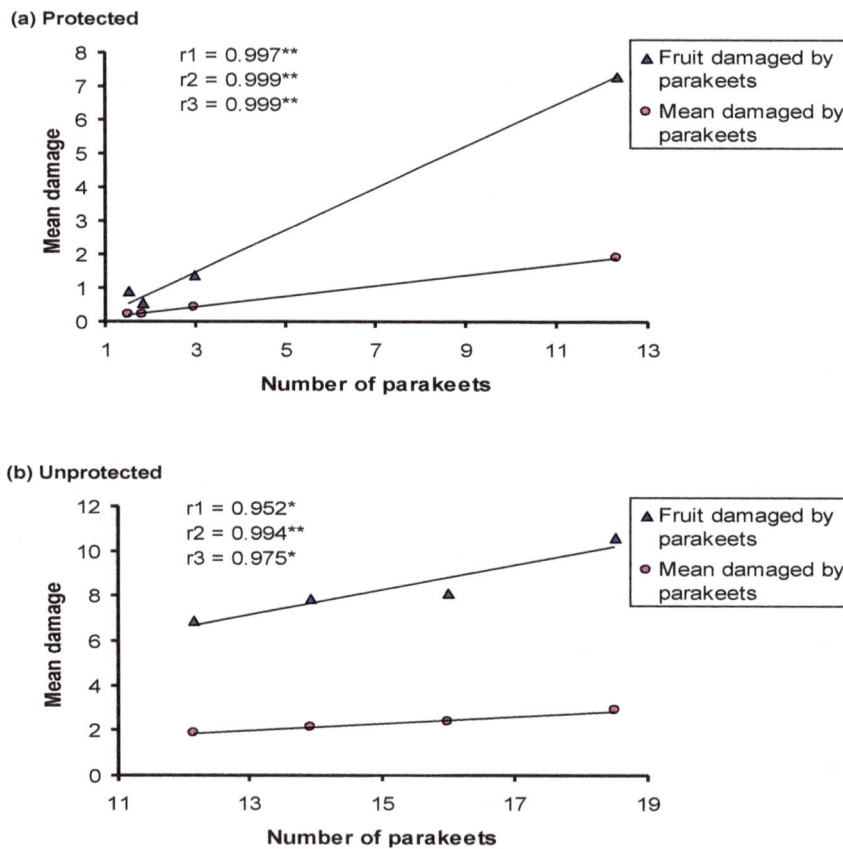

Figure 6. Mean correlation for unprotected and protected conditions of guava and mango in an orchard fruit farm. r1 = Correlation between Number of parakeets and fruit damaged by parakeets, r^2 = Correlation between Number of parakeets and mean damaged by parakeets, r3 = Correlation between mean damages and fruit damaged by parakeets.

implications of rose-ringed parakeets, have been also reported by (Anonymous, 2004a, b; Sushil and Kumar, 1994; Gupta et al., 1998).

An important aspect of this study was that the damage was apparently high in the evening with sufficiently large numbers of parakeet flocks inflicting the damage to both the fruits, to suffice their food requirements for spending the fasting night in their roost. In literature, similar results have also been described by (Dvir, 1985; Iqbal et al., 2001). Unquestionably, the wide feeding niche of the rose-ringed parakeet is mainly due to suitable ecological conditions here in the region of Central Punjab and,

Table 1. A Comparison between protected and unprotected conditions regarding number of rose- ringed parakeets (*Psittacula krameri*) visited.

Crops	Conditions	N	Mean	SD	SE	t-value	Prob.
Guava	Protected	108	7.620	11.186	1.076	-4.51**	0.000
	Unprotected	108	14.963	12.709	1.223		
Mango	Protected	108	1.639	1.743	0.168	-11.29**	0.000
	Unprotected	108	15.315	12.464	1.199		
All data	Protected	216	4.630	8.531	0.580	-10.17**	0.000
	Unprotected	216	15.139	12.559	0.855		

** Highly significant (P<0.01); SD = Standard deviation; SE = Standard error.

Table 2. A Comparison recorded between protected and unprotected conditions regarding number of damaged fruits.

Crops	Conditions	N	Mean	SD	SE	t-value	Prob.
Guava	Protected	108	4.343	6.715	0.646	-4.06**	0.000
	Unprotected	108	8.019	6.595	0.635		
Mango	Protected	108	0.731	1.056	0.102	-11.20	0.000
	Unprotected	108	8.769	7.381	0.710		
All data	Protected	216	2.537	5.125	0.349	-9.93	0.000
	Unprotected	216	8.394	6.993	0.476		

** Highly significant (P<0.01); SD = Standard deviation; SE = Standard error.

Table 3. An assessment between protected and unprotected conditions regarding mean damage.

Crops	Conditions	N	Mean	SD	SE	t-value	Prob.
Guava	Protected	108	1.172	1.756	0.169	-4.48**	0.000
	Unprotected	108	2.274	1.861	0.179		
Mango	Protected	108	0.236	0.259	0.025	-11.40**	0.000
	Unprotected	108	2.407	1.962	0.189		
All data	Protected	216	0.704	1.337	0.091	-10.32**	0.000
	Unprotected	216	2.341	1.909	0.130		

** Highly significant (P<0.01); SD = Standard deviation; SE = Standard error.

therefore, render it to be one of the worst vertebrate pests in the unguarded conditions, using certain trees as roosts and the hollows as nests (Shafi et al., 1986; Sarwar et al., 1989a, b; Khan and Beg, 1998). Throughout the region of Central Punjab, Pakistan, introduction of canal irrigation system to improve on agriculture more than a century ago, involved planting of trees along the canal rest houses of the main irrigating canals. Trees viz. *Salmalia malabarica*, *Dalbergia sissoo*, *Terminalia arjuna*, *Ficus benghalensis* and *Cedrella toona*, grown here, have become old and tall, providing many cavity nesting birds with safe roosts and nests, and therefore, augmenting their populations (Khan, 1999;

Ahmad et al., 2011). Seemingly with a shorter distance from their roosts, damage traveled by the rose-ringed parakeet and some allied birds on the cultivated and non-cultivated crops, and orchard fruits, damage remains unparalleled, causing substantial economic losses (Butler, 2003; Sarwar et al., 1989a, b; Roberts, 1991).

Use of the traditional methods for crop management against the damage of birds, has failed to deliver the required dividends to the farmers and stakeholders (Dechant et al., 2003). Present studies also report on the ecologically effective management measures like the use of avian repellents to inhibit their depredations. The sound player with play back noises of fearsome creatures helped to reduce the damage with minimized depredatory attacks on both guava and mango and that of an important oil-seed sunflower (Parwin, 1988) in the study site. This resulted in a better fruit crop production and also the least economic losses.

Conclusion

In the light of the present findings, it seems appropriate to incorporate few more such repellents viz. multi-mirror reflectors, reflecting ribbons, fire exploders and bird hawk eye rotators, to be tried under same conditions for varying crops, to substantially reduce the bird damage and also to gain maximum yields without exerting a fatal impact on the sustainability of the productive ecosystems. Similar eco-friendly management bird studies have remained wanting mainly due to the complexities involved in aerial mode of life. Present studies however, raise a useful anticipation regarding their efficient control using environmentally friendly methods, in contrast to the avicides to challenge the ecological safety issues.

ACKNOWLEDGEMENT

This work was financially supported by the funding from Agricultural Linkage Program, Pakistan Agricultural Research Council, Islamabad, Pakistan.

REFERENCES

Ahmad S, Khan HA, Javed, M, Rehman K (2012). Roost Composition and Damage Assessment of Rose-Ringed Parakeet (*Psittacula krameri*) on Maize and Sunflower in Agro-Ecosystem of Central Punjab, Pakistan. Int. J. Agric. Biol. 13:731-736.

Ahmad S, Khan HA, Javed M, Rehman, K (2011). An Estimation of Rose-Ringed Parakeet (*Psittacula krameri*) Depredations on Citrus, Guava and Mango in Orchard Fruit Farm. Int. J. Agric. Biol. 2:286-290.

Ahmad S, Khan HA, Javed M, Rehman K (2012). Management of Maize and Sunflower against the Depredations of Rose-ringed Parakeet (*Psittacula krameri*) Using Mechanical Repellents in an Agro-ecosystem. Int. J. Agric. Biol. 14:286–290.

Ali S, Ripley RD (1969). A Handbook of Birds of India and Pakistan. Vol. I, Oxford Univ. Press, London. P. 487.

Akande M (1986). The Economic Importance and Control of Vertebrate Pest of Graminacious Crops, Particularly the Rice (*Oryza sativa*) in Nigeria. Proc. 12th Vert. Pest Conf. pp. 302-305.

Anonymous (2004a). Agricultural Statistics of Pakistan, Ministry of Food, Agriculture and Livestock. Food, Agriculture and Livestock Division (Economics Wing), Islamabad, Government of Pakistan. P. 48.

Anonymous (2004b). Statistics of Fruits, Vegetables and Condiments, p: 52. Ministry of Food Agriculture and Livestock, Food, Agriculture and Livestock Division (Economics Wing), Islamabad, Government of Pakistan. P. 52.

Bruggers J, Matte J, Miskell W, Eriksson M, Jeager B, and Jumale Y. (1981). Reduction of Bird Damage to Field Crops in Eastern Africa with Methiocarb. Trop. Pest. Manage. 27:230-241.

Butler JN (2003). Population Biology of Introduced Rose-ringed Parakeet (*Psittacula krmaeri*: Scopoli) in the United Kingdom. Ph.D. Thesis, Depart. Zool. Univer. Oxford, UK P. 275.

Dechant JA, Sondreal, MA, Johnson, DH, Igl, LD, Goldade, CM, Rabie, PA, Euliss, BR (2003). Effects of Management Practices on Grassland Birds: Burrowing Owl. N. Prair. Wildl. Res. Centre, Nevada. P. 33.

De Grazio JF (1978). World Bird Damage Problems. 6th Proc. Vert. Pest Conf. Washington, United States. pp. 9-24

Dvir E (1985). Ring-necked Parakeets are Trying to make it in Israel. Nat. Sci., Israel Ltd. 10:115-120.

Forshaw JM (1989). Parrots of the World. Princeton, NJ: Princeton Press. P. 517.

Forshaw JM (2006). Parrots of the World: An Identification Guide. Princeton, NJ: Princeton Press, New York. P. 588.

Galle AS, George ML, Homan HJ, William JB (2009). Avian Use of Harvested Crop Fields in North Dakota During Spring Migration. West. N. Amer. Natural. 69:491-500.

Gilsdorf JM, Hygnstrom SE, Vercauteren KC (2002). Use of Frightening Devices in Wildlife Damage Management. *Integ.* Pest Manage. Rev. 7:29-45.

Gupta MK, Rajan B, Baruha R (1998). Parakeet Damage to Sugarcane. Ind. J. Sugar. 46:953–967.

Iqbal MT, Khan HA, Ahmad MH (2001). Feeding Regimens of the Rose-Ringed Parakeet on a Brassica and Sunflower in an Agro-ecosystems in Central Punjab, Pakistan. Pak. Vet. J. 4:111–115.

Khan HA (2002). Foraging, Feeding, Roosting and Nesting Behaviour of Rose-Ringed Parakeet (*Psittacula krameri*) in the cultivations of Central Punjab, Pakistan. *Ph.D. Thesis, Dept. Zool. And Fisheries*, Univ. Agric. Faisalabad. P. 152.

Khan HA, Beg MA (1998). Roosts and Roosting Habits of Rose-Ringed Parakeet (*Psittacula krameri*) in Central Punjab, Pakistan. Pak. J. Biol. Sci. 1:37-38.

Lambert MK, Massie G, Yoder CA, Cowman DP (2010). An Evaluation of *Diazacon* as a Potential Contraceptive in Non-Native Rose-Ringed Parakeets. J. Wildl. Manage. 74:573-581.

Manikowski S (1984) Birds Injurious to Crops in West Africa. Trop. Pest Manage. 30:379–387.

Malhi CS (2000). Timing of Operation to Control Damage by Rose-Ringed Parakeets to maturing sunflower crops and their relationship with sowing times. Int. J. Pest. Cont. 42:86–88.

Parwin A (1988). The Importance of Sunflower Pests in Iran. In: Proc. 12th Int. Sunflower Conf., Novi Sad, Yugoslavia, July 25-29.

Rao GS, Shivanarynan RS (1981). A Note on Feeding Habits of the Rose-ringed Parakeet in Hyderabad. PAVO 19:97-99.

Roberts TJ (1991). Birds of Pakistan. Oxford Univ. Press, London, England. PP. 740.

Sarwar M, Beg MA, Khan AA, Shahwar D (1989a). Breeding Behaviour and Reproduction in Rose-Ringed Parakeet. Pakistan J. Zool. 21:131–138.

Shafi MM, Khan AA, Hussain I (1986). Parakeet Damage to Citrus Fruit in Punjab. J. Bomb. Nat. His. Soc. 83:439-444.

Strubbe D, Mathysen E (2009). Experimental evidence for nest-site competition between invasive ring-necked parakeets (*Psittacula krameri*) and native nuthatches (*Sitta europaea*). Biol. Conser. 142:1588-1594.

Sushil K, Kumar S (1994). Seed Damage of Tree (*Terminalia arjuna*) by the Rose-Ringed Parakeet (*Psittacula krameri*). Indian J. For. 17:151–153.

Whistler NJ (1986). A Handbook of Birds of India, Natraj Publication Inc., New Delhi, India. P. 387.

Permissions

All chapters in this book were first published in AJAR, by Academic Journals; hereby published with permission under the Creative Commons Attribution License or equivalent. Every chapter published in this book has been scrutinized by our experts. Their significance has been extensively debated. The topics covered herein carry significant findings which will fuel the growth of the discipline. They may even be implemented as practical applications or may be referred to as a beginning point for another development.

The contributors of this book come from diverse backgrounds, making this book a truly international effort. This book will bring forth new frontiers with its revolutionizing research information and detailed analysis of the nascent developments around the world.

We would like to thank all the contributing authors for lending their expertise to make the book truly unique. They have played a crucial role in the development of this book. Without their invaluable contributions this book wouldn't have been possible. They have made vital efforts to compile up to date information on the varied aspects of this subject to make this book a valuable addition to the collection of many professionals and students.

This book was conceptualized with the vision of imparting up-to-date information and advanced data in this field. To ensure the same, a matchless editorial board was set up. Every individual on the board went through rigorous rounds of assessment to prove their worth. After which they invested a large part of their time researching and compiling the most relevant data for our readers.

The editorial board has been involved in producing this book since its inception. They have spent rigorous hours researching and exploring the diverse topics which have resulted in the successful publishing of this book. They have passed on their knowledge of decades through this book. To expedite this challenging task, the publisher supported the team at every step. A small team of assistant editors was also appointed to further simplify the editing procedure and attain best results for the readers.

Apart from the editorial board, the designing team has also invested a significant amount of their time in understanding the subject and creating the most relevant covers. They scrutinized every image to scout for the most suitable representation of the subject and create an appropriate cover for the book.

The publishing team has been an ardent support to the editorial, designing and production team. Their endless efforts to recruit the best for this project, has resulted in the accomplishment of this book. They are a veteran in the field of academics and their pool of knowledge is as vast as their experience in printing. Their expertise and guidance has proved useful at every step. Their uncompromising quality standards have made this book an exceptional effort. Their encouragement from time to time has been an inspiration for everyone.

The publisher and the editorial board hope that this book will prove to be a valuable piece of knowledge for researchers, students, practitioners and scholars across the globe.

List of Contributors

Angelo Maria Giuffrè
Università degli Studi Mediterranea di Reggio Calabria
– Dipartimento di Agraria

Victor Manuel Moo-Huchin
Laboratorio de Instrumentación Analítica. Instituto Tecnológico Superior de Calkiní. Av. Ah-Canul, C.P. 24900, Calkiní, Campeche, México

Iván Alfredo Estrada-Mota
Laboratorio de Instrumentación Analítica. Instituto Tecnológico Superior de Calkiní. Av. Ah-Canul, C.P. 24900, Calkiní, Campeche, México

Raciel Javier Estrada-Leon
Laboratorio de Instrumentación Analítica. Instituto Tecnológico Superior de Calkiní. Av. Ah-Canul, C.P. 24900, Calkiní, Campeche, México

Elizabeth Ortiz-Vazquez
Laboratorio de Ciencia y Tecnología de Alimentos. División de Estudios de Posgrado e Investigación. Instituto Tecnológico de Mérida, C.P. 97118, km 5 Mérida-Progreso, Mérida, Yucatán, México

Jorge Pino Alea
Instituto de Investigaciones para la Industria Alimentaria, Carretera al Guatao km 3½, La Habana, CP 19200, Cuba

Adriana Quintanar-Guzman
Adriana Consulting Services inc, Consultant Food Science and Technology, P. O. Box 6762, Siloam Springs, AR 72761, U.S.A

Luis Fernando Cuevas-Glory
Laboratorio de Ciencia y Tecnología de Alimentos. División de Estudios de Posgrado e Investigación. Instituto Tecnológico de Mérida, C.P. 97118, km 5 Mérida-Progreso, Mérida, Yucatán, México

Enrique Sauri-Duch
Laboratorio de Ciencia y Tecnología de Alimentos. División de Estudios de Posgrado e Investigación. Instituto Tecnológico de Mérida, C.P. 97118, km 5 Mérida-Progreso, Mérida, Yucatán, México

Zehra Tugba Abacı
Department of Food Engineering, Faculty of Engineering, Ardahan University, 75000 Ardahan-Turkey

Bayram Murat Asma
Department of Biology, Faculty of Science and Literature, Inonu University, 44069 Malatya-Turkey

Phatu W. Mashela
School of Agricultural and Environmental Sciences, University of Limpopo, Private Bag X1106, Sovenga 0727, South Africa

Kgabo M. Pofu
Agricultural Research Council, VOPI, Private Bag X293, Pretoria, 0001, South Africa

Bombiti Nzanza
School of Agricultural and Environmental Sciences, University of Limpopo, Private Bag X1106, Sovenga 0727, South Africa

B. Venudevan
Department of Seed Science and Technology, Tamil Nadu Agricultural University, Coimbatore-3, India

P. Srimathi
Seed Centre, Tamil Nadu Agricultural University, Coimbatore-3, India

N. Natarajan
Department of Nano Science and Technology, Tamil Nadu Agricultural University, Coimbatore-3, India

R. M. Vijayakumar
Department of Medicinal and Aromatic Crops, Tamil Nadu Agricultural University, Coimbatore-3, India

Zhang Yugang
College of Landscaping and Horticulture, Qingdao Agricultural University, Qingdao, Shandong Province 266109, China

Dai Hongyi
College of Landscaping and Horticulture, Qingdao Agricultural University, Qingdao, Shandong Province 266109, China

Imran Rauf
Nuclear Institute of Agriculture Tandojam - 70060, Pakistan

Nazir Ahmad
Nuclear Institute of Agriculture Tandojam - 70060, Pakistan

S. M. Masoom Shah Rashdi
Nuclear Institute of Agriculture Tandojam - 70060, Pakistan

Muhammad Ismail
Nuclear Institute of Agriculture Tandojam - 70060, Pakistan

M. Hamayoon Khan
Nuclear Institute of Agriculture Tandojam - 70060, Pakistan

Qingjiang Wei
College of Horticulture and Forestry, Huazhong Agricultural University, Shizishan Street No.1, Wuhan 430070, China

Yongzhong Liu
College of Horticulture and Forestry, Huazhong Agricultural University, Shizishan Street No.1, Wuhan 430070, China

Ou Sheng
College of Horticulture and Forestry, Huazhong Agricultural University, Shizishan Street No.1, Wuhan 430070, China
Institute of Fruit Tree Research, Guangdong Academy of Agricultural Science, Tianhe District, Guangzhou 510640, China

Jicui An
College of Horticulture and Forestry, Huazhong Agricultural University, Shizishan Street No.1, Wuhan 430070, China

Gaofeng Zhou
College of Horticulture and Forestry, Huazhong Agricultural University, Shizishan Street No.1, Wuhan 430070, China

Shu-ang Peng
College of Horticulture and Forestry, Huazhong Agricultural University, Shizishan Street No.1, Wuhan 430070, China

Olmar Baller Weber
Embrapa Tropical Agroindustry, Dra. Sara Mesquita, 2270, Pici, 60511-110, Fortaleza, Ceará, Brazil

Sandy Sampaio Videira
Embrapa Agrobiology, BR 465, km07, 23890-000, Seropédica, Rio de Janeiro, Brazil

Jean Luiz Simões de Araújo
Embrapa Agrobiology, BR 465, km07, 23890-000, Seropédica, Rio de Janeiro, Brazil

Manoj Kumar Mahawar
Department of Agricultural Engineering, IARI New Delhi, India

Anupama Singh
Department of PHPFE, GBPUAT Pantnagar, Uttarakhand, India

B. K. Kumbhar
Department of PHPFE, GBPUAT Pantnagar, Uttarakhand, India

Manvika Sehgal
Department of Microbiology, GBPUAT Pantnagar, Uttarakhand, India

K. Sunilkumar
Indian Council of Agricultural Research, Pedavegi–534 450 West Godavari District, Andhra Pradesh, India

D. S. Sparjan Babu
Indian Council of Agricultural Research, Pedavegi–534 450 West Godavari District, Andhra Pradesh, India

Savreet Sandhu
Punjab Agricultural University, Regional Research Station, Bathinda-151001, Punjab, India

J. S. Bal
Department of Agriculture, Khalsa College, Amritsar-143001, Punjab, India

J. Kubiriba
Kawanda Agricultural Research Institute-NARO, P. O. Box, 7065, Kampala, Uganda

W. K. Tushemereirwe
Kawanda Agricultural Research Institute-NARO, P. O. Box, 7065, Kampala, Uganda

L. Kenyon
Natural Resources Institute, University of Greenwich, Central Avenue, Chatham, Maritime Kent ME4 4TB, UK

T. C. B. Chancellor
Natural Resources Institute, University of Greenwich, Central Avenue, Chatham, Maritime Kent ME4 4TB, UK

Phebe Ding
Department of Crop Science, Faculty of Agriculture, Universiti Putra Malaysia, 43400 UPM Serdang, Selangor, Malaysia

Khairul Bariah Darduri
Department of Crop Science, Faculty of Agriculture, Universiti Putra Malaysia, 43400 UPM Serdang, Selangor, Malaysia

Satisha Jogaiah
National Research Centre for Grapes, P. B. No. 3, Manjri Farm, Solapur Road, Pune - 412 307, Maharashtra, India

Dasharath P. Oulkar
National Research Centre for Grapes, P. B. No. 3, Manjri Farm, Solapur Road, Pune - 412 307, Maharashtra, India

Amruta N. Vijapure
National Research Centre for Grapes, P. B. No. 3, Manjri Farm, Solapur Road, Pune - 412 307, Maharashtra, India

Smita R. Maske
National Research Centre for Grapes, P. B. No. 3, Manjri Farm, Solapur Road, Pune – 412 307, Maharashtra, India

Ajay Kumar Sharma
National Research Centre for Grapes, P. B. No. 3, Manjri Farm, Solapur Road, Pune – 412 307, Maharashtra, India

Ramhari G. Somkuwar
National Research Centre for Grapes, P. B. No. 3, Manjri Farm, Solapur Road, Pune – 412 307, Maharashtra, India

K. K. Srivastava
Section of Crop Improvement, Central Institute of Temperate Horticulture, Old Air Field, Srinagar– 190007, Jammu and Kashmir, India

N. Ahmad
Section of Crop Improvement, Central Institute of Temperate Horticulture, Old Air Field, Srinagar– 190007, Jammu and Kashmir, India

Dinesh Kumar
Section of Crop Improvement, Central Institute of Temperate Horticulture, Old Air Field, Srinagar– 190007, Jammu and Kashmir, India

Biswajit Das
Section of Crop Improvement, Central Institute of Temperate Horticulture, Old Air Field, Srinagar– 190007, Jammu and Kashmir, India

S. R. Singh
Section of Crop Improvement, Central Institute of Temperate Horticulture, Old Air Field, Srinagar– 190007, Jammu and Kashmir, India

Shiv Lal
Section of Crop Improvement, Central Institute of Temperate Horticulture, Old Air Field, Srinagar– 190007, Jammu and Kashmir, India

O. C. Sharma
Section of Crop Improvement, Central Institute of Temperate Horticulture, Old Air Field, Srinagar– 190007, Jammu and Kashmir, India

J. A. Rather
Section of Crop Improvement, Central Institute of Temperate Horticulture, Old Air Field, Srinagar– 190007, Jammu and Kashmir, India

S. K. Bhat
Section of Crop Improvement, Central Institute of Temperate Horticulture, Old Air Field, Srinagar– 190007, Jammu and Kashmir, India

Inuwa Shehu Usman
Department of Plant Science, Ahmadu Bello Univeristy, Samaru, Zaria, Nigeria

Maimuna Mohammed Abdulmalik
Department of Plant Science, Ahmadu Bello Univeristy, Samaru, Zaria, Nigeria

Lawan Abdu Sani
Department of Plant Science, Bayero Univeristy Kano, Kano, Nigeria

Ahmed Nasir Muhammad
School of Engeneering Science Technology, Federal Polytechnic, Kazuare, Jigawa state, Nigeria

Islam Saruhan
Department of Plant Protection, Faculty of Agriculture, Ondokuz Mayis University, Samsun, Turkey

Hüseyin Akyol
Bleak Sea Agricultural Research Institute Samsun, Turkey

Berhanu Megerssa
Department of Agricultural Economics, Rural Development and Value Chain Management, Jimma University College of Agriculture and Veterinary Medicine, P. O. Box 307, Jimma, Ethiopia

Arinaitwe Abel Byarugaba
Kachwekano Zonal Agricultural Research and Development Institute, P. O. Box 421, Kabale, Uganda

Turyamureeba Gard
Kachwekano Zonal Agricultural Research and Development Institute, P. O. Box 421, Kabale, Uganda

Imelda Night Kashaija
Kachwekano Zonal Agricultural Research and Development Institute, P. O. Box 421, Kabale, Uganda

Cissé Ibrahima
INHP, BP1313 Yamoussoukro, Côte d'ivoire

Montet Didier
CIRAD, UMR 95 Qualisud, TA B95/16, 34398 Montpellier cedex 5 – France

Reynes Max
CIRAD, UMR 95 Qualisud, TA B95/16, 34398 Montpellier cedex 5 – France

Danthu Pascal
3CIRAD, PD Forêts et Biodiversité, P. O. Box 853, Antananarivo, Madagascar and UR 105, Campus de Baillarguet, 34392 Montpellier Cedex 5, France

Yao Benjamin
INHP, BP1313 Yamoussoukro, Côte d'ivoire

Boulanger Renaud
CIRAD, UMR 95 Qualisud, TA B95/16, 34398 Montpellier cedex 5 – France

D. M. Modise
School of Agriculture and Life Sciences, University of South Africa (UNISA), P/Bag X6, Florida 1710, South Africa

Rita Narayanan
Department of Dairy Science, Madras Veterinary College, Chennai, TamilNadu, India

Jyothi Lingam
Department of Dairy Science, Madras Veterinary College, Chennai, TamilNadu, India

Mohammad Samir El- Habba
Economics of Date Palm Chair", King Faisal University, Alahssa, Saudi Arabia

Fahad Al- Mulhim
Economics of Date Palm Chair", King Faisal University, Alahssa, Saudi Arabia

Bruno Vinícius Castro Guimarães
Federal Institute of Education, Science and Technology of the Amazon, IFAM, Campus São Gabriel da Cachoeira, BR 307, km 03, Estrada do Aeroporto, Cachoeirinha, ZIP Code 69750-000, São Gabriel da Cachoeira, AM, Brazil

Sérgio Luiz Rodrigues Donato
Federal Institute of Education, Science and Technology of Bahia – Campus Guanambi, P. O. Box 09, Ceraima District, ZIP Code 46.430-000, Guanambi, BA, Brazil

Victor Martins Maia
Montes Claros State University– Campus Janaúba, Center for Exact Sciences and Technology, Department of Agricultural Sciences, 2.630 Reinaldo Viana Av., PO Box 91, Bico da Pedra, ZIP Code 39.440-000, Janaúba, MG, Brazil

Ignacio Aspiazú
Montes Claros State University– Campus Janaúba, Center for Exact Sciences and Technology, Department of Agricultural Sciences, 2.630 Reinaldo Viana Av., PO Box 91, Bico da Pedra, ZIP Code 39.440-000, Janaúba, MG, Brazil

Maria Geralda Vilela Rodrigues
Agricultural Research Company of Minas Gerais / Regional Unit Epamig Northern Minas Gerais, ZIP Code 39525-000, Nova Porteirinha, MG, Brazil

Pedro Ricardo Rocha Marques
Federal Institute of Education, Science and Technology of Bahia – Campus Guanambi, P. O. Box 09, Ceraima District, ZIP Code 46.430-000, Guanambi, BA, Brazil

M. U. Dimelu
Department of Agricultural Extension, Faculty of Agriculture, University of Nigeria, Nsukka, Enugu State, Nigeria

R. N. Odo
Department of Agricultural Extension, Faculty of Agriculture, University of Nigeria, Nsukka, Enugu State, Nigeria

Aiala Vieira Amorim
Universidade da Integração Internacional da Lusofonia Afro-brasileira, Avenida da Abolição, número 3, Centro, 62790-000, Redenção, CE - Brasil

Deborah dos Santos Garruti
Embrapa Agroindústria Tropical, Rua Dra. Sara Mesquita, 2270, 60511-110, Fortaleza, CE - Brasil

Claudivan Feitosa de Lacerda
Departamento de Engenharia Agrícola/UFC and INCTSal, Campus do Pici, bloco 804, 60455-760, Fortaleza, CE - Brasil

Carlos Farley Herbster Moura
Embrapa Agroindústria Tropical, Rua Dra. Sara Mesquita, 2270, 60511-110, Fortaleza, CE - Brasil

Enéas Gomes-Filho
Departamento de Bioquímica e Biologia Molecular/UFC and INCTSal, Campus do Pici, bloco 907, 60440-554, Fortaleza, CE - Brasil

Geoffrey O. Bedford
Department of Biological Sciences, Macquarie University, NSW 2109, Australia

Hammad Ahmad Khan
Department of Zoology and Fisheries, University of Agriculture, Faisalabad, Pakistan 38040

Muhammad Javed
Department of Zoology and Fisheries, University of Agriculture, Faisalabad, Pakistan 38040

Ammara Tahir
Department of Zoology and Fisheries, University of Agriculture, Faisalabad, Pakistan 38040

Madeeha Kanwal
Department of Zoology and Fisheries, University of Agriculture, Faisalabad, Pakistan 38040